Linear Algebra I

Frederick P. Greenleaf
Courant Institute

Sophie Marques
University of Cape Town

29 Linear Algebra I

Courant Institute of Mathematical Sciences
New York University
New York, New York

American Mathematical Society
Providence, Rhode Island

2010 *Mathematics Subject Classification.* Primary 97H60, 15-01, 15A04, 15A15, 15A16, 15A18, 15A21, 40A05, 40A25, 20B30.

For additional information and updates on this book, visit
www.ams.org/bookpages/cln-29

Library of Congress Cataloging-in-Publication Data

Names: Greenleaf, Frederick P., author. | Marques, Sophie, 1986- author.
Title: Linear algebra I / Frederick P. Greenleaf (Courant Institute, New York University), Sophie Marques (University of Cape Town, Capetown, South Africa).
Other titles: Linear algebra one
Description: Providence, RI : American Mathematical Society, [2019] | Series: Courant lecture notes ; volume 29 | "Courant Institute of Mathematical Sciences, New York University, New York, NY." | Includes index.
Identifiers: LCCN 2018047584 | ISBN 9781470448714 (alk. paper)
Subjects: LCSH: Algebras, Linear–Textbooks. | AMS: Mathematics education – Algebra – Linear algebra. msc | Linear and multilinear algebra; matrix theory – Instructional exposition (textbooks, tutorial papers, etc.). msc | Linear and multilinear algebra; matrix theory – Basic linear algebra – Linear transformations, semilinear transformations. msc | Linear and multilinear algebra; matrix theory – Basic linear algebra – Determinants, permanents, other special matrix functions. msc | Linear and multilinear algebra; matrix theory – Basic linear algebra – Matrix exponential and similar functions of matrices. msc | Linear and multilinear algebra; matrix theory – Basic linear algebra – Eigenvalues, singular values, and eigenvectors. msc | Linear and multilinear algebra; matrix theory – Basic linear algebra – Canonical forms, reductions, classification. msc | Sequences, series, summability – Convergence and divergence of infinite limiting processes – Convergence and divergence of series and sequences. msc | Sequences, series, summability – Convergence and divergence of infinite limiting processes – Approximation to limiting values (summation of series, etc.). msc | Group theory and generalizations – Permutation groups – Symmetric groups. msc
Classification: LCC QA184.2 .G7254 2019 | DDC 512/.5–dc23
LC record available at https://lccn.loc.gov/2018047584

∞ The paper used in this book is acid-free and falls within the guidelines established to ensure permanence and durability.
Visit the AMS home page at `https://www.ams.org/`

10 9 8 7 6 5 4 3 2 1 24 23 22 21 20 19

Contents

Preface

These notes have been used for several years, by various instructors, in a one-year course sequence, Linear Algebra I and Linear Algebra II, offered in the Mathematics Department at New York University's Courant Institute. These courses were developed with two audiences in mind:

- As Senior Honors courses for undergraduate mathematics majors seeking a stronger background in linear algebra.
- As part of a one-year course sequence for students in the Courant Master's Degree Program in Mathematics. This large and diverse audience includes:
    (i) ambitious undergraduate math majors seeking a deeper understanding of linear algebra,
    (ii) graduate students from other quantitative disciplines, particularly computer science and economics, intent on building a stronger math background,
    (iii) mathematically proficient students with undergraduate backgrounds in mathematics, the sciences, or engineering, who after working for a few years have developed a strong interest in Ph.D.-level studies in pure or applied mathematics.

A significant number of students in the last category have been accepted with advanced standing into the regular Courant Ph.D. program after completing their one-year master's program and passing written qualifying exams.

Accordingly, these notes are not a comprehensive introduction to linear algebra. Students were presumed to have had a prior course in linear algebra, a basic understanding of matrix algebra, and some proficiency with mathematical proofs. However, proofs are emphasized in this course sequence as a means of building students' skills at analyzing problems for themselves, anticipating the day when they will encounter questions that cannot be resolved by a Google search.

The overall emphasis in these notes is on understanding the structure of linear operators as the key to solving problems in which they arise.

Although these notes provide a self-contained tour of linear algebra, in practice they were often supplemented by two (optional) standard texts that provided alternative views of many topics as well as more extensive problem sets, and are well-aligned with the notes. These were:

1. *Linear Algebra* by Seymour Lipschutz, 2nd edition, Schaum's Outline Series, McGraw-Hill, 1991.
2. *Linear Algebra* by Stephen Friedberg, Arnold Insel, and Lawrence Spence, 4th edition, Prentice-Hall, 2003.

Students found the Lipschutz text particularly helpful for its clarity, well-chosen worked examples, extensive selection of exercises, and detailed discussion of proofs. The first four chapters also provide a concise and self-contained review of the topics that are prerequisites to this course: basic matrix algebra, row operations, and solution of systems of linear equations.

Organization of the Text

The handling of exercises is somewhat unconventional. Many are placed within the main text, as the topics they address first occur. Each chapter ends with an extensive set of section-by-section Additional Exercises. Some recap the main topics of of each section; others are longer and intended to be more challenging; each begins with a block of true/false questions, which students often find more challenging than you might expect.

There are a few unconventional notations, explained as they are introduced:

- We often write $|V|$ for the dimension $\dim(V)$, and $R(T), K(T)$ for the range and kernel of T, respectively, when this is convenient.
- A special symbol $\mathbb{1}$ is used to distinguish the constant function (or polynomial) everywhere equal to 1, from the scalar 1.

The symbol $\mathbb{K}$ indicates a generic ground field.

Acknowledgments

The present volume, Linear Algebra I, covers the first semester of the NYU course sequence Graduate Linear Algebra I–II. The companion volume, Linear Algebra II, used in the second semester, is currently being prepared for publication in the Courant Lecture Notes series.

These Notes were developed over several years, as handwritten lecture notes for Professor Greenleaf's courses. They were reproduced and distributed to students along with detailed problem sets (many with full solutions).

The typeset LaTeX version of these Notes might never have seen the light of day without the enthusiastic and tireless efforts of the second author, Dr. Sophie Marques, who came to Courant as a postdoc in 2014 and was assigned to teach the Linear Algebra I–II sequence. She quickly converted the existing notes into serviceable LaTeX code, and during the next two years actively helped reshape the final text based on her own experience teaching these courses. The final version owes a lot to those efforts, and her talent as a teacher.

CHAPTER 1

Vector Spaces over a Field $\mathbb{K}$

1.1. Vector Spaces

The objects of interest in this chapter will be vector spaces over arbitrary fields.

Basic Properties and Examples.

DEFINITION 1.1. A *vector space over a field* $\mathbb{K}$ is a set V equipped with two operations $(+)$ and $(\cdot)$ from $V \times V \to V$ and $\mathbb{K} \times V \to V$ having the following properties:

1. **Axioms for** $(+)$**:** For all $x, y \in V$ we have:
   COMMUTATIVE LAW: $x + y = y + x$
   ASSOCIATIVE LAW: $(x + y) + z = x + (y + z)$
   ZERO ELEMENT: There exists an element "0" in V such that $0 + v = v$ for all v.
   ADDITIVE INVERSE: For every $x \in V$, there is an element $v \in V$ such that $v + x = 0$.
2. **Axioms for** $(\cdot)$**:** For all $v, w \in V$ and $a, b \in \mathbb{K}$ we have:
   IDENTITY LAW: $1 \cdot v = v$ ($1 =$ the multiplicative identity in $\mathbb{K}$)
   ASSOCIATIVE LAW: $(a \cdot b) \cdot v = a \cdot (b \cdot v)$
   DISTRIBUTIVE LAW: $a \cdot (v + w) = (a \cdot v) + (a \cdot w)$
   DISTRIBUTIVE LAW: $(a + b) \cdot v = (a \cdot v) + (b \cdot v)$

NOTATION: We generally write "0" for both the zero vector in V and the scalar "0" in the field $\mathbb{K}$. Usually this causes no trouble, but when it becomes necessary to distinguish the two we will write 0_V for the zero element in V. The "additive inverse" of $x \in V$ is written "$-x$."

As a consequence we have:

LEMMA 1.2. *The zero element is unique: if* $0, 0' \in V$ *are elements such that* $0 + v = v$ *and* $0' + v = v$*, for all* $v \in V$*, then* $0' = 0$.

PROOF. $0 + 0' = 0'$ and $0 + 0' = 0$, so $0' = 0$. □

LEMMA 1.3. *The additive inverse is unique. That is, given* $v \in V$ *there is just one element* $u \in V$ *such that* $u + v = 0$.

PROOF. Suppose $v \in V$ and we are given u and u' with $u + v = 0$ and $u' + v = 0$. Look at the combination $u + v + u'$; by associativity we get

$$u' = 0 + u' = (u + v) + u' = u + (v + u') = u + 0 = u \ ,$$

so that $u' = u$. □

EXERCISE 1.4. From the axioms and previous results prove:

(a) $0 \cdot v = 0_V$
(b) $\lambda \cdot 0_V = 0_V$
(c) $\lambda \cdot v = v$ and $v \neq 0_V \Rightarrow \lambda = 1$

for all $v \in V$ and $\lambda \in \mathbb{K}$, where "0" is the zero element in the field of scalars $\mathbb{K}$.

EXERCISE 1.5. Prove that $-v = (-1) \cdot v$ where -1 is the negative of $1 \in \mathbb{K}$.

HINT: $1+(-1) = 0$ in $\mathbb{K}$ and $0 \cdot v = 0_V$. Remember: "$-v$" is the unique element that added to v is 0_V; prove that $(-1) \cdot v$ has this property and conclude that $(-1) \cdot v = -v$ by uniqueness of the additive inverse.

EXERCISE 1.6. Prove that $-(-v) = v$, for all $v \in V$.

HINT: Same as the previous exercise: what identity does $-(-v)$ satisfy?

EXERCISE 1.7 (Cancellation Laws). If $a + v = a + w$ for $a, v, w \in V$, prove that $v = w$. Then use this to prove:

(a) $\lambda \cdot v = 0_V$ and $v \neq 0_V$ implies that $\lambda = 0$ in $\mathbb{K}$.
(b) $\lambda \cdot v = v$ and $v \neq 0_V$ implies $\lambda = 1$.

DEFINITION 1.8. *Coordinate space* over the field $\mathbb{K}$ consists of all ordered n-tuples $\mathbb{K}^n = \{\mathbf{x} = (x_1, \dots, x_n) : x_k \in \mathbb{K}\}$, equipped with the usual $(+)$ and $(\cdot)$ operations:

1. $(x_1, \dots, x_n) + (y_1, \dots, y_n) = (x_1 + y_1, \dots, x_n + y_n)$,
2. $\lambda \cdot (x_1, \dots, x_n) = (\lambda \cdot x_1, \dots, \lambda \cdot x_n)$ for $\lambda \in \mathbb{K}$.

EXERCISE 1.9. explain why $(+)$ in $\mathbb{R}^2$ is described geometrically by the "parallelogram law" for vector addition shown in Figure 1.1.

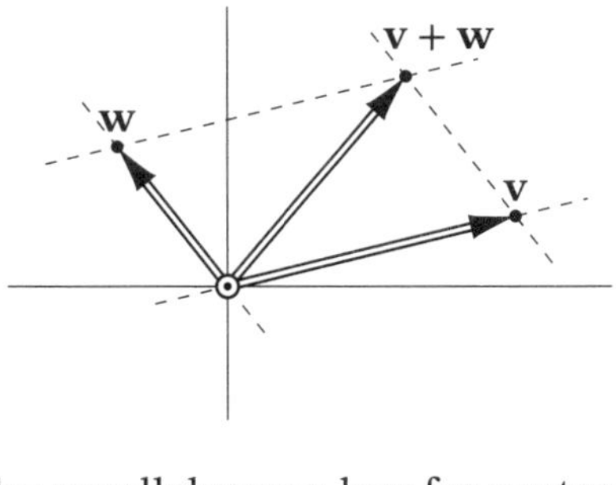

FIGURE 1.1. The parallelogram law for vector addition, illustrated in $\mathbb{R}^2$.

EXAMPLE 1.10. The space $\mathrm{M}(n \times m, \mathbb{K})$ of $n \times m$ matrices with entries in $\mathbb{K}$ becomes a vector space when equipped with the operations

ADDITION: $(A + B)_{ij} = A_{ij} + B_{ij}$
SCALING: $(\lambda \cdot A)_{ij} = \lambda A_{ij}$

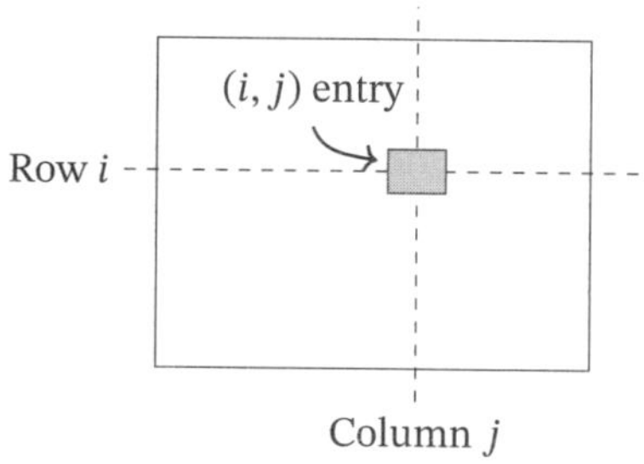

FIGURE 1.2. The entry in a matrix array with "address" (i, j) is the one in Row i and Column j.

The space of square matrices, with $m = n$, is denoted $\mathrm{M}(n, \mathbb{K})$.
NOTE: Matrix entry A_{ij} is the one in the ith row and j^{th} column. The pair (i, j) is referred to as its "address"; see Figure 1.2.

There is also a matrix multiplication that makes $\mathrm{M}(n, \mathbb{K})$ an associative-multiplication that makes $\mathrm{M}(n, \mathbb{K})$ an associative algebra with identity, but the matrix product AB can be defined more generally for nonsquare matrices as long as they are "compatible," with the number of columns in A equal to the number of rows in B. Thus if A is $m \times q$ and B is $q \times n$ we get an $m \times n$ matrix AB with entries

$$(AB)_{ij} = \sum_{k=1}^{q} A_{ik} B_{kj}.$$

The algebra $\mathrm{M}(n, \mathbb{K})$ of square matrices is not commutative unless $n = 1$.

EXAMPLE 1.11 (polynomial ring $\mathbb{K}[x]$). The set $\mathbb{K}[x]$ consists of all finite "formal sums" $a_0 + a_1 x + \cdots + a_n x^n + \cdots = \sum_{k \ge 0} a_k x^k$ with $a_i \in \mathbb{K}$, and $a_i = 0$ for all but a finite number of indices. These sums can have arbitrary length. They include the "constant polynomials," which have the form $c \cdot \mathbb{1}$ with $c \in \mathbb{K}$, where $\mathbb{1}$ is the particular constant polynomial $1 + 0 \cdot x + 0 \cdot x^2 + \cdots$; the zero polynomial $0 \cdot \mathbb{1}$ is written as "0", which might get confusing.

The algebraic operations in $\mathbb{K}[x]$ are

1. ADDITION:
$$\Big(\sum_{k \ge 0} a_k x^k\Big) + \Big(\sum_{k \ge 0} b_k x^k\Big) = \sum_{k \ge 0} (a_k + b_k) x^k$$

2. SCALING:
$$\lambda \cdot \Big(\sum_{k \ge 0} a_k x^k\Big) = \sum_{k \ge 0} (\lambda a_k) x^k$$

There is also a multiplication operation, obtained by multiplying terms in the formal sums and gathering together those of the same degree:

3. PRODUCT:
$$\Big(\sum_{k \ge 0} a_k x^k\Big) \times \Big(\sum_{l \ge 0} b_l x^l\Big) = \sum_{k,l \ge 0} a_k b_l \, x^{k+l} = \sum_{r \ge 0} \Big(\sum_{\substack{k,l \ge 0 \\ k+l=r}} a_k b_l \Big) \cdot x^r$$

(the sum being finite for each r). This makes $\mathbb{K}[x]$ into a commutative associative algebra over $\mathbb{K}$ with $\mathbb{1}$ as its multiplicative identity.

All information about a polynomial resides in the symbol string $(a_0, a_1, a_2, \dots)$ of coefficients, and the algebraic operations on $\mathbb{K}[x]$ can be performed as operations on symbol strings; the zero polynomial is represented by $(0, 0, \dots)$, the identity by $\mathbb{1} = (1, 0, \dots, 0)$, and x by $x = (0, 1, 0, \dots.)$, etc.

EXERCISE 1.12. If $f(x) = 3 + 3x + x^2$ and $g(x) = 4x^2 - 2x^3 + x^5$, compute the sum $f + g$ and product $f \cdot g$.

The *degree* $\deg(f)$ of $f = \sum_{k \geq 0} a_k x^k$ is n if $a_n \neq 0$ and $a_k = 0$ for all $k > n$. The degree of a constant polynomial $c\mathbb{1}$ is zero, except that no "degree" can be assigned to the zero polynomial 0. (For various reasons, the only possible assignment would be "$-\infty$".)

EXERCISE 1.13. If $f, g \neq 0$ in $\mathbb{K}[x]$, prove that $fg \neq 0$ and $\deg(fg) = \deg(f) + \deg(g)$.

EXERCISE 1.14. If $f, g \neq 0$ in $\mathbb{K}[x]$, what (if anything) can you say about $\deg(f + g)$?

Multivariate Polynomials and Multi-Index Notation.

EXAMPLE 1.15. (Polynomials in several unknowns). The polynomial ring $\mathbb{K}[\mathbf{x}] = \mathbb{K}[x_1, \dots, x_n]$is handled using very efficient "multi-index notation." A *multi-index* is an element $\alpha = (\alpha_1, \dots, \alpha_n)$ of the Cartesian product set $\mathbb{Z}_+^n = \mathbb{Z}_+ \times \cdots \times \mathbb{Z}_+$ (n factors). Each α determines a *monomial* $x^\alpha = x_1^{\alpha_1} \cdot \dots \cdot x_n^{\alpha_n}$, in which we interpret $x_k^0 = 1$. Elements of $\mathbb{K}[x_1, \dots, x_n]$ are finite formal linear combinations of monomials

$$f(x_1, \dots, x_n) = \sum_{\alpha \in \mathbb{Z}_+^n} c_\alpha x^\alpha \quad (c_\alpha \in \mathbb{K}).$$

The monomial $x^{(0,\dots,0)}$ is the constant polynomial $\mathbb{1}$ in $\mathbb{K}[x_1, \dots, x_n]$. With these ideas in mind we define the following:

1. The *total degree* of a multi-index is $|\alpha| = \alpha_1 + \cdots + \alpha_n$, and the degree of the corresponding monomial is $\deg(x^\alpha) = |\alpha|$. Many monomials can have the same total degree, for example x^2y and xy^2.
2. The *degree of a polynomial* $f \in \mathbb{K}[\mathbf{x}]$ is

$$\deg(f) = \max\{|\alpha| : c_\alpha \neq 0\}.$$

Nonzero constant polynomials $c\mathbb{1}$ have degree zero; if f is the zero polynomial (all coefficients $c_\alpha = 0$) $\deg(f)$ cannot be defined. The generators $f_k(\mathbf{x}) = x_k$ of the polynomial ring all have degree 1.

The following operations make $V = \mathbb{K}[x_1, \dots, x_n]$ a vector space and a commutative associative algebra with identity $\mathbb{1} = x^{(0,\dots,0)}$.

1. SUM: $f + g = \sum_\alpha (a_\alpha + b_\alpha) x^\alpha$
2. SCALING: $\lambda \cdot f = \sum_\alpha (\lambda a_\alpha) x^\alpha$

3. PRODUCT OPERATION:

$$f \cdot g = \Big(\sum_\alpha a_\alpha x^\alpha\Big) \cdot \Big(\sum_\beta b_\beta x^\beta\Big) = \sum_{\alpha,\beta \in \mathbb{Z}_+^n} a_\alpha b_\beta x^{\alpha+\beta}$$
$$= \sum_{\gamma \in \mathbb{Z}_+^n} \Big(\sum_{\alpha+\beta=\gamma} a_\alpha \cdot b_\beta \Big) \cdot x^\gamma$$

where we define $\alpha + \beta = (\alpha_1 + \beta_1, \dots, \alpha_n + \beta_n)$.

As an example, the monomials of degree 2 in $\mathbb{K}[x_1, x_2, x_3]$ are

multi-index α	monomial x^α	dictionary word
$(0, 0, 2)$	x_3^2	AAC
$(0, 1, 1)$	$x_2 x_3$	ABA
$(0, 2, 0)$	x_2^2	ACA
$(1, 0, 1)$	$x_1 x_3$	BAB
$(1, 1, 0)$	$x_1 x_2$	BBA
$(2, 0, 0)$	x_1^2	CAA

Here we have lined up the monomials in "lexicographic"or "dictionary" order (taking $A = 0$, $B = 1$, $C = 2$, ...), which is a useful way to manage them. This is a strict *linear* ordering of monomials; they are only partially ordered by their "total degree" $\deg(x^\alpha) = |\alpha|$. The system $\mathbb{K}[x_1, \dots, x_n]$ is a commutative associative algebra with identity element $\mathbb{1}$. Its properties are quite a bit more complicated than those of polynomials $\mathbb{K}[x]$ in one unknown, but they do share two crucial algebraic properties.

EXERCISE 1.16. (Hard, but try it.) If $f, g \neq 0$ in $\mathbb{K}[x_1, \dots, x_n]$ prove that:

1. DEGREE FORMULA: $\deg(f \cdot g) = \deg(f) + \deg(g)$ for all $f, g \neq 0$ in $\mathbb{K}[x_i, \dots, x_n]$.
2. NO ZERO DIVISORS: $f, g \neq 0$ in $\mathbb{K}[x_1, \dots, x_n] \Rightarrow f \cdot g \neq 0$. This implies we can perform "cancellation": if $f \neq 0$ and $f \cdot h_1 = f \cdot h_2$, then $h_1 = h_2$.

HINT: Try it first for $n = 1$. For $n = 2$ try lexicographic ordering of monomials in $\mathbb{K}[x, y]$.

NOTE: The maximum possible degree for a nonzero monomial in the product fg is obviously $d = \deg(f) + \deg(g)$. The problem is that the coefficient c_γ of such a monomial will be a sum of products $(\sum_{\alpha+\beta=\gamma} a_\alpha b_\beta)$, and not a simple product as it is when there is just one variable. Such sums could equal zero even if all terms are nonzero, so why couldn't these coefficients (sums) be zero for *all* monomials with the maximum possible degree d, making $\deg(fg) < \deg(f) + \deg(g)$?

A more complete discussion of the degree formula for $n \geq 2$, and especially its proof using lexicographic ordering of monomials, is provided in the appendix of this chapter. I-1.17

EXAMPLE 1.17 (Function Spaces). If S is a set, $\mathcal{C}(S)$ = all scalar-valued functions $f : S \to \mathbb{K}$ becomes a vector space under the usual operations

$$(f+g)(x) = f(x) + g(x), \quad (\lambda \cdot f)(x) = \lambda f(x) \quad \forall x \in S.$$

There is also a pointwise multiplication operation

$$(f \cdot g)(x) = f(x) \cdot g(x),$$

which makes $\mathcal{C}(S)$ a commutative associative algebra over $\mathbb{K}$ with identity element $\mathbb{1}(x) = 1$ for all x and zero element $0(x) = 0$ for $x \in S$.

EXAMPLE 1.18 (Polynomial Functions vs. Formal Sums). The *polynomial functions* $\mathcal{P}_{\mathbb{K}}$ with values in $\mathbb{K}$ are the functions $\phi_f : \mathbb{K} \to \mathbb{K}$ of the form

$$\phi_f(t) = [f(x)|_{x=t}] = \sum_{k \geq 0} a_k\, t^k \quad (t \in \mathbb{K})$$

for some $f(x) \in \mathbb{K}[x]$. (Thus, $\phi(t) = \sin(t)$ is not a polynomial function on $\mathbb{K}$.) Note carefully that the elements of $\mathcal{P}_{\mathbb{K}}$ are functions, while $\mathbb{K}[x]$ is made up of symbol strings or formal sums. They are not the same thing, though there is a close relation between them implemented by the surjective (= "onto") mapping $\Phi : \mathbb{K}[x] \to \mathcal{P}_{\mathbb{K}}$ such that

$$\Phi f(t) = \sum_{k \geq 0} a_k t^k \quad (t \in \mathbb{K})$$

if $f(x) = \sum_{k \geq 0} a_k x^k$ in $\mathbb{K}[x]$. This surjective map is a *homomorphism*: it preserves, or "intertwines," the algebraic operations in $\mathbb{K}[x]$ and in the "target space" $\mathcal{P}_{\mathbb{K}}$, so that

$$\Phi(\lambda \cdot f) = \lambda \cdot \Phi(f), \quad \Phi(f+g) = \Phi(f) + \Phi(g), \quad \Phi(f \cdot g) = \Phi(f) \cdot \Phi(g).$$

EXERCISE 1.19. If $\mathbb{K} = \mathbb{R}$ or $\mathbb{C}$, explain why Φ is a bijection, hence an "isomorphism" between commutative associative algebras. In fact, prove that this is so for polynomials over any *infinite* field $\mathbb{K}$.

HINT: Φ is linear, hence being one-to-one is equivalent to saying that $\Phi(f) = 0 \Rightarrow f = 0$ in $\mathbb{K}[x]$. If f is nonzero in $\mathbb{K}[x]$, the corresponding polynomial function $\Phi(f) : \mathbb{K} \to \mathbb{K}$ can take on the value zero at no more than $n = \deg(f)$ points—i.e., the number of roots in $\mathbb{K}$ cannot exceed $\deg(f)$. Since $\mathbb{R}$ and $\mathbb{C}$ (and even $\mathbb{Q}$) are infinite, we cannot have $\Phi(f) \equiv 0$ on these fields unless f is the zero polynomial.

The finite fields $\mathbb{Z}_p$ (p a prime) are widely used in number theory, cryptography, image processing, etc. The one-to-one correspondence between formal polynomials and scalar-valued functions $\Phi f(t)$ defined on $\mathbb{K}$ breaks down for these fields. For example, if $\mathbb{K} = \mathbb{Z}_p$, the nonzero polynomial $f = x^p - x$ has value zero for every choice of $x \in \mathbb{Z}_p$, so f has $p = \deg(f)$ distinct roots in $\mathbb{Z}_p$. (By a theorem of Fermat, if p is a prime, then $t^{p-1} = 1$ for all nonzero t in $\mathbb{Z}_p$. Thus $t^p - t$ is zero at *every* $t \in \mathbb{Z}_p$ and the function $\Phi(f)$ is the zero function on $\mathbb{Z}_p$.)

EXERCISE 1.20. For $p = 3$, verify that $t^3 - t = 0$ for each of the three elements $t = [0], [1], [2]$ in $\mathbb{Z}_3$. But the corresponding element of $\mathbb{Z}_3[x]$ is $f(x) = x^3 - x$, whose symbol string $(0, -1, 0, 1, 0, 0, \dots)$ differs from that of the zero polynomial in $\mathbb{Z}_3[x]$.

Row Space and Column Space of a Matrix.

EXAMPLE 1.21 (Row Space and Column Space). Two vector spaces can be associated with any $n \times m$ matrix A having entries in $\mathbb{K}$:

1. ROW SPACE: $\text{RowSpace}(A) = \{\text{linear combinations } c_1R_1(A) + \cdots + c_nR_n(A) : c_i \in \mathbb{K}\}$ of the rows $R_i(A) = (a_{i1}, \dots, a_{im})$, regarded as vectors in $\mathbb{K}^m$.
2. COLUMN SPACE: $\text{ColSpace}(A) = \{\text{linear combinations } c_1C_1(A) + \cdots + c_mC_m(A) : c_i \in \mathbb{K}\}$ of the columns $C_i(A) = \text{col}(a_{1i}, \dots, a_{ni})$, regarded as vectors in $\mathbb{K}^n$.

These will play important roles in our study of matrix operations.

1.2. Vector Subspaces

DEFINITION 1.22. A nonempty subset W of a vector space V is a *vector subspace* if

1. W is *closed under* $(+)$: $W + W \subseteq W$, so $w_1, w_2 \in W \Rightarrow w_1 + w_2 \in W$.
2. W is *closed under* $(\cdot)$: $\mathbb{K} \cdot W \subseteq W$, so $\lambda \in \mathbb{K}, w \in W \Rightarrow \lambda \cdot w \in W$.

The zero vector 0_V then lies in W, for if $w \in W$ then $-w = (-1) \cdot w \in W$ and then $0_V = w + (-w) \in W$. Thus W becomes a vector space over $\mathbb{K}$ in its own right under the $(+)$ and $(\cdot)$ operations applied to elements of W.

Subspaces of V include the trivial examples $W = (0)$ and $W = V$; all others are "proper" subspaces of V.

DEFINITION 1.23. Given a nonempty set S of vectors in V, its *linear span* $\langle S \rangle = \mathbb{K}\text{-span}(S)$ is the smallest subspace $W \subseteq V$ such that W contains S.

EXERCISE 1.24. If $\{W_\alpha : \alpha \in I\}$ is any family of subspaces in V, prove that their intersection $W = \bigcap_{\alpha \in I} W_\alpha$ is also a subspace.

Thus Definition 1.23 makes sense: Given S there is at least one subspace containing S, namely V. If E = intersection of all subspaces W that contain S, then E is a subspace and is obviously the smallest subspace containing S. Thus $\mathbb{K}$-span(S) exists, even if V is "infinite dimensional," for instance, if $V = \mathbb{K}[x]$.

This "top-down"definition has its uses, but an equivalent "bottom-up" version is often more informative.

LEMMA 1.25. *If $S \neq \varnothing$ in V, its linear span is the set of finite sums*

$$\left\{ \sum_{i=1}^{n} a_i v_i : a_i \in \mathbb{K},\ v_i \in S,\ n < \infty \right\}.$$

PROOF. Let $E = \{\sum_{i=1}^{n} c_i v_i : n < \infty,\ c_i \in \mathbb{K},\ v_i \in S\}$. Since $S \subseteq \mathbb{K}$-span(S), every finite sum lies in this span, proving $E \subseteq \mathbb{K}$-span(S). For $(\supseteq)$, it is clear that the family E of finite linear combinations is closed under $(+)$ and $(\cdot)$ operations because a linear combination of linear combinations is just one big linear combination of elements of S. It is a subspace of V, and it contains S because $1 \cdot s = s$ is a (trivial) linear combination. On the other hand, every subspace $W \supseteq S$ must contain all these linear sums, so $S \subseteq E \subseteq W$. Hence E is the smallest subspace containing S and $E = K$-span(S). □

EXERCISE 1.26. If $K = \mathbb{R}$ and $V = \mathbb{R}^3$ show that

$$W = \{x \in \mathbb{R}^3 : 3x_1 + 2x_2 - x_3 = 0\}$$

is a subspace and $W' = \{x \in \mathbb{R}^3 : 3x_1 + 2x_2 - x_3 = 1\}$ is not a subspace.

HINT: For one thing the zero vector $0 = (0,0,0)$ is not in W'. The situation is shown in Figure 1.3.

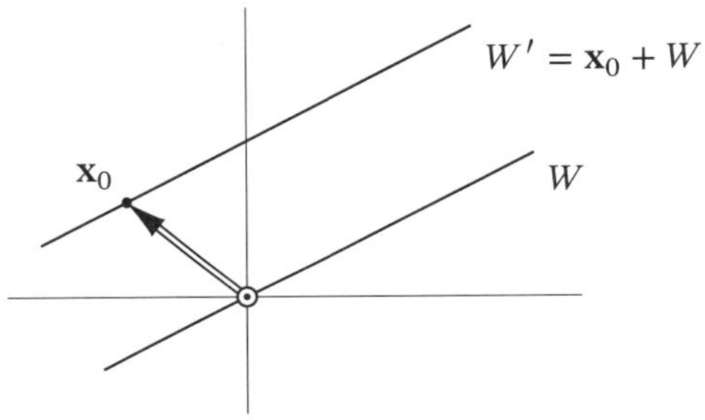

FIGURE 1.3. The subspace W in Exercise 1.26 and a translate $W' = \mathbf{x}_0 + W$ by some $\mathbf{x}_0 \in V$ such that $3x_1^0 + 2x_2^0 - x_3^0 = 1$, for instance, $\mathbf{x}_0 = (0,1,1)$. The set W' is not a subspace.

Systems of Linear Equations and Their Solutions. Systems of n linear equations in m unknowns are of two general types:

$$\text{Homogeneous}\begin{cases} a_{11}x_1 + \cdots + a_{1m}x_m = 0 \\ \qquad\vdots \\ a_{n1}x_1 + \cdots + a_{nm}x_m = 0 \end{cases}$$

$$\text{Inhomogeneous}\begin{cases} a_{11}x_1 + \cdots + a_{1m}x_m = b_1 \\ \qquad\vdots \\ a_{n1}x_1 + \cdots + a_{nm}x_m = b_n \end{cases}$$

with a_{ij} and b_k in $\mathbb{K}$.

EXERCISE 1.27. Verify that the solutions $\mathbf{x} = (x_1, \dots, x_m)$ of the homogeneous system form a vector subspace of $\mathbb{K}^m$. Explain why the solution set of an inhomogeneous system cannot be a vector subspace unless $\mathbf{b} = (b_1, \dots, b_n) = \mathbf{0}$ in $\mathbb{K}^n$.

If we regard vectors $\mathbf{x} = (x_1, \dots, x_m) \in \mathbb{K}^m$ as $m \times 1$ column matrices,

$$\mathbf{x} = \text{col}(x_1, \dots, x_m) = \begin{pmatrix} x_1 \\ \vdots \\ x_m \end{pmatrix},$$

you will recognize that the solutions $\mathbf{x} \in \mathbb{K}^m$ of the homogeneous system of equations are precisely the solutions of the matrix equation

$$A\mathbf{x} = \mathbf{0} \quad \text{where the zero vector is} \quad \mathbf{0} = \begin{pmatrix} 0 \\ \vdots \\ 0 \end{pmatrix}_{n\times 1},$$

and for inhomogeneous systems we must solve

$$A\mathbf{x} = \mathbf{b} \quad \text{where } \mathbf{b} = \begin{pmatrix} b_1 \\ \vdots \\ b_n \end{pmatrix}_{n\times 1} \in \mathbb{K}^n.$$

The homogeneous system always has the zero vector $\mathbf{0} \in \mathbb{K}^m$ as a solution, and the solution set $\{\mathbf{x} \in \mathbb{K}^m : A\mathbf{x} = \mathbf{0}\}$ is a vector subspace in $\mathbb{K}^m$. If $\mathbb{K} = \mathbb{R}$ or $\mathbb{C}$, then the number of solutions is either 1 or ∞ for this system. An inhomogeneous system might not have any solutions at all; otherwise, it has just one solution or infinitely many.

If A is an $n \times m$ matrix with entries in $\mathbb{K}$, we will find it useful to let A act by left multiplication as an operator $L_A : \mathbb{K}^m \to \mathbb{K}^n$ on column vectors

$$\mathbf{y} = L_A(\mathbf{x}) = A \cdot \mathbf{x} \quad (\text{an } (n \times m) \cdot (m \times 1) \text{ matrix product})$$

for $\mathbf{x} \in \mathbb{K}^m$. This is a *linear operator* in the sense that

$$L_A(\mathbf{x} + \mathbf{y}) = L_A(\mathbf{x}) + L_A(\mathbf{y}) \quad \text{and} \quad L_A(\lambda \cdot \mathbf{x}) = \lambda \cdot L_A(\mathbf{x})$$

for $\mathbf{x}, \mathbf{y} \in \mathbb{K}^m$ and $\lambda \in \mathbb{K}$. Solving a system of linear equations is then equivalent to finding solutions of $L_A(\mathbf{x}) = \mathbf{0}$ or $L_A(\mathbf{x}) = \mathbf{b}$ for $\mathbf{x} \in \mathbb{K}^m$. From this point of view, $A\mathbf{x} = \mathbf{b}$ has solutions if and only if $\mathbf{b}$ lies in the range $R(L_A) = \{A\mathbf{x} : \mathbf{x} \in \mathbb{K}^m\}$, which is a vector subspace in $\mathbb{K}^n$. If $\mathbf{b} = \mathbf{0}$ the "homogeneous" equation $A\mathbf{x} = \mathbf{0}$ always has the trivial solution $\mathbf{x} = \mathbf{0}$.

EXERCISE 1.28. If A is an $n \times m$ matrix and $L_A : \mathbb{K}^m \to \mathbb{K}^n$ is defined as above, verify the following:

(a) The *range* $R(L_A) = L_A(\mathbb{K}^m) = \{A \cdot \mathbf{x} : \mathbf{x} \in \mathbb{K}^m\}$ of the operator L_A is a vector subspace in $\mathbb{K}^n$.

(b) The *kernel* of L_A,

$$K(L_A) = \ker(L_A) = \{\mathbf{x} \in \mathbb{K}^m : L_A(\mathbf{x}) = A \cdot \mathbf{x} = \mathbf{0} \text{ in } \mathbb{K}^n\},$$

is a vector subspace in $\mathbb{K}^m$.

EXERCISE 1.29 (Inhomogeneous vs. Homogeneous Systems). Given a particular solution $\mathbf{x}_0$ of $A\mathbf{x} = \mathbf{b}$, the full solution set of this equation consists of the vectors $W_\mathbf{b} = \mathbf{x}_0 + W$, where $W = \{x \in \mathbb{K}^m : A\mathbf{x} = \mathbf{0}\}$ is the solution set for the homogeneous system $A \cdot \mathbf{x} = \mathbf{0}$. W is a vector subspace of $\mathbb{K}^m$ because

$A \cdot \mathbf{x}_1 = A \cdot \mathbf{x}_2 = \mathbf{0}$ implies $A \cdot (\mathbf{x}_1 + \mathbf{x}_2) = A \cdot \mathbf{x}_1 + A \cdot \mathbf{x}_2 = \mathbf{0} + \mathbf{0} = \mathbf{0}$ and $A(\lambda \cdot \mathbf{x}) = \lambda A \cdot \mathbf{x} = \lambda \cdot \mathbf{0} = \mathbf{0}$. But $\mathbf{x}_0 + W$ is not a subspace unless $\mathbf{x}_0 \in W$, because it does not contain $\mathbf{0}$.

NOTE: The converse is also true: in $\mathbb{K}^m$ every vector subspace is the solution set of some homogeneous system of linear equations $A\mathbf{x} = \mathbf{0}$, but we are not ready to prove that yet. The situation is shown in Figure 1.4.

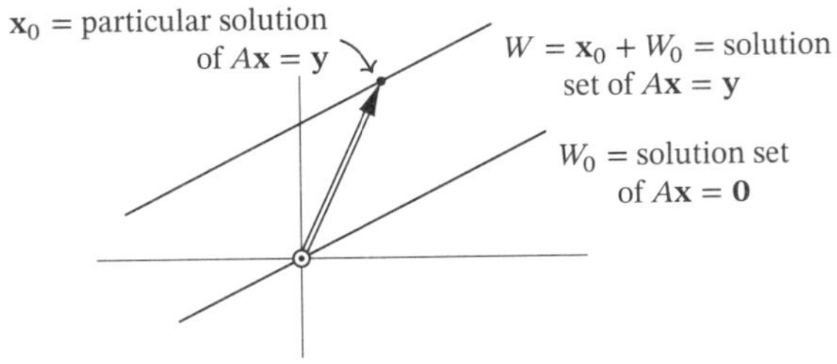

FIGURE 1.4. The subspace W_0 is the solution set for a homogeneous equation $A\mathbf{x} = \mathbf{0}$. If the inhomogeneous equation $A\mathbf{x} = \mathbf{b}$ has solutions and if $\mathbf{x}_0$ is a particular solution, so $A\mathbf{x}_0 = \mathbf{b}$, the full solution set $W = \{\mathbf{x} : A\mathbf{x} = \mathbf{b}\}$ is the translate $W' = \mathbf{x}_0 + W$ of W_0.

This of course presumes that $A\mathbf{x} = \mathbf{b}$ has any solutions at all; if it does not, we say that the system is *inconsistent*. Geometrically, that means $\mathbf{b}$ does not lie in the range $R(L_A)$. Here is an example of an inconsistent inhomogeneous system:

$$\begin{pmatrix} 1 & 0 \\ 2 & 0 \end{pmatrix} \mathbf{x} = \begin{pmatrix} 0 \\ 1 \end{pmatrix}.$$

The corresponding system of linear equations

$$\begin{cases} x_1 + 0 \cdot x_2 = 0, \\ 2x_1 + 0 \cdot x_2 = 1, \end{cases}$$

implies that $x_1 = 0$ and $2x_1 = 1$, an obvious impossibility.

We will continue discussion of linear systems and their solutions via elementary row operations on A, or on the $n \times (m+1)$ augmented matrix $[A : \mathbf{b}]$, where $\mathbf{b}$ regarded as an $n \times 1$ column matrix $\mathbf{b} = \text{col}(b_1, \dots, b_n)$. But first we provide a few more examples of vector spaces we will encounter from time to time.

EXAMPLE 1.30 (Sequence Space ℓ^∞). Let ℓ^∞ = all bounded sequences of complex numbers $a = (a_1, a_2, \dots)$ with $a + b = (a_1 + b_1, a_2 + b_2, \dots)$ and $\lambda \cdot a = (\lambda a_1, \lambda a_2, \dots)$ for $\lambda \in \mathbb{C}$. This infinite-dimensional space over $\mathbb{C}$ has the following subspaces:

1. W_0 = {sequences such that $a_n \to 0$ as $n \to \infty$}.
2. W_n = {all sequences of the form $(a_1, \dots, a_n, 0, 0, \dots)$}.
3. $\ell^1 = \{a : \sum_{n=1}^{\infty} |a_n| < \infty\}$.

M(n, ℝ) and Other Spaces of Matrices.

EXAMPLE 1.31. Matrix space $\mathrm{M}(n,\mathbb{K})$ contains a few significant subspaces.

1. SYMMETRIC MATRICES:

$$A^{\mathsf{T}} = A \quad \text{where } A^{\mathsf{T}} = (\text{transpose of } A) \text{ with } A^{\mathsf{T}}_{ij} = A_{ji}.$$

2. DIAGONAL MATRICES:

$$D = \begin{pmatrix} d_1 & & & 0 \\ & \cdot & & \\ & & \cdot & \\ & & & \cdot \\ 0 & & & d_n \end{pmatrix}$$

3. BLOCK DIAGONAL MATRICES:

$$D_{m_1,\dots,m_r} = \begin{pmatrix} \boxed{B_{m_1\times m_1}} & & & 0 \\ & \cdot & & \\ & & \cdot & \\ & & & \cdot \\ 0 & & & \boxed{B_{m_r\times m_r}} \end{pmatrix}$$

for fixed indices $m_1, \dots, m_r \geq 1$. (The square "blocks" are allowed to have arbitrary entries; all other entries are zero and $m_1 + \cdots + m_r = n$.)

4. UPPER TRIANGULAR and STRICTLY UPPER TRIANGULAR MATRICES:

$$\begin{pmatrix} * & & & * \\ & \cdot & & \\ & & \cdot & \\ & & & \cdot \\ 0 & & & * \end{pmatrix} \quad \text{and} \quad \begin{pmatrix} 0 & * & & * \\ & \cdot & * & \\ & & \cdot & * \\ & & & \cdot \quad * \\ 0 & & & 0 \end{pmatrix} \quad \text{with } * \text{ in } \mathbb{K}.$$

EXERCISE 1.32. Which of these four subspaces, if any, are closed under matrix multiplication as well as $(+)$?

EXERCISE 1.33. Show that the vector subspaces of upper triangular and strictly upper triangular matrices are closed under formation of the matrix product AB.

EXERCISE 1.34. Show that the vector subspaces of upper triangular (or of strictly upper triangular) matrices are *Lie algebras*: all *commutators* $[A,B] = AB - BA$ are (strictly) upper triangular if A, B are.

EXERCISE 1.35. If an $n \times n$ matrix A has the strictly upper triangular form shown in (a), prove that A^2 has the form in (b).

$$\text{(a) } A = \begin{pmatrix} 0 & * & & & * \\ & 0 & * & & * \\ & & \ddots & & \\ & & & 0 & * \\ 0 & & & & 0 \end{pmatrix} \qquad \text{(b) } A^2 = \begin{pmatrix} 0 & 0 & * & & & & * \\ & 0 & 0 & * & & & * \\ & & 0 & 0 & * & & * \\ & & & \ddots & \ddots & & \\ & & & & \ddots & \ddots & * \\ & & & & & 0 & 0 \\ 0 & & & & & & 0 \end{pmatrix}$$

NOTE: Further computations show that A^3 has three diagonal files of zeros so that A is a *nilpotent matrix*, with $A^n = 0_{n \times n}$.

1.3. Solving Matrix Equation $A\mathbf{x} = \mathbf{b}$

We have seen that solving a system of n linear equations in m unknowns is equivalent to solving the matrix equation $A\mathbf{x} = \mathbf{b}$ where A is the $n \times m$ array of coefficients in the system. Given such a system several questions arise:

1. *For which vectors* $\mathbf{b}$ *does the system have any solutions at all? This is equivalent to determining the range of the linear map* $L_A(\mathbf{x}) = A\mathbf{x}$ *from* $\mathbb{K}^m \to \mathbb{K}^n$.
2. *When the system* $A\mathbf{x} = \mathbf{b}$ *is consistent, so there are solutions, how can we find them all? What is the connection between the systems* $A\mathbf{x} = \mathbf{b}$ *and* $A\mathbf{x} = \mathbf{0}$?

Row operations on matrices are the main tool for resolving such questions. These can also be applied to solve the basic problems of interest in studying vector spaces.

3. *Given vectors* $\{u_1, \dots, u_r\} \subseteq V$ *and* $b \in V$, *decide whether there exist* $c_1, \dots, c_r \in \mathbb{K}$ *such that* $b = \sum_{i=1}^{r} c_i u_i$*—i.e., decide which* b *are in the linear span* $\mathbb{K}\text{-span}\{u_1, \dots, u_r\}$.
4. *Are the vectors* $u_1, \dots, u_r$ *independent, or is there some dependency between them that would allow us to write one of them as a linear combination of the others?*

This last question is equivalent to asking whether some nontrivial linear combination is zero:

$$4'. \quad c_1 u_1 + \cdots + c_r u_r = 0_V \quad (\text{not all } c_i = 0)$$

In fact, if $c_i \neq 0$ in $(4')$ there is a nontrivial dependency $u_i = \sum_{j \neq i} -(c_j/c_i)\, v_j$; the converse is trivial.

Questions 3 and 4, posed for an arbitrary vector space V, can often be reduced to computations in coordinate space, in which $V = \mathbb{K}^m$ and $\mathbf{b} \in \mathbb{K}^n$, so we will focus on this special case.

Elementary Operations and Echelon Form of $A\mathbf{x} = \mathbf{b}$. Before proceeding farther we recall some basic facts about solving matrix equations using elementary row operations. These are based on observations with which you should already be familiar.[1] The simple (but important) verifications are left as exercises.

PROPOSITION 1.36. *The following elementary operations on the rows $R_i(A)$ of an $n \times m$ matrix A do not change the set of solutions $\mathbf{x}$ of the homogeneous equation $A\mathbf{x} = \mathbf{0}$.*

1. $R_i \leftrightarrow R_j$: *switch two rows*
2. $R_i \to \lambda R_i$: *scale (Row i) by some $\lambda \neq 0$ in $\mathbb{K}$*
3. $R_i \to R_i + \lambda R_j$: *for $i \neq j$ add any scalar multiple of (Row j) to (Row i), leaving (Row j) unaltered.*

Furthermore, if row operations on a matrix A yield the matrix A', then:

4. *Each row $R_i(A')$ is a linear combination of the rows of A, viewed as vectors in $\mathbb{K}^m$.*
5. *The linear span of the rows of A' is the same as the linear span of the rows of A.*

The same is obviously true if A' results from a sequence of elementary row operations.

SKETCH OF PROOF. This follows because each row operation is reversible, with $R_i \to R_i - \lambda R_j$ the inverse of $R_i \to R_i + \lambda R_j$. Applied to the $n \times (m+1)$ "augmented matrix" $[A : \mathbf{b}]$ associated with an inhomogeneous system $A\mathbf{x} = \mathbf{b}$, the matrix equation $A'\mathbf{x} = \mathbf{b}'$ corresponding to the modified matrix $[A' : \mathbf{b}']$ has the same solution set as $A\mathbf{x} = \mathbf{b}$. □

Although row operations do not change the solution set they can greatly simplify the system of equations to be solved, or the task of determining the linear span of some set of vectors in $\mathbb{K}^m$, leading to easy systematic solution of matrix equations. For instance, when $\mathbb{K} = \mathbb{Q}, \mathbb{R}, \mathbb{C}$ it is always possible to find a sequence of row operations that put A into *echelon form* A',

$$\text{ECHELON FORM:}\quad A' = \left(\begin{array}{cccccccccc} \boxed{1} & * & \cdot & \cdot & \cdot & & & & & * \\ & & & \boxed{1} & * & \cdot & \cdot & \cdot & \cdot & * \\ 0 & & & & & & \boxed{1} & * & \cdot \; \cdot & * \\ \hline & & & & & \mathbf{0} & & & & \end{array}\right).$$

[1]See chapters 1–4 of *Linear Algebra*, by Seymour Lipschutz, Schaum's Outline Series, 1991.

The same moves put the augmented matrix $[A : \mathbf{b}]$ into a similar form:

$$
(1.1)\qquad [A' : \mathbf{b}'] = \left(\begin{array}{cccccccccc|c}
\boxed{1} & * & . & 0 & * & 0 & & & & * & b'_1 \\
 & & & \boxed{1} & * & 0 & . & . & . & * & \vdots \\
0 & & & & & \boxed{1} & * & . & . & * & b'_r \\
\hline
 & & & & & & & & & & b'_{r+1} \\
 & & & & & \mathbf{0} & & & & & \vdots \\
 & & & & & & & & & & b'_m
\end{array}\right).
$$

However, when we apply the row operations that put A in echelon form A' to the augmented matrix $[A : \mathbf{b}]$, the last column of $[A' : \mathbf{b}']$ need not terminate in a string of zeros.

Solutions of the systems $A'\mathbf{x} = \mathbf{0}$, $A'\mathbf{x} = \mathbf{b}'$ are quickly found by "backsolving" (illustrated below). One could go further, forcing A into even simpler form by knocking out all terms $*$ above the "step corners." These additional operations would of course affect $\mathbf{b}'$ in the augmented matrix yielding the *reduced echelon form*

$$
[A'' : \mathbf{b}''] = \left(\begin{array}{cccccccccc|c}
\boxed{1} & * & . & 0 & * & 0 & & & & * & b''_1 \\
 & & & \boxed{1} & * & 0 & . & . & . & * & \vdots \\
0 & & & & & \boxed{1} & * & . & . & * & b''_r \\
\hline
 & & & & & & & & & & b''_{r+1} \\
 & & & & & \mathbf{0} & & & & & \vdots \\
 & & & & & & & & & & b''_m
\end{array}\right).
$$

The step corners appearing in the echelon displays A' and A'' are often referred to as *pivots*, and the columns in which they occur are the *pivot columns*.

The systems $A\mathbf{x} = \mathbf{b}$ and $A'\mathbf{x} = \mathbf{b}'$ have the same solutions. But solutions of $A'\mathbf{x} = \mathbf{b}'$ exist if and only if $b'_{r+1} = \cdots = b'_n = 0$, because the rows lying below the last step corner in the echelon form $[A' : \mathbf{b}']$ correspond to equations

$$b'_{r+1} = 0, \dots, b'_n = 0,$$

which cannot be satisfied if any of the right-hand entries $b'_{r+1}, \dots, b'_m$ in 1.1 are nonzero. (The variables $x_1, \dots, x_m$ don't appear!) To summarize, we have established the following:

> CONSISTENCY CONDITION: *A matrix equation* $A\mathbf{x} = \mathbf{b}$ *has solutions if and only if the echelon form* $[A' : \mathbf{b}']$ *of the augmented matrix* $[A : \mathbf{b}]$ *has only rows of zeros below the last step corner.*

Once the consistency conditions $b'_{r+1} = \cdots = b'_n = 0$ have been met (without which there are no solutions at all), the columns $C_i(A')$ in the echelon form of A that DO NOT include a step corner correspond to "free variables" or "pivot variables" x_i in the solutions of the equation $A'\mathbf{x} = \mathbf{0}$; they are also free variables in solutions of $A'\mathbf{x} = \mathbf{b}'$. If $I = \{1 \le i_1 < \cdots < i_r \le m\}$ are the indices labeling the pivot columns, the REMAINING indices correspond to free variables

x_k $(k \notin I)$ in the solutions to $A\mathbf{x} = \mathbf{b}$. Once values have been specified for the free variables, backsolving yields the values of the remaining "dependent" variables x_k $(k \in I)$. We get a unique solution of $A'\mathbf{x} = \mathbf{b}'$ for *every* choice of the free variables $(k \notin I)$; different choices yield different solutions and all solutions are accounted for. By Proposition 1.36 these are also the solutions of the original equation $A\mathbf{x} = \mathbf{b}$. This concludes our review of row operations.

EXAMPLE 1.37 (A Case Study). Consider the vectors in $\mathbb{K}^3$

$$\mathbf{u}_1 = \begin{pmatrix}1\\2\\1\end{pmatrix},\ \mathbf{u}_2 = \begin{pmatrix}-2\\-4\\-2\end{pmatrix},\ \mathbf{u}_3 = \begin{pmatrix}0\\2\\3\end{pmatrix},\ \mathbf{u}_4 = \begin{pmatrix}2\\0\\-3\end{pmatrix},\ \mathbf{u}_5 = \begin{pmatrix}-3\\8\\16\end{pmatrix},$$

and let A be the matrix with these vectors as its columns

$$A = \begin{pmatrix}1 & -2 & 0 & 2 & -3\\ 2 & -4 & 2 & 0 & 8\\ 1 & -2 & 3 & -3 & 16\end{pmatrix}.$$

We will show how to solve $A\mathbf{x} = \mathbf{0}$ and $A\mathbf{x} = \mathbf{b}$.

DISCUSSION: By definition of row/column matrix multiplication, finding a solution $\mathbf{x} = \mathrm{col}(x_1, \dots, x_5) \in \mathbb{R}^5$ of the matrix equation $A\mathbf{x} = \mathbf{b}$ for an arbitrary vector $\mathbf{b} = \mathrm{col}(b_1, b_2, b_3)$ in $\mathbb{R}^3$ is the same as finding coefficients $x_1, \dots, x_5$ such that

$$\mathbf{b} = \sum_{i=1}^{5} x_i\mathbf{u}_i = x_1\begin{pmatrix}1\\2\\1\end{pmatrix} + x_2\begin{pmatrix}-2\\-4\\-2\end{pmatrix} + x_3\begin{pmatrix}0\\2\\3\end{pmatrix} + x_4\begin{pmatrix}2\\0\\-3\end{pmatrix} + x_5\begin{pmatrix}-3\\8\\16\end{pmatrix}$$

$$= \begin{pmatrix}1 & -2 & 0 & 2 & -3\\ 2 & -4 & 2 & 0 & 8\\ 1 & -2 & 3 & -3 & 16\end{pmatrix}\begin{pmatrix}x_1\\x_2\\x_3\\x_4\\x_5\end{pmatrix} = A\mathbf{x}.$$

Clearly, $A\mathbf{x} = \mathbf{b}$ has solutions if and only if $\mathbf{b}$ is a linear combination of the columns of A. A similar calculation provides a solution to $A\mathbf{x} = \mathbf{b}$ for any finite set of vectors $\mathbf{u}_1, \dots, \mathbf{u}_m$ in coordinate space $\mathbb{K}^n$. Viewing the $\mathbf{u}_i$ as $n \times 1$ column vectors we may assemble them into an $n \times m$ matrix $A = [\mathbf{u}_1; \dots; \mathbf{u}_n]$ and apply the following general result, whose proof is a simple variant of the preceding discussion.

LEMMA 1.38. *If A is any $n \times m$ matrix and $\mathbf{b} \in \mathbb{K}^n$, the system $A\mathbf{x} = \mathbf{b}$ has solutions if and only if $\mathbf{b}$ is in the linear span of the columns $\mathbf{u}_1, \dots, \mathbf{u}_m$ of A, regarded as vectors in $\mathbb{K}^n$.*

In particular, a system $A\mathbf{x} = \mathbf{b}$ is consistent precisely when $\mathbf{b}$ belongs to the column space $\mathrm{ColSpace}(A)$, so it is in the range of $L_A : \mathbb{K}^m \to \mathbb{K}^n$.

EXERCISE 1.39. Fill in the details needed to verify the claim in Lemma 1.38.

Continuing the discussion of Example 1.37, we now allow arbitrary $\mathbf{b} = (b_1, b_2, b_3) \in \mathbb{K}^3$ and will determine whether the system $A\mathbf{x} = \mathbf{b}$ has solutions $\mathbf{x} = \mathrm{col}(x_1, \dots, x_5)$, and when it does we will find them all. By allowing undetermined coefficients in $\mathbf{b}$ we will also be determining the linear span of $\mathbf{u}_1, \dots, \mathbf{u}_5$, which is just the set of $\mathbf{b}$ for which the system $A\mathbf{x} = \mathbf{b}$ is consistent.

The Homogeneous Equation $A\mathbf{x} = \mathbf{0}$. Start by putting the system $A\mathbf{x} = \mathbf{0}$ into echelon form by applying row operations to the augmented matrix

$$[A : \mathbf{0}] = \left(\begin{array}{ccccc|c} 1 & -2 & 0 & 2 & -3 & 0 \\ 2 & -4 & 2 & 0 & 8 & 0 \\ 1 & -2 & 3 & -3 & 16 & 0 \end{array}\right).$$

By applying $R_2 \to R_2 - 2R_1$ and $R_3 \to R_3 - R_1$, this becomes

$$\left(\begin{array}{ccccc|c} \boxed{1} & -2 & 0 & 2 & -3 & 0 \\ 0 & 0 & 2 & -4 & 14 & 0 \\ 0 & 0 & 3 & -5 & 19 & 0 \end{array}\right).$$

Now apply $R_3 \to R_3 - \frac{3}{2}R_2$, $R_2 \to \frac{1}{2}R_2$, and then $R_3 \to R_3 - 3R_2$ to get

$$[A' : \mathbf{0}] = \left(\begin{array}{ccccc|c} \boxed{1} & -2 & 0 & 2 & -3 & 0 \\ 0 & 0 & \boxed{1} & -2 & 7 & 0 \\ 0 & 0 & 0 & \boxed{1} & -2 & 0 \end{array}\right).$$

This is the desired echelon form. The system is consistent since there are no nonzero rows below the last step corner. Some additional work, needless for most purposes, would yield the *reduced echelon form*,

$$A'' = \left(\begin{array}{ccccc|c} \boxed{1} & -2 & 0 & 0 & 1 & 0 \\ 0 & 0 & \boxed{1} & 0 & 3 & 0 \\ 0 & 0 & 0 & \boxed{1} & -2 & 0 \end{array}\right).$$

By recursively backsolving the corresponding system of linear equations $A'\mathbf{x} = \mathbf{0}$ we find that

1. x_2, x_5 are free variables;
2. $x_4 - 2x_5 = 0 \Rightarrow x_4 = 2x_5$;
3. $x_3 - 2x_4 + 7x_5 = 0 \Rightarrow x_3 = -7x_5 + 2(2x_5) = -3x_5$;
4. $x_1 - 2x_2 + 2x_4 - 3x_5 = 0 \Rightarrow x_1 = 2x_2 - 2(2x_5) + 3x_5 = 2x_2 - x_5$.

The solutions of $A'\mathbf{x} = \mathbf{0}$ (which are also the solutions of $A\mathbf{x} = \mathbf{0}$) form a vector subspace in $\mathbb{R}^5$, each of whose points is uniquely labeled (parametrized) by the choice of the free variables x_2, x_5. Relabeling $x_2 = s$, $x_5 = t$ ($s, t \in \mathbb{K}$) we find that the solution set $W = \{\mathbf{x} \in \mathbb{K}^5 : A\mathbf{x} = \mathbf{0}\} = \{\mathbf{x} \in \mathbb{K}^5 : A'\mathbf{x} = \mathbf{0}\}$ is equal to

$$W = \left\{ \begin{pmatrix} 2x_2 - x_5 \\ x_2 \\ -3x_5 \\ 2x_5 \\ x_5 \end{pmatrix} : x_2, x_5 \in \mathbb{K} \right\} = \left\{ \begin{pmatrix} 2s - t \\ s \\ -3t \\ 2t \\ t \end{pmatrix} : s, t \in \mathbb{K} \right\}.$$

These solutions can be rewritten in a more instructive form

$$\mathbf{x} = \begin{pmatrix} 2s-t \\ s \\ -3t \\ 2t \\ t \end{pmatrix} = s\begin{pmatrix} 2 \\ 1 \\ 0 \\ 0 \\ 0 \end{pmatrix} + t\begin{pmatrix} -1 \\ 0 \\ -3 \\ 2 \\ 1 \end{pmatrix} = s\mathbf{w}_1 + t\mathbf{w}_2 \quad (s, t \in \mathbb{K}),$$

which shows that every solution of $A\mathbf{x} = \mathbf{0}$ is a linear combination of two basic solutions

$$\mathbf{w}_1 = \begin{pmatrix} 2 \\ 1 \\ 0 \\ 0 \\ 0 \end{pmatrix} \quad \text{and} \quad \mathbf{w}_2 = \begin{pmatrix} -1 \\ 0 \\ -3 \\ 2 \\ 1 \end{pmatrix}. \tag{1.2}$$

Thus the solution set of $A\mathbf{x} = \mathbf{0}$ is the linear span $W = \mathbb{K}\text{-span}\{\mathbf{w}_1, \mathbf{w}_2\}$ of a set of generators $\{\mathbf{w}_1, \mathbf{w}_2\}$. We will later observe that these vectors are a "basis" for the solution set W.

Solving Inhomogeneous Equations $A\mathbf{x} = \mathbf{b}$. The same elementary row operations that put A into echelon form may be applied to the augmented matrix $[A : \mathbf{b}]$ where $\mathbf{b} = \text{col}(b_1, b_2, b_3)$ has indeterminate coefficients. We already know what happens to A; applying the same moves to $\mathbf{b}$, the operations $R_2 \to R_2 - 2R_1$ and $R_3 \to R_3 - R_1$ transform

$$\mathbf{b} = \begin{pmatrix} b_1 \\ b_2 \\ b_3 \end{pmatrix} \to \begin{pmatrix} b_1 \\ b_2 - 2b_1 \\ b_3 - b_1 \end{pmatrix}.$$

Then $R_3 \to R_3 - \frac{3}{2}R_2$, $R_2 \to \frac{1}{2}R_2$, and $R_3 \to R_3 - 3R_2$ yield

$$\to \begin{pmatrix} b_1 \\ \frac{1}{2}b_2 - b_1 \\ b_3 - b_1 - \frac{3}{2}(b_2 - 2b_1) \end{pmatrix} = \begin{pmatrix} b_1 \\ \frac{1}{2}b_2 - b_1 \\ b_3 - \frac{3}{2}b_2 + 2b_1 \end{pmatrix}$$

so the augmented matrix becomes

$$[A : \mathbf{b}] \to [A' : \mathbf{b}'] = \left(\begin{array}{ccccc|c} \boxed{1} & -2 & 0 & 2 & -3 & b_1 \\ 0 & 0 & \boxed{1} & -2 & 7 & \frac{1}{2}b_2 - b_1 \\ 0 & 0 & 0 & \boxed{1} & -2 & b_3 - \frac{3}{2}b_2 + 2b_1 \end{array}\right). \tag{1.3}$$

Again, x_2, x_5 are free variables. The general solution $\mathbf{x} = \text{col}(x_1, x_2, x_3, x_4, x_5)$ of $A\mathbf{x} = \mathbf{b}$ can be found by backsolving. We already know the general solutions of $A\mathbf{x} = \mathbf{0}$, so all we need is one particular solution $\mathbf{x_b}$, and the simplest way to find one is to set the free variables $x_2 = x_5 = 0$ and backsolve to get

$$x_2, x_5 = 0, \qquad x_4 = b_3 - \frac{3}{2}b_2 + 2b_1,$$

$$
\begin{aligned}
x_3 - 2x_4 = \frac{1}{2}b_2 - b_1 \Rightarrow x_3 &= 2\left(b_3 - \frac{3}{2}b_2 + 2b_1\right) + \frac{1}{2}b_2 - b_1 \\
&= 2b_3 - \frac{5}{2}b_2 + 3b_1,
\end{aligned}
$$

$$
x_1 - 2 \cdot 0 + 0 + 2x_4 + 0 = b_1 \Rightarrow
$$

$$
x_1 = b_1 - 2x_4 = -2\left(b_3 - \frac{3}{2}b_2 + 2b_1\right) + b_1 = -2b_3 + 3b_2 - 3b_1.
$$

So

$$
\mathbf{x_b} = \mathrm{col}\left(-2b_3 + 3b_2 - 3b_1, 0, 2b_3 - \frac{5}{2}b_2 + 3b_1, b_3 - \frac{3}{2}b_2 + 2b_1, 0\right)
$$

is a particular solution and the full solution set is

$$
\begin{aligned}
W_\mathbf{b} = \{\mathbf{x} : A\mathbf{x} = \mathbf{b}\} &= \begin{pmatrix} -2b_3 + 3b_2 - 3b_1 \\ 0 \\ 2b_3 - \frac{5}{2}b_2 + 3b_1 \\ b_3 - \frac{3}{2}b_2 + 2b_1 \\ 0 \end{pmatrix} + \mathbb{K}\mathbf{w}_1 + \mathbb{K}\mathbf{w}_2 \\
&= \mathbf{x_b} + \mathbb{K}\mathbf{w}_1 + \mathbb{K}\mathbf{w}_2
\end{aligned}
$$

where $\mathbf{w}_1$ and $\mathbf{w}_2$ are the basis vectors for the space $W = \{x : A\mathbf{x} = \mathbf{0}\}$ of homogeneous solutions determined previously. Writing $x_2 = s$, $x_5 = t$ for the variable attached to $\mathbf{w}_1, \mathbf{w}_2$ we have found a *parametric description* of the solution set, in which each point in $W_\mathbf{b}$ is tagged by a unique pair (s, t) in the parameter space $\mathbb{K}^2$.

If $\mathbf{b} = \mathrm{col}(2, 6, 8)$, a particular solution is $\mathbf{x_b} = \mathrm{col}(-4, 0, 7, 3, 0)$, and the full solution set for $A\mathbf{x} = \mathbf{b}$ is

$$
W_\mathbf{b} = \begin{pmatrix} -4 + 2s - t \\ s \\ 7 - 3t \\ 3 + 2t \\ t \end{pmatrix} = \mathbf{x_b} + \mathbb{K}\mathbf{w}_1 + \mathbb{K}\mathbf{w}_2.
$$

Inspection of the echelon form (1.3) reveals that there is no obstruction to backsolving to get valid solutions of $A'\mathbf{x} = \mathbf{b}'$ (and $A\mathbf{x} = \mathbf{b}$) no matter which $\mathbf{b} = (b_1, b_2, b_3) \in \mathbb{K}^3$ we consider. In fact, there will be a two-parameter family of solutions for any given $\mathbf{b}$ because two columns in A' fail to meet a step corner.

That concludes the present discussion. We will have more to say about these issues in Section 1.4.

Determining the Linear Span of a Set of Vectors. The linear span $E = \mathbb{K}\text{-span}(S)$ of a set of m column vectors $S = \mathbf{u}_1, \dots, \mathbf{u}_m$ in $\mathbb{K}^n$ consists of all finite linear combinations $\sum_{i=1}^m c_i\mathbf{u}_i$. In Lemma 1.38 we showed that E is precisely the set of vectors $\mathbf{b} \in \mathbb{K}^n$ such that the linear system $A\mathbf{x} = \mathbf{b}$ has solutions, where A is the $n \times m$ matrix $A = [\mathbf{u}_1; \dots; \mathbf{u}_m]$.

Implicit and Parametric Descriptions of Vector Subspaces. Although this characterization is useful in theoretical discussions, one would also like to have more explicit descriptions of the linear span E. There are two approaches.

1. IMPLICIT DESCRIPTION. Describe the vectors in E as the solution set of some new homogeneous system of linear equations $C\mathbf{x} = \mathbf{0}$ where $\mathbf{x} \in \mathbb{K}^n$. This tells us how the linear span is carved out of coordinate space $\mathbb{K}^n$ in which the given vectors $\mathbf{u}_i$ live.
2. PARAMETRIC DESCRIPTION. Determine a *basis* for E, a set of vectors $\mathbf{b}_1, \dots, \mathbf{b}_r$ such that every $\mathbf{b} \in E$ is a unique linear combination of these vectors.

In Section 1.4 we will show how such descriptions can be obtained, but for the moment we simply show how basis vectors for E can be found using row operations.

EXAMPLE 1.40. Given n vectors $\mathbf{v}_1, \dots, \mathbf{v}_n$ in $\mathbb{K}^m$ we can find a set of basis vectors for their linear span W using row operations.

DISCUSSION: Let us regard the $\mathbf{v}_i$ as the *rows* of an $n \times m$ matrix

$$B = \begin{pmatrix} \mathbf{v}_1 \\ \vdots \\ \mathbf{v}_n \end{pmatrix}_{m \times n} .$$

Now $E = \mathbb{K}\text{-span}\{\mathbf{v}_1, \dots, \mathbf{v_n}\}$ coincides with the row space RowSpace(B). Applying row operations to put B into echelon form

$$B' = \left(\begin{array}{cccccccccc} \boxed{1} & * & . & . & . & & & & & * \\ & & & \boxed{1} & * & . & . & . & . & * \\ 0 & & & & & \boxed{1} & * & . & . & * \\ \hline & & & & & & & & & \\ & & & & \mathbf{0} & & & & & \end{array}\right),$$

we know that the span of the rows is the same for B and B', and in particular the *nonzero* rows $R_1(B'), \dots, R_r(B')$ of B' span E. They also have the desired independence property. In fact, if some vector $\mathbf{v} \in E$ had two different expansions

$$\mathbf{v} = \sum_{i=1}^{r} a_i R_i(B') = \sum_{i=1}^{r} b_i R_i(B'),$$

their difference, the zero vector $\mathbf{0} = \mathbf{v} - \mathbf{v}$, would have a nontrivial expansion

$$\mathbf{0} = \sum_{i=1}^{r} (a_i - b_i) R_i(B') \quad \text{(not all coefficients} = 0).$$

The echelon form of B' prevents this from happening. In fact, if any coefficient in this sum is nonzero, let $R_i(B')$ be the first row vector $(1 \le i \le r)$ for which this happens. The previous rows $(j < i)$ cannot undo this because they all get multiplied by zero, and the later rows $(j > i)$ can't either because they all have zeros in the column that contains the step corner of $R_i(B')$. Therefore we have $a_i = b_i$ in the preceding sum and the coefficients are unique.

More on Elementary Row and Column Operations. Elementary row operations on a matrix A do not change the linear span RowSpace(A) of its rows. Note carefully, however, that row operations will mess up column space!

As for columns, there is an obvious family of *elementary column operations* on any matrix A.

1. $C_i \leftrightarrow C_j$.
2. $C_i \to \lambda C_i$ for $\lambda \neq 0$ in $\mathbb{K}$.
3. $C_i \to C_i + \lambda C_j$ for $i \neq j$ where λ is any element in $\mathbb{K}$.

Column operations on A have properties completely analogous to those listed for row operations in Proposition 1.36. In particular, column operations preserve the linear span of the columns of A. We won't spell all this out since we will work mostly with row operations. The properties of column operations can of course be verified by direct calculation, but they also follow by observing that row and column operations are related via the transpose operation $A \to A^{\mathrm{T}}$on matrices given by

$$(A^{\mathrm{T}})_{ij} = A_{ji}$$

(see Figure 1.5). Note that $(A^{\mathrm{T}})^{\mathrm{T}} = A$.

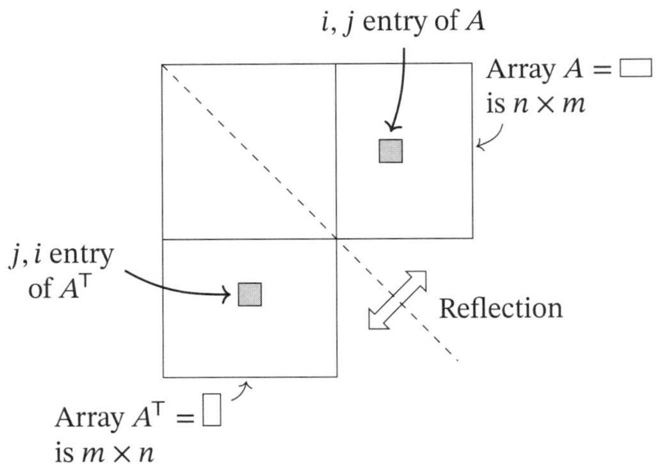

FIGURE 1.5. A matrix A and its transpose A^{T} are related by a reflection that sends rows in A to columns in A^{T}, and columns to rows.

Obviously transposition takes rows of A to columns of A^{T} and vice versa. Furthermore, elementary row operations on A become the corresponding elementary operations on columns of A^{T}, so that ColSpace(A) = RowSpace(A^{T}). Invariance of ColSpace(A) under column operations follows from invariance of RowSpace(A^{T}) under row operations.

EXERCISE 1.41. Verify that the transpose operation $A \to A^{\mathsf{T}}$ on matrices of any shape has the following algebraic properties:

(a) $(A^{\mathsf{T}})^{\mathsf{T}} = A$.
(b) If A is square and I is the $n \times n$ identity matrix, then $I^{\mathsf{T}} = I$.
(c) $(c \cdot A)^{\mathsf{T}} = c \cdot A^{\mathsf{T}}$ for any scalar $c \in \mathbb{K}$.
(d) $(A + B)^{\mathsf{T}} = A^{\mathsf{T}} + B^{\mathsf{T}}$.

If A is $m \times k$ and B is $k \times n$ they have compatible shapes and their matrix product AB is a well-defined $m \times n$ matrix.

(e) Prove that $(AB)^{\mathsf{T}} = B^{\mathsf{T}} \cdot A^{\mathsf{T}}$ for compatible matrices.

Note the reversal of order in (e).

EXERCISE 1.42. If A is an $n \times m$ matrix, consider an elementary row operation T_R, and let T_C be the corresponding column operation on $m \times n$ matrices. (For instance, $T_R : R_i \leftrightarrow R_j$ in A corresponds to $T_C : C_i \leftrightarrow C_j$ in A^{T}.) Verify that

$$\left(T_R(A)\right)^{\mathsf{T}} = T_C(A^{\mathsf{T}}) \quad \text{or equivalently} \quad \left(T_C(A^{\mathsf{T}})\right)^{\mathsf{T}} = T_R(A)$$

for all $n \times m$ matrices A.

EXERCISE 1.43. Use Exercise 1.42 to verify the following:

LEMMA 1.44. *If a row operation T_R on $\mathrm{M}(n, \mathbb{K})$ corresponds to left multiplication $T_R : A \to EA$ by the elementary matrix E, the corresponding column operation is $T_C : A \to AE^{\mathsf{T}}$.*

1.4. Linear Span, Independence, and Bases

DEFINITION 1.45. A set of vectors $S = \{v_1, \dots, v_r\}$ in a vector space V *spans* a subspace W if

$$W = \mathbb{K}\text{-span}\{S\} = \left\{\sum_{i=1}^{r} c_i v_i \ : \ c_i \in \mathbb{K}\right\}.$$

The vectors are *linearly independent* if the only linear combination $\sum_i c_i v_i = \mathbf{0}$ adding up to zero in V is the trivial combination with $c_1 = \cdots = c_r = 0$. The vectors are a *basis* for W if they span W and are independent, so every $w \in W$ has a unique representation as $\sum_{i=1}^{n} \lambda_i v_i$ ($\lambda_i \in \mathbb{K}$).

EXERCISE 1.46. If $\mathfrak{X} = \{v_1, \dots, v_n\}$ span V and are independent, explain why every $v \in V$ has a unique representation as $\sum_{i=1}^{n} \lambda_i v_i$ ($\lambda_i \in K$), so $\mathfrak{X}$ is a basis for V.

Existence and Construction of Bases. The next result exhibits two ways to construct a basis in a vector space. One starts with a spanning set and then "prunes" it, deleting redundant vectors until we arrive at an independent subset with the same span as the original vectors. This yields a basis for V. The other constructs a basis recursively by adjoining "outside vectors" to an initial family of independent vectors in V. The initial family might consist of a single nonzero vector (obviously an independent set).

PROPOSITION 1.47. *Every finite set $\{v_1, \dots, v_n\}$ that spans a vector space V can be made into a basis by deleting suitably chosen entries from the list.*

PROOF. We argue by induction on $n = \#$(vectors in list). There is nothing to prove if $n = 1$; then $V = \mathbb{K} \cdot v_1$ and $\{v_1\}$ is already a basis. The induction hypotheses (one for each index $n = 1, 2, \dots$) are:

> HYPOTHESIS $P(n)$: *If a set of n vectors spans a vector space V, we can delete vectors from the list to get a basis for V.*

We have proved this for $n = 1$. It will be true for all $n \in \mathbb{N}$ if we can prove $P(n+1)$ is true, using only the information that $P(n)$ is true—i.e., if we can verify that

$$P(n) \text{ true} \Rightarrow P(n+1) \text{ true} \qquad \text{for all } n = 1, 2, \dots .$$

(Remember: This is a *conditional* statement owing to the presence of the word "*if*". By itself, it does not assert that either statement $P(n)$ or $P(n+1)$ is actually true.)

So, assuming $P(n)$ true, consider a spanning set $\mathfrak{X} = \{v_1, \dots, v_n, v_{n+1}\}$ in V. If these vectors are already independent (which could be checked using row operations if $V = \mathbb{K}^m$), we already have a basis for V without deleting any vectors. If $\mathfrak{X}$ is not independent there must be coefficients $c_1, \dots, c_{n+1} \in \mathbb{K}$ (not all equal to 0) such that $\sum_{i=1}^{n+1} c_i v_i = \mathbf{0}$. Relabeling, we may assume $c_{n+1} \neq 0$, and then ($\mathbb{K}$ being a field)

$$-c_{n+1}v_{n+1} = \sum_{i=1}^{n} c_i v_i \quad \text{and} \quad v_{n+1} = \sum_{i=1}^{n} -(c_i/c_{n+1}) \cdot v_i.$$

Thus $v_{n+1} \in \mathbb{K}\text{-span}\{v_1, \dots, v_n\}$ and $\mathbb{K}\text{-span}\{v_1, \dots, v_{n+1}\} = \mathbb{K}\text{-span}\{v_1, \dots, v_n\}$ is all of V. By the induction hypotheses we may thin out $\{v_1, \dots, v_{n+1}\}$ to get a basis for V. □

PROPOSITION 1.48. *If $\{v_1, \dots, v_n\}$ are independent in a vector space V, and v_{n+1} is a vector not in $W_0 = \mathbb{K}\text{-span}\{v_1, \dots, v_n\}$, then:*

1. *$\{v_1, \dots, v_n, v_{n+1}\}$ are independent.*
2. *$W_0 \subsetneqq W_1 = \mathbb{K}\text{-span}\{v_1, \dots, v_n, v_{n+1}\}$.*
3. *$\{v_1, \dots, v_n, v_{n+1}\}$ is a basis for W_1.*

PROOF. If $v_1 \dots, v_{n+1}$ are not independent there would be $c_i \in \mathbb{K}$ (not all zero) such that $\sum_{i=1}^{n+1} c_i v_i = \mathbf{0}$. We can't have $c_{n+1} = 0$; otherwise $\sum_{i=1}^{n} c_i v_i = \mathbf{0}$ contrary to the assumed independence of $\{v_1, \dots, v_n\}$. Thus

$$v_{n+1} = \sum_{i=1}^{n} -(c_i/c_{n+1}) \cdot v_i$$

is in W_0, which contradicts the assumption $v_{n+1} \notin W_0$. We then conclude that $v_1, \dots, v_{n+1}$ are independent. It follows immediately that $\{v_1, \dots, v_{n+1}\}$ is a basis for $W_1 = \mathbb{K}\text{-span}\{v_1, \dots, v_{n+1}\}$.

NOTE: This is an example of a "proof by contradiction," in which the assumption that $v_1, \dots, v_n$ are *not* independent leads to an impossible conclusion. Thus the statement "$v_1, \dots, v_n$ are independent" must be true. □

IMPORTANT REMARK: This process of "adjoining an outside vector" can be iterated to construct larger and larger independent sets and subspaces

$$W_0 = \mathbb{K}\text{-span}\{v_1, \dots, v_n\} \subsetneqq W_1 = \mathbb{K}\text{-span}\{v_1, \dots, v_n, v_{n+1}\}$$
$$\subsetneqq \cdots \subsetneqq W_r = \mathbb{K}\text{-span}\{v_1, \dots, v_{n+r}\}.$$

Since $\{v_1, \dots, v_n\}$ are independent, they are a basis for the initial space W_0, and by Proposition 1.48, $v_1, \dots, v_n, \dots, v_{n+r}$ will be a basis for W_r. If this process stops in finitely many steps (because $W_r = V$ and we can no longer find a vector outside W_r), we have produced a basis for V. If the process never stops, no finite subset of vectors can span V and in this case we say V is *infinite dimensional.* To begin the process we need an initial set of independent vectors, but if $V \neq (0)$ we could start with any $v_1 \neq 0$ and $W_0 = \mathbb{K} \cdot v_1$. Then apply Proposition 1.48 recursively as above.

DEFINITION 1.49. A vector space V is *finite dimensional* if there is a finite set of vectors $S = \{v_1, \dots, v_n\}$ that span V. Otherwise V is *infinite dimensional* which we indicate by writing $\dim(V) = \infty$.

Coordinate space $\mathbb{K}^n$ and matrix spaces $\mathrm{M}(m \times n, \mathbb{K})$ are finite dimensional; the spaces of polynomials $\mathbb{K}[x]$ and $\mathbb{K}[x_1, \dots, x_n]$ are infinite dimensional.

EXAMPLE 1.50. Coordinate space $\mathbb{K}^n$ is finite dimensional and is spanned by the *standard basis vectors* $\mathfrak{X} = \{\mathbf{e}_1, \dots, \mathbf{e}_n\}$

$$\mathbf{e}_1 = (1, 0, \dots, 0), \quad \mathbf{e}_2 = (0, 1, 0, \dots, 0), \quad \dots \quad , \mathbf{e}_n = (0, \dots, 0, 1).$$

In fact, $\mathfrak{X}$ is a basis for $\mathbb{K}^n$.

EXAMPLE 1.51. Polynomial space $\mathbb{K}[x]$ is infinite dimensional. Given any finite set of nonzero vectors $\mathfrak{X} = \{f_1, \dots, f_r\}$, let $d_i = \deg(f_i)$. All coefficients of terms $c_j x^j$ in f_i are zero if $j > N = \max\{d_1, \dots, d_r\}$, and the same is true for all linear combinations $\sum_{i=1}^r c_i f_i$. But then $\mathfrak{X}$ cannot span $\mathbb{K}[x]$ because x^{N+1} is not in $\mathbb{K}\text{-span}\{f_1, \dots, f_r\}$.

However, the vectors $f_0 = 1, f_1 = x, f_2 = x^2, \dots$ *are* a basis for $\mathbb{K}[x]$. This (infinite) set of vectors clearly spans $\mathbb{K}[x]$, but it is also independent, for if $\sum_{i=0}^r c_i f_i = 0$ that means $c_0 + c_1 x + \cdots + c_r x^r = 0$ as a polynomial, so the symbol string $(c_0, \dots, c_r, 0, 0, \dots)$ is equal to $(0, 0, 0, \dots)$.

LEMMA 1.52. *Every finite-dimensional vector space has a basis.*

PROOF. If $\{v_1, \dots, v_r\}$ span V, then by Proposition 1.47 we may delete some of the vectors to get an independent set with the same linear span. □

PROPOSITION 1.53. *If $S \subseteq V$ is an independent set of vectors in a finite-dimensional vector space V, and T a finite set of vectors that span V, we can adjoin certain vectors from T to S to get a basis for V that contains the original set of independent vectors S.*

PROOF. Let $W = \mathbb{K}\text{-span}\{S\}$. If $W = V$, S is already a basis. If $W \neq V$, there exists some $v_1 \in T$ such that $v_1 \notin W$ and then $S \cup \{v_1\}$ is an independent set, a basis for the larger space $W_1 = \mathbb{K}\text{-span}\{S \cup \{v_1\}\} \supsetneqq W$. Continuing, we get vectors $v_1, \dots, v_r$ in T such that $W \subsetneqq W_1 \subsetneqq W_2 \subsetneqq \cdots \subsetneqq W_r$ for $0 \leq i \leq r$, where $W_i = \mathbb{K}\text{-span}\{v_1, \dots, v_i\}$. The process must terminate when no vector $v_{r+1} \in T$ can be found outside of W_r. Then $T \subseteq W_r$, so $\mathbb{K}\text{-span}\{T\} = V \subseteq W_r$ and $W_r = V$. Therefore $S \cup \{v_1, \dots, v_r\}$ is a basis for $V = W_r$ (and $S \cup \{v_1, \dots, v_k\}$ is a basis for W_k for each $1 \leq k \leq r$). □

The Dimension $\dim(V)$ of a Vector Space.

THEOREM 1.54. *All bases in a finite-dimensional vector space have the same cardinality. More generally, if V is finite dimensional, and S is a finite spanning set (with $|S| = n$), every independent set of vectors $L \subseteq V$ has cardinality $|L| \leq |S|$. In other words, the size of any independent set is always less than or equal to that of any spanning set.*

PROOF. We can eliminate vectors from S to get an independent spanning set $S' \subseteq S$, which is then a basis for V, and will show that $|L| \leq |S'| \leq |S|$. Let $S' = \{u_1, \dots, u_n\}$ and $L = \{v_1, \dots, v_m\}$. Every $v_i \in L$ can be written $v_i = \sum_{i=1}^{n} a_{ji} u_j$ since the $u_i \in S'$ are a basis for V. On the other hand, if $c_1, \dots, c_m$ are scalars such that $0 = \sum_{j=1}^{m} c_j v_j$, we must have $c_1 = \cdots = c_m = 0$ because the v_j are independent. But the identity $\sum_{j=1}^{m} c_j v_j = 0$ can be written another way, as

$$0 = \sum_{i=1}^{m} c_i \Big(\sum_{j=1}^{n} a_{ji} u_j \Big) = \sum_{j=1}^{n} \Big(\sum_{i=1}^{m} a_{ji} c_i \Big) u_j .$$

Since the $u_j \in S'$ span V and are independent, each sum $(\cdots)$ is equal to 0 so the coefficients $c_1, \dots, c_m$ satisfy the system of n equations in m unknowns

$$\sum_{i=1}^{m} a_{ji} c_i = 0 \quad \text{for } 1 \leq j \leq n. \tag{1.4}$$

Hence the $m \times 1$ matrix $\mathbf{c} = \text{col}(c_1, \dots, c_m)$ is a solution of the matrix equation $A\mathbf{c} = \mathbf{0}$.

A linear system such as (1.4) always has nontrivial solutions if the number of unknowns $m = |L|$ exceeds the number of equations $n = |S'|$, and from this it will follow that $|L| \leq |S'|$, as claimed. In fact, row operations on the coefficient matrix A yield an echelon form A' shown below. There are just n columns in A

and A', hence at most n step corners in A', and if $m > n$ there must be at least one column that fails to meet a step corner:

Echelon form when $m > n$:

$$A' = \begin{pmatrix} \boxed{1} & * & \cdot & \cdot & \cdot & & & & * \\ & & & \cdot & \cdot & \cdot & \cdot & \cdot & \cdot \\ & & & \cdot & \cdot & \cdot & \cdot & \cdot & \cdot \\ & & & \boxed{1} & * & \cdot & \cdot & \cdot & * \\ 0 & & & & & \boxed{1} & * & \cdot\ \cdot & * \\ \hline 0 & \cdot & & \cdot & & \cdot & & \cdot\ \cdot & 0 \end{pmatrix}_{n\times m}$$

Hence there is at least one free variable and the system $A\mathbf{c} = \mathbf{0}$ has nontrivial solutions. But we showed above that $\mathbf{c} = \mathbf{0}$ is the only solution, so we obtain a contradiction unless $|L| \le |S'| \le |S|$. The theorem is proved. □

COROLLARY 1.55. *In a finite-dimensional vector space, all bases have the same cardinality, which we refer to hereafter as the dimension* $\dim_{\mathbb{K}}(V)$.

NOTATION: We will often simplify notation when the underlying ground field $\mathbb{K}$ is understood by writing $\dim(V)$ or even $|V|$ for the dimension of V.

EXAMPLE 1.56. We have already seen that $\dim_{\mathbb{K}}(\mathbb{K}^n) = n$, with the standard basis vectors $\mathbf{e}_1 = (1, 0, \dots, 0), \dots, \mathbf{e}_n = (0, \dots, 0, 1)$. We may view $\mathbb{C}^n$ (or any vector space over $\mathbb{C}$) as a vector space over $\mathbb{R}$ by restricting scalars in $\lambda \cdot v$ to be real. As a vector space over $\mathbb{C}$ we have $\dim_{\mathbb{C}}(V) = n$, but as a vector space over $\mathbb{R}$ we have $\dim_{\mathbb{R}}(V) = 2n$.

DISCUSSION: In fact, any $v \in \mathbb{C}^n$ can be written as a sum with complex coefficients $v = \sum_{j=1}^{n} z_j \mathbf{e}_j$, and if $z_j = x_j + iy_j$ this becomes

$$v = x_1\mathbf{e}_1 + \cdots + x_n\mathbf{e}_n + y_1(i\mathbf{e}_1) + \cdots + y_n(i\mathbf{e}_n) \quad \text{with } x_i, y_j \in \mathbb{R}.$$

Thus the vectors $\{\mathbf{e}_1, \dots, \mathbf{e}_n, i\mathbf{e}_1, \dots, i\mathbf{e}_n\} \subseteq \mathbb{C}^n$ span $\mathbb{C}^n$ as a vector space over $\mathbb{R}$. They are also independent over $\mathbb{R}$, for if $a_i, b_j \in \mathbb{R}$ and

$$\mathbf{0} = \sum a_j\mathbf{e}_j + \sum b_j(i\mathbf{e}_j) = \sum (a_j + ib_j)\mathbf{e}_j\,,$$

we must have $a_j + ib_j = 0$ and $a_j = b_j = 0$ because $\{\mathbf{e}_j\}$ is a basis over $\mathbb{C}$. Since elements of $\mathbb{C}$ are $\mathbb{R}$-linear combinations $z = a + ib = a1 + ib$, the elements $1, i$ are a basis for $\mathbb{C}$ when we regard it as a vector space over $\mathbb{K} = \mathbb{R}$. Thus $\dim_{\mathbb{R}}(\mathbb{C}) = 2$ while $\dim_{\mathbb{C}}(\mathbb{C}) = 1$.

EXERCISE 1.57. If V is a finite-dimensional vector space and $W \subseteq V$ a subspace, explain why W must also be finite dimensional.

NOTE: When $\mathbb{C}$ is regarded as a vector space over $\mathbb{K} = \mathbb{R}$ then $\dim_{\mathbb{R}}(\mathbb{C}) = 2$ and the elements $v_1 = 1, v_2 = i$ are convenient basis vectors over $\mathbb{R}$ because $z = a + ib = a \cdot 1 + b \cdot i$.

EXERCISE 1.58. If V_1, V_2 are finite-dimensional vector spaces, prove the following:

(a) If $V_1 \subseteq V_2$ then $\dim(V_1) \leq \dim(V_2)$.
(b) If $\dim(V_1) = \dim(V_2)$ and $V_1 \subseteq V_2$, then $V_1 = V_2$ as sets.

EXERCISE 1.59. Explain why $W \subseteq V \Rightarrow \dim(W) \leq \dim(V)$, even if one or both of these spaces is infinite dimensional.

Implicit and Parametric Description of Subspaces (Revisited). How can a subspace W in a vector space be specified? Every V of dimension n can be identified in a natural way with $\mathbb{K}^n$ once a basis $\{f_1, \dots, f_n\}$ in V has been determined, so we may as well restrict attention to describing subspaces W of coordinate space $\mathbb{K}^n$. Given a basis $\mathfrak{X} = \{f_i\}$ in V, the map $j_{\mathfrak{X}} : \mathbb{K}^n \to V$ given by

$$\mathbf{x} = (x_1, \dots, x_n) \mapsto j_{\mathfrak{X}}(\mathbf{x}) = \sum_{i=1}^{n} x_i f_i$$

is a bijection that respects all vector space operations in the sense that

$$j_{\mathfrak{X}}(\lambda \cdot \mathbf{x}) = \lambda \cdot j_{\mathfrak{X}}(\mathbf{x}) \quad \text{and} \quad j_{\mathfrak{X}}(\mathbf{x} + \mathbf{y}) = j_{\mathfrak{X}}(\mathbf{x}) + j_{\mathfrak{X}}(\mathbf{y}).$$

It is an *isomorphism* between $\mathbb{K}^n$ and V, by which properties of one space can be matched with those of the other.

DEFINITION 1.60. Subspaces $W \subseteq \mathbb{K}^n$ can be described in two ways.

1. By exhibiting a basis $\mathfrak{X} = \{f_1, \dots, f_r\}$ in W, so $W = \mathbb{K}\text{-span}\{\mathfrak{X}\}$ and $\dim_{\mathbb{K}}(W) = r$. This is a "parametric description" of W since each w in W is labeled by a unique coordinate r-tuple $\mathbf{c} = (c_1, \dots, c_r)$.
2. By finding a set of linear equations

$$\begin{aligned} a_{11}x_1 + \cdots + a_{1m}x_m &= 0 \\ &\vdots \\ a_{n1}x_1 + \cdots + a_{nm}x_m &= 0 \end{aligned}$$

described by a matrix equation $A\mathbf{x} = \mathbf{0}$ (where $A = n \times m$, $\mathbf{0} = n \times 1$, $\mathbf{x} = m \times 1$) with $A_{ij} = a_{ij}$, whose solution set $\{\mathbf{x} \in \mathbb{K}^m : A\mathbf{x} = \mathbf{0}\}$ is equal to W. Such an "implicit description" may include redundant equations. When there are no redundant equations we will see that $W = \{\mathbf{x} \in \mathbb{K}^n : A\mathbf{x} = \mathbf{0}\}$ has dimension

$$\dim(W) = m - n = \dim(V) - \#(\textit{constraint equations})$$

We illustrate this with some computational examples.

EXAMPLE 1.61. Find a basis for the subspace $W = \mathbb{R}\text{-span}\{\mathbf{u}_1, \mathbf{u}_2, \mathbf{u}_3\}$ in $\mathbb{R}^3$ if

$$\mathbf{u}_1 = \begin{pmatrix} 1 \\ 2 \\ 3 \end{pmatrix}, \quad \mathbf{u}_2 = \begin{pmatrix} 2 \\ 3 \\ 4 \end{pmatrix}, \quad \mathbf{u}_3 = \begin{pmatrix} 3 \\ 4 \\ 5 \end{pmatrix},$$

and determine $\dim_{\mathbb{R}}(W)$. Then describe W as the solution set of a system of linear equations

$$\begin{aligned} a_{11}x_1 + a_{12}x_2 + a_{13}x_3 &= 0 \\ &\vdots \\ a_{31}x_1 + a_{32}x_2 + a_{33}x_3 &= 0 \end{aligned}$$

where $\mathbf{x} = (x_1, x_2, x_3) \in \mathbb{R}^3$.

SOLUTION: Write the vectors as the *rows* of the 3×3 matrix

$$A = \begin{pmatrix} \mathbf{u}_1 \\ \mathbf{u}_2 \\ \mathbf{u}_3 \end{pmatrix} = \begin{pmatrix} 1 & 2 & 3 \\ 2 & 3 & 4 \\ 3 & 4 & 5 \end{pmatrix}.$$

The span RowSpace(A) of the rows is unaffected by elementary row operations, which yield the echelon form

$$A \to \begin{pmatrix} 1 & 2 & 3 \\ 0 & -1 & -2 \\ 0 & -2 & -4 \end{pmatrix} \to \begin{pmatrix} \boxed{1} & 2 & 3 \\ 0 & \boxed{1} & 2 \\ 0 & 0 & 0 \end{pmatrix}.$$

Therefore $\mathbf{w}_1 = (1, 2, 3)$ and $\mathbf{w}_2 = (0, 1, 2)$ span W; they are also independent because $0 = c_1\mathbf{w}_1 + c_2\mathbf{w}_2 = (c_1, 2c_1 + c_2, 3c_1 + 2c_2)$ implies

$$\begin{cases} c_1 = 0 \\ 2c_1 + c_2 = 0 \\ 3c_1 + 2c_2 = 0 \end{cases} \Rightarrow \ c_1 = c_2 = 0.$$

Thus $\mathfrak{X} = \{\mathbf{w}_1, \mathbf{w}_2\}$ is a basis and $\dim(W) = 2$. A typical vector in W can be written (uniquely) as

$$s\mathbf{w}_1 + t\mathbf{w}_2 = (s, 2s + t, 3s + 2t) = (x_1, x_2, x_3) \quad \text{with } s, t \in \mathbb{R}.$$

To describe W as the solution set of a system of equations in x_1, x_2, x_3 we need to "eliminate" s and t from this parametric description of W. This can be done by writing x_1, x_2, x_3 in terms of s, t

$$\begin{cases} x_1 = s & \Rightarrow s = x_1, \\ x_2 = 2s + t & \Rightarrow x_2 = 2s + t = 2x_1 + t \Rightarrow t = x_2 - 2x_1, \\ x_3 = 3s + 2t. \end{cases}$$

The last equation yields the "constraint" identity that determines W:

$$x_3 = 3s + 2t = 3x_1 + 2(x_2 - 2x_1) = -x_1 + 2x_2$$

or $x_1 - 2x_2 + x_3 = 0$ (1 equation in 3 unknowns). Thus $W = \{\mathbf{x} \in \mathbb{R}^3 : x_1 - 2x_2 + x_3 = 0\}$, which has dimension $\dim(\mathbb{R}^3) - 1 = 2$.

EXAMPLE 1.62. Let $W \subseteq \mathbb{R}^4$ be the solution set of the system of linear equations

$$\begin{cases} x_1 + x_2 - x_3 + 2x_4 = 0, \\ 3x_1 - x_2 + x_4 = 0, \end{cases}$$

so $A\mathbf{x} = \mathbf{0}$ ($\mathbf{x} \in \mathbb{R}^4$) where

$$A = \begin{pmatrix} 1 & 1 & -1 & 2 \\ 3 & -1 & 0 & 1 \end{pmatrix}_{2\times 4}.$$

Find a basis for W and determine $\dim_{\mathbb{R}}(W)$. Do the answers change if we replace $\mathbb{R}$ by $\mathbb{Q}$ or $\mathbb{C}$?

SOLUTION: Elementary row operations yield

$$A \to \begin{pmatrix} 1 & 1 & -1 & 2 \\ 0 & -4 & 3 & -6 \end{pmatrix} \to \begin{pmatrix} \boxed{1} & 1 & -1 & 2 \\ 0 & \boxed{1} & -\frac{3}{4} & -\frac{3}{2} \end{pmatrix},$$

and for any solution of $A\mathbf{x} = 0$, $\mathbf{x} = \mathrm{col}(x_1, x_2, x_3, x_4)$ has x_3, x_4 as free variables. Backsolving yields the dependent variables

$$x_2 = \frac{3}{4}x_3 - \frac{3}{2}x_4,$$

$$x_1 = -x_2 + x_3 - 2x_4 = \left(-\frac{3}{4}x_3 + \frac{3}{2}x_4\right) + x_3 - 2x_4 = \frac{1}{4}x_3 - \frac{1}{2}x_4.$$

Thus solutions have the form

$$\mathbf{x} = \begin{pmatrix} \frac{1}{4}x_3 - \frac{1}{2}x_4 \\ \frac{3}{4}x_3 - \frac{3}{2}x_4 \\ x_3 \\ x_4 \end{pmatrix} = x_3 \begin{pmatrix} \frac{1}{4} \\ \frac{3}{4} \\ 1 \\ 0 \end{pmatrix} + x_4 \begin{pmatrix} -\frac{1}{2} \\ -\frac{3}{2} \\ 0 \\ 1 \end{pmatrix} = x_3\mathbf{f}_1 + x_4\mathbf{f}_2$$

for every $x_3, x_4 \in \mathbb{K}$. After rescaling basis vectors to clear denominators, we see that the solution set is

$$\mathbb{R}\text{-span}\{\mathrm{col}(1,3,4,0), \mathrm{col}(3,1,0,-2)\} = \mathbb{R}\text{-span}\{\mathbf{f}_1, \mathbf{f}_2\}.$$

The vectors $\mathbf{f}_1$, $\mathbf{f}_2$ span the solution set W but are also independent because

$$c_1(1,3,4,0) + c_2(3,1,0,-2) = (c_1 + 3c_2, 3c_1 + c_2, 4c_1, -2c_2) = (0,0,0,0)$$

implies that $c_1 = c_2 = 0$. Thus $\{\mathbf{f}_1, \mathbf{f}_2\}$ is a basis and $\dim_{\mathbb{R}}(W) = 2$. The result is the same if we replace the ground field $\mathbb{R}$ with $\mathbb{Q}$ or $\mathbb{C}$. As a rule of thumb, each constraint equation $a_{i1}x_1 + \cdots + a_{im}x_m = 0$ on $\mathbb{K}^m$ reduces the dimension of the solution set $W = \{\mathbf{x} \in \mathbb{K}^m : A\mathbf{x} = 0\}$ by 1, but this is not always the case.

EXERCISE 1.63. Consider the special case of one constraint equation,

$$W = \left\{\mathbf{x} : \sum_{i=1}^{n} c_i x_i = 0\right\} \quad \text{with } c_1, \dots, c_n \in \mathbb{K}.$$

(a) Under what condition on $\{c_1, \dots, c_n\}$ do we have $\dim_{\mathbb{K}}(W) = n - 1$?
(b) Explain why $\dim(W) < n - 1$ is impossible.

EXERCISE 1.64. Given vectors in $\mathbb{R}^4$

$$\mathbf{v}_1 = (1,0,-1,2), \quad \mathbf{v}_2 = (2,1,1,-3), \quad \mathbf{v}_3 = (-2,1,0,-2),$$

determine whether they are linearly independent, and then decide whether the vector $\mathbf{w} = (1, 2, 3, -4)$ lies in their linear span.

EXERCISE 1.65. Find an explicit basis for the subspace in $\mathbb{R}^n$ determined by the single constraint equation

$$x_1 + \cdots + x_n = 0.$$

What is its dimension?

The Lagrange Interpolation Formula.

EXAMPLE 1.66 (Lagrange Interpolation Formula). For any infinite field, such as $\mathbb{K} = \mathbb{Q}, \mathbb{R}, \mathbb{C}$, the problem of finding a polynomial $f \in \mathbb{K}[x]$ having specified values $f(p_j) = \lambda_j$ at a finite set of distinct points $p_1, \dots, p_n$ in $\mathbb{K}$ always has a solution. The solution is nonunique (the problem is underdetermined) unless we require that $\deg(f) = n-1$; there may be no solution if $\deg(f) < n-1$.

DISCUSSION: The product $h(x) = \prod_{j=1}^{n}(x - p_j)$ has degree equal to n and is zero at each p_j (and zero nowhere else), so the solution to the interpolation problem cannot be unique without restrictions on $f(x)$: one can add h (or any scalar multiple thereof) to any proposed solution f. It is reasonable to ask for a solution $f(x)$ of minimal degree to reduce the ambiguity. The polynomial

$$f(x) = \sum_{i=1}^{n} \lambda_i \cdot \frac{\prod_{j \neq i}(x - p_j)}{\prod_{j \neq i}(p_i - p_j)} \tag{1.5}$$

has nonzero denominator, is equal to λ_i at p_i for each i, and has $\deg(f) = n-1$.

This is the *Lagrange interpolation formula*, determined by direct methods. It is a bit complicated to rewrite this sum of products in the form $f = c_0 + c_1 x + \cdots + c_{n-1}x^{n-1}$. But the coefficients $c_0, \dots, c_{n-1}$ can also be found directly as the solution of a system of linear equations

$$\lambda_j = f(p_j) = \sum_{k=0}^{n-1} p_j^k c_k \quad \text{for } 1 \le j \le n-1,$$

which is equivalent to the matrix equation $A\mathbf{c} = \lambda$ in which

$$A = \begin{pmatrix} p_1^0 & \cdots & p_1^{n-1} \\ \vdots & \ddots & \vdots \\ p_n^0 & \cdots & p_n^{n-1} \end{pmatrix}_{n\times n} \quad \text{and} \quad \mathbf{c} = \begin{pmatrix} c_0 \\ \vdots \\ c_{n-1} \end{pmatrix}_{n\times 1}, \quad \lambda = \begin{pmatrix} \lambda_0 \\ \vdots \\ \lambda_{n-1} \end{pmatrix}_{n\times 1}.$$

EXERCISE 1.67. (Another Dimension Formula) If E, F are subspaces in a finite-dimensional vector space V and $E + F = \{e + f : e \in E,\ f \in F\}$ is their linear span, prove that

$$\dim(E + F) = \dim(E) + \dim(F) - \dim(E \cap F).$$

HINT: Choose appropriate bases related to E, F and $E \cap F$.

Rank of a Matrix: Row Rank vs. Column Rank. If we view vectors in $\mathbb{K}^m$ as $m \times 1$ column vectors, an $n \times m$ matrix A determines a map $L_A : \mathbb{K}^m \to \mathbb{K}^n$,

$$\mathbf{y} = L_A(\mathbf{x}) = A \cdot \mathbf{x} \quad \text{(matrix product)}$$

As an immediate consequence of this definition, L_A is a *linear operator* from $\mathbb{K}^m$ to $\mathbb{K}^n$, one that preserves linear combinations of vectors in the sense that

$$L_A\Big(\sum_{i=1}^{r} \lambda_i \mathbf{x}_i\Big) = \sum_{i=1}^{r} \lambda_i \cdot L_A(\mathbf{x}_i) \quad (\mathbf{x}_i \in \mathbb{R}^m).$$

It follows that the *range* of the map L_A,

$$R(L_A) = L_A(\mathbb{K}^m) = \{A \cdot \mathbf{x} : \mathbf{x} \in \mathbb{K}^m\},$$

is a vector subspace in $\mathbb{K}^n$.

DEFINITION 1.68. We have defined the *rank* of a matrix A to be the dimension $\dim R(L_A)$ of the range of this operator,

$$\text{RANK OF A MATRIX:} \quad \operatorname{rank}(A) = \dim(R(L_A)).$$

The problem of finding a basis for $R(L_A)$, and $\dim R(L_A)$, arises quite often in linear algebra. If $\mathbf{e}_1, \dots, \mathbf{e}_m$ is the standard basis in $\mathbb{K}^m$, viewed as $m \times 1$ column vectors, then by definition of matrix multiplication the image vector $L_A(\mathbf{e}_i) = A \cdot \mathbf{e}_i$ is precisely the ith column $C_i(A)$. The linearity property of L_A then implies that $R(L_A)$ is the linear span $\mathbb{K}\text{-span}\{C_1, \dots, C_m\}$ of the columns of A.

DEFINITION 1.69. (Column Space/Row Space) The linear span of the rows of an $n \times m$ matrix A is the *row space* of A, denoted RowSpace(A), which can be viewed as a subspace of $\mathbb{K}^m$. Similarly, the linear span of the columns is the *column space* of A, ColSpace(A), which can be viewed as a vector subspace in $\mathbb{K}^n$. It is precisely the range of the map $L_A : \mathbb{K}^m \to \mathbb{K}^n$.

We now cite two important results addressing these issues, but defer the proofs until Chapter 4 since they involve *determinants*, which will be discussed in that chapter.

THEOREM 1.70. *If row reduction of an $n \times m$ matrix A yields an echelon form E, the columns $C_{j_1}(A), \dots, C_{j_r}(A)$ of A that correspond to the pivot columns $C_{j_1}(E), \dots, C_{j_r}(E)$ of E (those columns in E that contain a step corner) are a basis for* ColSpace(A) *and the rank of A is the number r of pivot columns in E.*

This is unexpected because row operations do not interact well with column operations; for instance, if A' is obtained from A by an elementary operation on its rows, the column spaces of A and A' need not be the same subspace in $\mathbb{K}^n$. Nevertheless, a basis for ColSpace(A), which would normally be determined by performing elementary *column operations* on A, can be found by inspection of the echelon matrix E determined by performing *row operations* on A.

There is another important consequence of Theorem 1.70. As noted above, the rank $r = \dim(L_A(\mathbb{K}^m))$ of an $n \times m$ matrix is equal to its column rank,

$$\begin{aligned}\mathrm{ColRank(A)} &= \dim\big(\mathbb{K}\text{-span}\{C_1(A), \dots, C_m(A)\}\big)\\ &= \#(\text{linearly independent columns in } A).\end{aligned}$$

The row rank of A is defined similarly, as the dimension of the subspace in $\mathbb{K}^m$ spanned by the rows of A,

$$\begin{aligned}\mathrm{RowRank}(A) &= \dim(\mathbb{K}\text{-span}\{R_1(A), \dots, R_n(A)\})\\ &= \#(\text{linearly independent rows in } A).\end{aligned}$$

These turn out to be equal.

COROLLARY 1.71. RowRank(A) = ColRank(A), *whether or not A is a square matrix.*

EXERCISE 1.72. Prove Corollary 1.71 is a consequence of Theorem 1.70.

1.5. Quotient Spaces V/W

If V is a vector space and W a subspace, the *additive cosets* of a subspace W are translates of W by vectors in V. They are the subsets $x + W = \{x + w : w \in W\}$ for some $x \in V$, which we shall often denote by $[x]$ when the subspace W is understood. In particular, W itself is the "zero coset": $[0] = 0 + W = W$. The key observation is that the whole space V gets partitioned into disjoint cosets that fill V. The collection of all cosets $[x]$ is the *quotient space* V/W. Observe that *points* in the space V/W are at the same time *subsets* in V.

LEMMA 1.73. *If W is a subspace in V and $x, y \in V$, then:*

1. *Two cosets $x + W$ and $y + W$ either coincide or are disjoint; hence the distinct cosets of W partition the space V.*
2. *An additive coset can have various representatives $x \in V$, so $y + W = x + W \Leftrightarrow$ there is some $w \in W$ such that $y = x + w$ (or $y - x \in W$).*
3. *If $y \in x + W$, then $y + W = x + W$.*

PROOF. We start by defining sums $A + B = \{a + b : a \in A, b \in B\}$ of sets $A, B \subseteq V$ that will be invoked repeatedly in what follows.

EXERCISE 1.74. If W is a subspace of a vector space V and $w \in W$, prove that:

(a) $w + W = W$ for all $w \in W$ and $(x + W) \cap W = \emptyset$ if $x \notin W$;
(b) $W + W = \{w_1 + w_2 : w_1, w_2 \in W\}$ is equal to W;
(c) $W - W = W$.

Resuming the proof of Lemma 1.73, if cosets $x + W$ and $y + W$ have a point p in common, there are $w_1, w_2 \in W$ such that $x + w_1 = p = y + w_2$; hence $y = x + (w_1 - w_2)$. By Exercise 1.74 the cosets are equal:

$$y + W = (x + (w_1 - w_2)) + W = x + ((w_1 - w_2) + W) = x + W$$

proving 1.73(a).

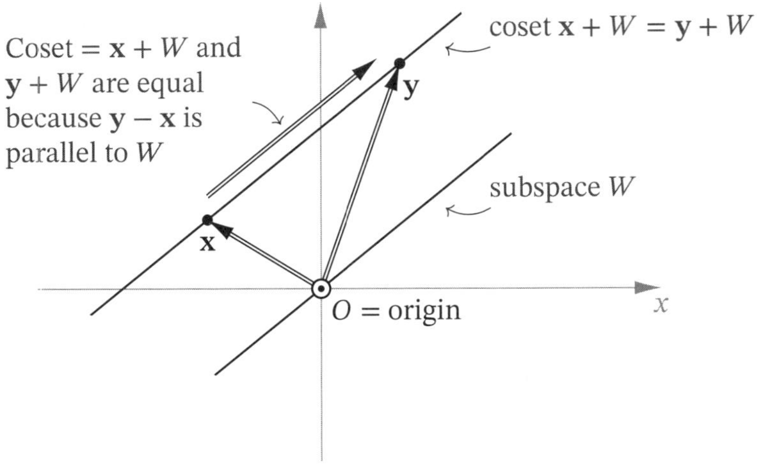

FIGURE 1.6. Additive cosets $\mathbf{x} + W$ of a subspace W are a family of parallel "hyperplanes" in a vector space V. When $V = \mathbb{R}^2$ and W is a line through the origin, *all* lines parallel to W are cosets. Two vectors $\mathbf{x}, \mathbf{y}$ in the same coset yield the same translate of W: $\mathbf{x} + W = \mathbf{y} + W$ because $\mathbf{y} - \mathbf{x}$ is parallel to the subspace W.

For (b), $x + W = y + W \Rightarrow y = y + 0 = x + w$ for some $w \in W$. Conversely, if $y = x + w$ for $w \in W$, then $y + W = x + (w + W) = x + W$ (again by Exercise 1.74). For (c), it follows from part (1.) in Lemma 1.73 that $y \in x + W \Rightarrow (y + W) \cap (x + W) \neq \varnothing \Rightarrow y + W = x + W$. □

As an example, if $V = \mathbb{R}^2$ and $W = \{(x, y) \in \mathbb{R}^2 : x = y\}$, the cosets of W are precisely the distinct lines in the plane that make an angle of 45° with the positive x-axis. These lines are the "points" in the quotient space V/W; see Figure 1.6.

DEFINITION 1.75. There is a natural surjective *quotient map* $\pi : V \to V/W$, given by

$$\pi(x) = [x] = x + W.$$

If C is a coset, any point $v \in C$ such that $C = [v] = v + W$ is called a *representative of the coset*. Part (b) of Lemma 1.73 tells us when two vectors x, y represent the same coset.

Algebraic Structure in V/W. The quotient space V/W inherits natural sum ($\oplus$) and scaling ($\odot$) operations from the overlying vector space V.

DEFINITION 1.76. For any $x, y \in V$ and $\lambda \in \mathbb{K}$ we define the following operations in V/W:

1. ADDITION: $[x] \oplus [y] = [x + y]$
2. SCALAR MULTIPLICATION: $\lambda \odot [x] = [\lambda \cdot x]$

To spell out what is involved, this definition tells us how to form the sum $X \oplus Y$ of two cosets $X, Y \in V/W$ via the following algorithm:

1. Pick representatives $x, y \in V$ such that $X = [x],\ Y = [y]$.
2. Add the representatives to get $x + y \in V$.
3. Form the coset $[x + y] = (x + y) + W$ and report the output: $X \oplus Y = [x + y]$.

But why should this make sense? The outcome depends on a choice of representatives for each coset X, Y, and if different choices yield different outputs, everything written above is nonsense. Fortuately, the outcome is independent of the choice of representatives and the operation $(\oplus)$ is *well-defined*. In fact, if $[x] = [x']$ and $[y] = [y']$ there must exist $w_1, w_2 \in W$ such that $x' = x + w_1,\ y' = y + w_2$, and then

$$\begin{aligned}[x' + y'] = (x' + y') + W &= (x + y) + ((w_1 + w_2) + W) \\ &= (x + y) + W = [x + y].\end{aligned}$$

Similarly, the scaling operation is well-defined: if $[x'] = [x]$ we have $x' = x + w$ for some $w \in W$, and then

$$[\lambda \cdot x'] = (\lambda \cdot x') + W = (\lambda \cdot x) + (\lambda w + W) = (\lambda \cdot x) + W = [\lambda \cdot x].$$

Once we know the operations $(\oplus)$ and $(\odot)$ make sense, direct calculations involving representatives show that all vector space axioms are satisfied by the system $(V/W, \oplus, \odot)$. For instance,

1. Associativity of $(\oplus)$ on V/W follows from associativity of $(+)$ on V: since $x + (y + z) = (x + y) + z$ in V we get

$$\begin{aligned}[x] \oplus ([y] \oplus [z]) &= [x] \oplus [y + z] = [x + (y + z)] \\ &= [(x + y) + z] = [x + y] \oplus [z] = ([x] \oplus [y]) \oplus [z].\end{aligned}$$

2. The zero element is $[0] = 0 + W = W$ because $[0] \oplus [x] = [0 + x] = [x]$.
3. The additive inverse $-[x]$ of $[x] = x + W$ is $[-x] = (-x) + W$ since $[x] \oplus [-x] = [x + (-x)] = [0]$.

EXERCISE 1.77. Verify the remaining vector space axioms for $(V/W, \oplus, \odot)$. Then show that the quotient map $\pi : V \to V/W$ with $\pi(x) = [x] = x + W$ "*intertwines*" the algebraic operations in $(V, +, \cdot)$ with those in $(V/W, \oplus, \odot)$ in the sense that, for any $v_1, v_2 \in V$ and $\lambda \in \mathbb{K}$, we have

(a) $\pi(v_1 + v_2) = \pi(v_1) \oplus \pi(v_2)$,
(b) $\pi(\lambda \cdot v_1) = \lambda \odot \pi(v_1)$.

Thus $\pi : V \to V/W$ is a *linear operator* that maps V onto V/W.

When $W = (0)$ elements of the quotient are single points $[v] = v + W = \{v\}$, and V/W has a natural identification with V under the quotient map that is now a bijection. When $W = V$ there is just one coset, $v + W = v + V = V$; the quotient space reduces to a single point, the zero element $[0] = 0 + V = V$.

EXERCISE 1.78. Let $V = \mathbb{R}^3$ and $W = \{(x_1, x_2, x_3) : x_3 = 0\}$ = the xy-plane in 3-dimensional space. The cosets in V/W are the distinct planes parallel to the xy-plane: if $v = (v_1, v_2, v_3)$, then

$$\begin{aligned} v + W &= \{v + w : w \in W\} \\ &= \{(v_1, v_2, v_3) + (w_1, w_2, 0) : w_1, w_2 \in \mathbb{R}\} \\ &= \{(v_1 + s, v_2 + t, v_3) : s, t \in \mathbb{R}\} \\ &= \{(x_1, x_2, x_3) : x_1, x_2 \in \mathbb{R}, x_3 = v_3\} \end{aligned}$$

(the plane parallel to W passing through $(0, 0, v_3)$). Each value of $v_3 \in \mathbb{R}$ gives a different coset.

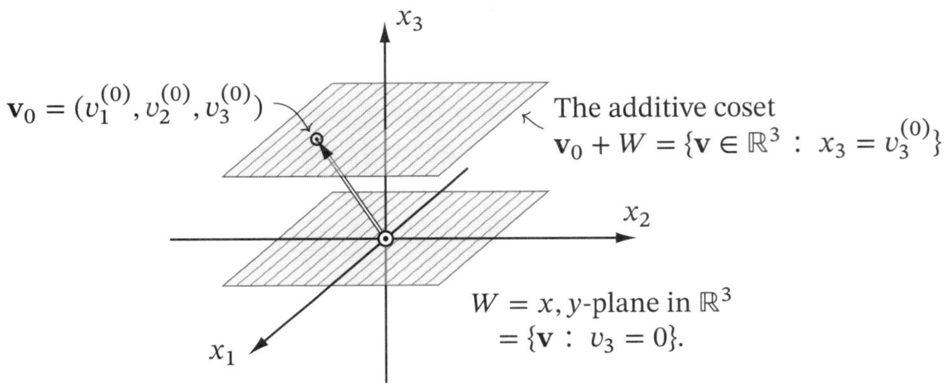

FIGURE 1.7. Additive cosets of $W = \{\mathbf{v} \in \mathbb{R}^3 : v_3 = 0\}$ are planes parallel to W in $\mathbb{R}^3$. A typical coset $\mathbf{v}_0 + W$ is shown. These planes are the elements of the quotient space $\mathbb{R}^3/W$.

One important viewpoint is to think of the quotient map $\pi : v \to V/W$ as "erasing" inessential aspects of the original vector space, retaining only those relevant to the problem at hand. Whole "bunches" of vectors in V, the cosets $v + W$, collapse to single points in the target space V/W (the planes in the last example become points in V/W). A lot of detail is lost in this collapse, but if W is suitably chosen the quotient space will retain information that is buried in superfluous detail when we look at what is happening in the larger space V. We will soon encounter many examples of this once we start looking at the structure of "linear operators" between vector spaces. For the moment we assemble a few more basic facts about quotients of vector spaces.

Finding Bases in V/W and the Dimension Formula.

THEOREM 1.79 (Dimension Theorem for Quotients). *If V is finite dimensional and W is a subspace. Then:*

1. $\dim(V/W) \leq \dim(V) < \infty$,
2. $\dim(W) \leq \dim(V) < \infty$,

and

$$\dim(V) = \dim(W) + \dim(V/W). \tag{1.6}$$

By our notational convention of writing $\dim_{\mathbb{K}}(V) = |V|$, *this identity can also be written in the abbreviated form* $|V| = |W| + |V/W|$.

PROOF. The quotient map $\pi : V \to V/W$ sends V onto V/W and preserves linear combinations in the sense that

$$\pi\Big(\sum_{i=1}^{m} \lambda_i v_i\Big) = \sum_{i=1}^{m} \lambda_i \pi(v_i).$$

(recall Exercise 1.77). Furthermore, $\pi(V) = V/W$, so if vectors $\{v_i\}$ span V their images $\bar{v}_i = \pi(v_i)$ span V/W. That proves

$$\dim(V/W) \leq \#\{\bar{v}_i\} \leq \#\{v_i\} = \dim(V) < \infty$$

as claimed in (1.).

As for item (2.), we know $\dim(V) < \infty$ but have no a priori information about W; however, we showed earlier that no independent set in V can have more than $\dim(V)$ elements, and a basis for W would be such a set.

The identity (1.6) is proved by constructing a basis in V/W aligned with a specially chosen basis in V. Since $\dim(W) < \infty$ there is a basis $\{w_1, \dots, w_m\}$ in W. If $W = V$ then V/W is trivial and there is nothing more to do, but otherwise we can find an "outside vector" $v_{m+1} \notin W$ such that the larger set $\{w_1, \dots, w_m, v_{m+1}\}$ is independent, and hence a basis for

$$W_1 = \mathbb{K}\text{-span}\{w_1, \dots, w_m, v_{m+1}\} \supsetneqq W_0 = W.$$

If $W_1 \neq V$, we can adjoin one more vector $v_{m+2} \notin W_1$ to get an independent set $\{w_1, \dots, w_m, v_{m+1}, v_{m+2}\}$ with

$$W_0 \subsetneqq W_1 \subsetneqq W_2 = \mathbb{K}\text{-span}\{w_1, \dots, w_m, v_{m+1}, v_{m+2}\}.$$

This process must terminate; otherwise we would have arbitrary large independent sets in the finite-dimensional space V. When the construction terminates we get an independent spanning set $\{w_1, \dots, w_m, v_{m+1}, \dots, v_{m+k}\}$ in $W_k = V$. This is a basis for V so $\dim(V) = m + k = \dim(W) + k$.

To conclude the proof we demonstrate that $k = \#(\text{outside vectors})$ is equal to $\dim(V/W)$ by showing that the π-images $\overline{v}_{m+1}, \dots, \overline{v}_{m+k} \in V/W$ of the outside vectors are a basis for V/W. Since π is surjective the images $\pi(w_1), \dots,$ $\pi(v_{m+k})$ span V/W. But π "kills" all vectors in W, so

$$\pi(w_1) = \cdots = \pi(w_m) = [0] \quad \text{in } V/W,$$

and the remaining images $\overline{v}_{k+i} = \pi(v_{m+i})$ span V/W. They are also linearly independent, because if some linear combination $\sum_{i=1}^{k} c_{m+i}\overline{v}_{m+i} = [0]$ in V/W, then by linearity of the quotient map π we get

$$[0] = \sum_{j=1}^{k} c_{m+j}\pi(v_{m+j}) = \pi\Big(\sum_{j=1}^{k} c_{m+j} v_{m+j}\Big).$$

But $\pi(v) = [0]$ for a vector $v \in V \Leftrightarrow [v] = v + W$ is equal to the zero coset $[0] = W$. Since $v + W = W \Leftrightarrow v \in W$, we can find coefficients $c_1, \dots, c_m$ such that

$$\sum_{i=1}^{m} c_i w_i = v = \sum_{j=1}^{k} c_{m+j} v_{m+j}$$

or

$$0 = \sum_{i=1}^{m} c_i w_i + \sum_{j=1}^{k} (-1) c_{m+j} v_{m+j} \quad \text{in } V.$$

Since $w_1, \dots, w_m, v_{m+1}, \dots v_{m+k}$ is a basis for V, this can only happen if all coefficients in this sum are zero, and in particular $c_{m+1}, \dots, c_{m+k} = 0$. Thus the $\{\overline{v}_i\}$ are independent and a basis for V/W, so $\dim(V/W) = k$ is equal to $\dim(V) - \dim(W)$. □

REMARK 1.80. The construction developed in proving Theorem 1.79 shows how to find bases in a quotient space V/W and perform effective calculations with them. The key was to find representatives v_i back in V in order to transfer calculations involving cosets in V/W to calculations involving actual vectors v_i in V. This was achieved by finding a basis $\mathfrak{X} = \{w_1, \dots, w_r; u_{r+1}, \dots, u_n\}$ in V such that $w_1, \dots, w_r$ are a basis for W. As a corollary" we can read the following observations out of that discussion.

COROLLARY 1.81. *If W is a subspace in a finite-dimensional vector space V, and if $\mathfrak{X} = \{w_1, \dots, w_r; u_{r+1}, \dots, u_n\}$ is a basis in V such that $\{w_1, \dots, w_r\}$ is a basis for W, then*:

1. *The cosets $v_{r+1} + W, \dots, v_n + W$ in V/W corresponding to the "outside vectors" in $\mathfrak{X}$ are a basis for the quotient space V/W.*
2. *The linear span $H = \mathbb{K}\text{-span}\{v_{r+1}, \dots, v_n\}$ in V is a subspace of V that "cross-sections" the cosets in V/W: each coset $x + W$ meets H in exactly one point $y \in V$ so that $H \cap (x + W) = \{y\}$.*

Thus each point in V/W is labeled by a unique point in H. The first statement can be read directly out of the proof of Theorem 1.79. *Proof of* (2.) *is left as an exercise.*

EXERCISE 1.82. If H is the linear span of the outside vectors $u_{r+1}, \dots, u_n$ in Corollary 1.81, explain why:

(a) Each coset $x + W \in V/W$ meets H in a single point $y \in H$.
(b) The quotient map $\pi : V \to V/W$ restricts to a bijection $\pi|_H$ between H and V/W.

EXERCISE 1.83. Consider the subspace $W = \{\mathbf{x} \in \mathbb{R}^n : x_1 + \cdots + x_n = 0\}$ in $V = \mathbb{R}^n$.

(a) What is its dimension?
(b) Prove that the following vectors are a basis for W:

$$(1, -1, 0 \dots, 0),\ (0, 1, -1, 0 \dots 0),\ \dots\ ,(0, \dots, 1, -1),\ (-1, 0, \dots, 0, 1)$$

(c) Find a vector subspace $H \subseteq \mathbb{R}^n$ that cross-sections the cosets $\mathbf{x}+W$ in the quotient space V/W. Explain why H must be a one-dimensional subspace in V.

EXERCISE 1.84. Find explicit bases for the following quotient spaces V/W.

(a) $V = \mathbb{R}^3$, $W = \mathbb{R}\mathbf{e}_1 + \mathbb{R}\mathbf{e}_2$

(b) $V = \mathbb{R}^3$, $W = \mathbb{R}\text{-span}\{\mathbf{w}_1 = (1,2,3), \mathbf{w}_2 = (0,1,-1)\}$

(c) $V = \mathbb{C}^4$,

$$W = \mathbb{C}\text{-span}\{\mathbf{z}_1 = (1, 1+i, 3-2i, -i),\ \mathbf{z}_2 = (4-i, 0, -1, 1+i)\}$$

(d) $V = \mathbb{R}^4$,

$$W = \{\mathbf{x} \in \mathbb{R}^4 : x_1 + x_2 - x_3 + x_4 = 0 \text{ and } 4x_1 - 3x_2 + 2x_3 + x_4 = 0\}.$$

Here is a simple example involving bases in a quotient space V/W.

EXAMPLE 1.85. Let $V = \mathbb{R}^4$ and $W = \{\mathbf{x} \in \mathbb{R}^4 : 2x_1 - x_2 + x_4 = 0\}$. The subspace W is the solution set of the matrix equation

$$A\mathbf{x} = \mathbf{0} \quad \text{where} \quad A = [\,2, -1, 0, 1\,]_{1\times 4}$$

that imposes a single linear constraint on $\mathbb{R}^4$. Find a basis for V/W.

SOLUTION: Row operations yield

$$A \to A' = \left[\,\boxed{1}, -\frac{1}{2}, 0, \frac{1}{2}\,\right]$$

The free variables are x_2, x_3, x_4 and $x_1 = \frac{1}{2}x_2 - \frac{1}{2}x_4$, so the solutions have the form

$$\mathbf{x} = \begin{pmatrix} \frac{1}{2}x_2 - \frac{1}{2}x_4 \\ x_2 \\ x_3 \\ x_4 \end{pmatrix} = x_2\begin{pmatrix} \frac{1}{2} \\ 1 \\ 0 \\ 0 \end{pmatrix} + x_3\begin{pmatrix} 0 \\ 0 \\ 1 \\ 0 \end{pmatrix} + x_4\begin{pmatrix} -\frac{1}{2} \\ 0 \\ 0 \\ 1 \end{pmatrix}$$

for $x_2, x_3, x_4 \in \mathbb{R}$. Thus the solution set for $A\mathbf{x} = \mathbf{0}$ is the linear span of the column vectors

$$\mathbf{u}_1 = \text{col}(1,2,0,0), \quad \mathbf{u}_2 = \text{col}(0,0,1,0), \quad \mathbf{u}_3 = \text{col}(-1,0,0,2),$$

which are a basis for W since they are easily seen to be linearly independent by row reducing the 3×4 matrix M that has these vectors as its rows

$$M = \begin{pmatrix} 1 & 2 & 0 & 0 \\ 0 & 0 & 1 & 0 \\ -1 & 0 & 0 & 2 \end{pmatrix}$$

to see if you get a row of zeros. You do not. Therefore $\dim(\text{RowSpace}(M)) = 3$ and the vectors are independent.

Since $\dim(V) = \dim(W) + \dim(V/W)$ and $\dim(W) = 3$, we need only find one outside vector $\mathbf{u}_4 \notin W$ to complete a basis for $V = \mathbb{R}^4$; then the coset $\pi(\mathbf{u}_4) = \mathbf{u}_4 + W$ will be nonzero, and a basis vector for the 1-dimensional quotient space. The vector $\mathbf{u}_4 = \mathbf{e}_4 = (0,0,0,1)$ is not in W because it fails to satisfy the constraint equation $2x_1 - x_2 + x_4 = 0$. Thus the single vector $[\mathbf{e}_4] = \pi(\mathbf{e}_4) = \mathbf{e}_4 + W$ is a basis for V/W, and $\dim(V/W) = 1$.

EXERCISE 1.86. Let W be the vector subspace in $\mathbb{R}^4$ determined by the equations

$$\begin{aligned} x_1 - \ x_2 + x_3 + x_4 &= 0 \\ 4x_1 - 3x_2 \qquad - x_4 &= 0 \end{aligned}$$

(a) Find a basis $\{\mathbf{u}_1, \ldots, \mathbf{u}_r\}$ for W. What are the dimensions of W and V/W?

(b) Find additional vectors $\{\mathbf{v}_{r+1}, \ldots, \mathbf{v}_4\}$ lying outside of W such that

$$\mathbf{u}_1, \ldots, \mathbf{u}_r, \mathbf{v}_{r+1}, \ldots, \mathbf{v}_4 \text{ is a basis for } \mathbb{R}^4.$$

(c) Find a system $A\mathbf{x} = \mathbf{0}$ of $4 - r$ linear equations in the unknowns $x_1, \ldots, x_4$ whose solution set is $H = \mathbb{R}\text{-span}\{\mathbf{u}_{r+1}, \ldots, \mathbf{u}_4\}$, the linear span of the outside vectors found in (b).

HINT: Some trial-and-error may be needed to find the outside vectors requested in (b). A vector lies outside of W if it fails to satisfy the equations that define W. The outside vectors you seek are not unique.

Thus the points in H uniquely label the cosets in V/W.

Additional Exercises

Section 1.1. Vector Spaces

1. TRUE/FALSE QUESTIONS: ("True" if the statement is *always* true.)

(a) In any vector space $c \cdot x = c \cdot y$ implies $x = y$.

(b) In any vector space $c \cdot x = c' \cdot x$ implies $c' = c$.

(c) If W is a subspace of a finite-dimensional vector space V and $\dim V = \dim W$, then $W = V$.

(d) $\deg(f + h) = \deg(f) + \deg(h)$ for nonzero polynomials f, h in $\mathbb{K}[x]$.

(e) $\deg(f + h) = \max\{\deg(f), \deg(h)\}$ for nonzero polynomials f, h in $\mathbb{K}[x]$,

(f) For $f \in \mathbb{K}[x]$ and $\lambda \in \mathbb{K}$, $\deg(\lambda \cdot f) = \deg(f)$.

(g) A "constant polynomial" $c\mathbb{1}$ in $\mathbb{K}[x]$ has degree zero.

(h) $\deg(f \cdot h) = \deg(f) + \deg(h)$ for all nonzero polynomials f, h in $\mathbb{K}[x]$.

(i) $\deg(f \cdot h) = \deg(f) + \deg(h)$ for all nonzero polynomials f, h in $\mathbb{K}[x_1, \ldots, x_n]$.

2. For $\mathbf{a} = (a_1, a_2)$ and $\mathbf{b} = (b_1, b_2)$ in $\mathbb{R}^2$ define operations $(*)$ and $(+)$, letting

$$\text{(i)}\ \ \mathbf{a} + \mathbf{b} = (a_1 + b_1, a_2 + b_2),$$

$$\text{(ii)}\ \ \lambda * \mathbf{a} = \begin{cases} (0, 0) & \text{if } \lambda = 0, \\ \left(\lambda a_1, \frac{1}{\lambda} a_2\right) & \text{if } \lambda \neq 0. \end{cases}$$

Is V a vector space when equipped with these operations? If not, exhibit a vector space axiom not satisfied by the system $(V, +, *)$.

3. In $\mathbb{R}^2$ the parallelogram law for addition of two vectors $\mathbf{a} = (a_1, a_2)$ and $\mathbf{b} = (b_1, b_2)$ based at the origin was illustrated in Figure 1.3. Explain why this geomeangle/tric procedure for adding vectors is equivalent to the algebraic rule

$$\mathbf{a} + \mathbf{b} = (a_1 + b_1, a_2 + b_2)$$

when vector addition is described in Cartesian coordinates.

Section 1.2. Vector Subspaces

1. TRUE/FALSE QUESTIONS: ("True" if the statement is *always* true.)

(a) The set $W = \{(x_1, x_2, x_3) : x_1 = x_3 + 2 \text{ and } x_1 - x_2 + x_3 = 0\}$ is a vector subspace in $\mathbb{R}^3$.

(b) A nonzero vector space V contains a "proper" subspace W (one such that $\{0\} \subsetneqq W \subsetneqq V$.)

(c) The union $V \cup W$ of two vector subspaces of V is a subspace.

(d) The set of odd polynomials (coefficients $c_n = 0$ for all even n) is a vector subspace of $\mathbb{K}[x]$.

(e) The set of polynomials with coefficients $c_n = 0$ for all indices $n \geq$ some cutoff N is a vector subspace of $\mathbb{K}[x]$.

(f) The number of nonzero entries in an upper-triangular $n \times n$ matrix is $\leq \frac{1}{2}n^2$.

(g) The set W consisting of the zero vector $\mathbf{0}$ and all *nonconstant* polynomials is a vector subspace of $\mathbb{K}[x]$.

2. Fix some integer $r \geq 1$ and let E_r be all polynomials of the form $f = \sum_{n \geq r} c_n x^n$ (finite sums). Show that:

(a) E_r is a vector subspace of $\mathbb{K}[x]$.

(b) E_r is also a *subalgebra* of $\mathbb{K}[x]$, so $f, h \in E \Rightarrow f \cdot h$ is also in E.

NOTE: E_r consists of all polynomials that have a "*zero of order at least* $(r-1)$ *at the origin.*"

3. If W_1 and W_2 are subspaces of a vector space V, prove that $S = W_1 \cup W_2$ is a subspace $\Leftrightarrow W_1 \subseteq W_2$ or $W_2 \subseteq W_1$.

4. The *sum* $S_1 + S_2$ of two nonempty subsets in a vector space V is defined to be

$$S_1 + S_2 = \{a + b : a \in S_1 \text{ and } b \in S_2\}$$

(a) If S_1 and S_2 are both vector subspaces, prove that their sum $S_1 + S_2$ is also a vector subspace.

(b) This is no longer true if just one of the sets is a vector subspace. In $\mathbb{R}^2$ describe the sum of sets $S_1 + S_2$ when

(i) $S_1 = \{(x_1, x_2) : x_1 - x_2 = 0\}$ (a line through the origin).

(ii) S_2 is the open disc $S_2 = \{(x_1, x_2) : x_1^2 + x_2^2 < 1\}$.

Make a sketch of the set $S_1 + S_2$.

5. The transpose A^{T} of a matrix A is obtained by reflecting all entries in the array across the diagonal, so that $(A^{\mathrm{T}})_{ij} = A_{ji}$. Obviously $(A^{\mathrm{T}})^{\mathrm{T}} = A$. A matrix in $\mathrm{M}(n, \mathbb{K})$ is *symmetric* if $A_{ji} = A_{ij}$ for all indices $1 \leq i, j \leq n$ and is *antisymmetric* if $A_{ji} = -A_{ij}$.
   (a) Prove that the sets

$$\mathcal{S}_n = \text{(symmetric matrices)}$$
$$\mathcal{A}_n = \text{(antisymmetric matrices)}$$

   are both vector subspaces of matrix space $\mathrm{M}(n, \mathbb{K})$.
   (b) If $\mathbb{K}$ is a field in which $1_{\mathbb{K}} + 1_{\mathbb{K}} \neq 0_{\mathbb{K}}$, explain why every square matrix A is a sum $A = A_s + A_a$ of a symmetric and an antisymmetric matrix, so the sum of sets $\mathcal{S}_n + \mathcal{A}_n$ is all of $\mathrm{M}(m, \mathbb{K})$.
   (c) If $\mathbb{K}$ is a field in which $1_{\mathbb{K}} + 1_{\mathbb{K}} \neq 0_{\mathbb{K}}$, prove that the decomposition $A = A_s + A_a$ is *unique*.

   HINTS: If the decomposition in (c) fails to be unique for some matrix A there would be a related matrix B that is both symmetric and antisymmetric. That is impossible unless we have $1 + 1 = 0$ in the ground field. (This can happen, for instance, when $\mathbb{K} = \mathbb{Z}_2$, but not in fields such as $\mathbb{Q}$, $\mathbb{R}$, and $\mathbb{C}$.)

Section 1.3. Solving Systems of Linear Equations

1. TRUE/FALSE QUESTIONS: ("True" if the statement is *always* true.)
   (a) The linear span of the empty set $\varnothing$ is the empty set.
   (b) The linear span of a nonempty subset S in a vector space V is the intersection of all vector subspaces $W \subseteq V$ that contain S.
   (c) In solving a system of linear equations, the solution set is unaffected if you multiply all terms in one of the equations by the same constant $c \in \mathbb{K}$.
   (d) In solving a system of linear equations, the solution set is unaffected if you multiply each equation by a different randomly chosen nonzero constant.
   (e) In solving a system of linear equations, the solution set is unaffected if you add a multiple of one equation to a different equation.
   (f) Every system of linear equations in n unknowns has at least one solution.
   (g) If A is an $n \times m$ matrix, $\mathbf{x} \in \mathbb{K}^m$, and $\mathbf{b} \in \mathbb{K}^n$, the solution set of the matrix equation $A\mathbf{x} = \mathbf{b}$ is a vector subspace of $\mathbb{K}^n$.

2. A homogeneous system $A\mathbf{x} = \mathbf{0}$ of n linear equations in m unknowns $x_1, \dots, x_m$ is *overdetermined* if there are more constraint equations than unknowns (so $n > m$). Do such systems have $x_1 = \cdots = x_n = 0$ as their only solution? Explain.

3. Solve the following systems of equations using row operations:

(a) $\begin{cases} x_1 + 2x_2 + 2x_3 & = 2 \\ x_1 + 2x_2 + 8x_3 + 5x_4 & = -16 \\ x_1 + x_2 + 5x_3 + 5x_4 & = 3 \end{cases}$ (b) $\begin{cases} x_1 + 2x_2 + 6x_3 = -1 \\ 2x_1 + x_2 + x_3 = 8 \\ 3x_1 + x_2 - x_3 = 15 \\ x_1 + 3x_2 + 10x_3 = -5 \end{cases}$

4. Determine the intersection of the following subspaces in $\mathbb{R}^3$:

(i) $W_1 = \{(a_1, a_2, a_3) : a_1 = 3a_2 \text{ and } a_3 = -a_2\}$
(ii) $W_2 = \{(a_1, a_2, a_3) : 2a_1 - 7a_2 + a_3 = 0\}$

by finding a set of basis vectors for $W_1 \cap W_2$. What is the dimension of this intersection?

HINT: The intersection is the solution set of all the equations defining W_1 and W_2.

5. For each list of vectors given below decide whether the first vector can be written as a linear combination of the other two.

(a) In $\mathbb{R}^3$: $(1, 2, 3)$; $(-3, 2, 1), (2, -1, -1)$
(b) In $\mathbb{K}[x]$: $-2x^3 - 11x^2 + 3x + 2$; $x^3 - 2x^2 + 3x - 1$, $2x^3 + x^2 + 3x - 2$

6. If $\mathbf{b} \in \mathbb{R}^n$, A is a matrix in $\mathrm{M}(n, \mathbb{R})$, and $\widetilde{A} = \lambda \cdot A$ is a scalar multiple with $\lambda \neq 0$, how are the solution sets of the following matrix equations related?

(a) $A\mathbf{x} = \mathbf{0}$ and $\widetilde{A}(\mathbf{x}) = \mathbf{0}$
(b) $A\mathbf{x} = \mathbf{b}$ and $\widetilde{A}(\mathbf{x}) = \lambda \cdot \mathbf{b}$

If $H = \{\mathbf{x} : A\mathbf{x} = \mathbf{b}\}$ and $\widetilde{H} = \{\mathbf{x} : \widetilde{A}\mathbf{x} = \mathbf{b}\}$,

(c) show that the solution set $\widetilde{H}$ is equal to $\frac{1}{\lambda} \cdot H$.

7. A "line" in $\mathbb{R}^3$ is any translate $L = \mathbf{x}_0 + E$ of some 1-dimensional subspace E; a "hyperplane" is any translate $H = \mathbf{x}_0 + E$ of some 2-dimensional vector subspace E. Answer the questions below about the following set of vectors in $\mathbb{R}^3$:

$$\mathbf{v}_1 = (2, -5, 1), \quad \mathbf{v}_2 = (0, 4, 6), \quad \mathbf{v}_3 = (-3, 7, 1)$$

(a) Are these vectors *collinear*? Do they lie in some line $L \subseteq \mathbb{R}^3$ (not necessarily one passing through the origin)?
(b) Prove that they lie in some hyperplane H by exhibiting (i) a 2-dimensional subspace E (determined by exhibiting basis vectors $\mathbf{f}_1, \mathbf{f}_2$), and (ii) a vector $\mathbf{x}_0 \in \mathbb{R}^3$ such that the given vectors all lie in the hyperplane $H = \mathbf{x}_0 + E$.
(c) Use (b) to provide an explicit "parametric" description of H as the range of a map $(x_1, x_2, x_3) = \phi(s_1, s_2)$ from $\mathbb{R}^2$ into $\mathbb{R}^3$ that sends $\mathbb{R}^2$ bijectively to the hyperplane H.
(d) Find a 3×2 matrix A and a vector $\mathbf{b} \in \mathbb{R}^3$ such that H is the solution set of the inhomogeneous linear system $A\mathbf{x} = \mathbf{b}$.

HINT: Start by considering some difference vectors $\mathbf{v}_i - \mathbf{v}_j$.

NOTE: The vector subspace E is uniquely determined, but the translation vector $\mathbf{x}_0$ is not, since it can be replaced by $\mathbf{y}_0 = \mathbf{x}_0 + \mathbf{e}$ where $\mathbf{e} \in E$. However, all such choices of the translation vector yield the same hyperplane $\mathbf{x}_0 + E$ in $\mathbb{R}^3$ because $\mathbf{e} + E = E$ when $\mathbf{e} \in E$ (Figure 1.4 suggests what is happening.)

Section 1.4. Linear Span, Independence, and Bases

1. TRUE/FALSE QUESTIONS: ("True" if the statement is *always* true.)

(a) If $S = \{v_1, \dots, v_r\}$ is a linearly dependent set containing at least two vectors, then v_1 is a linear combination of the *other* vectors in S.

(b) A set S that contains the zero vector cannot be independent.

(c) A nonempty subset of a set S whose vectors are linearly independent is itself linearly independent.

(d) A nonempty subset of a set S whose vectors are linearly dependent is always linearly dependent.

(e) If $c_1v_1 + \cdots + c_mv_m = 0$ in a vector space V and the v_i are independent, then all coefficients c_i must be zero.

(f) A nonempty set S of linearly dependent vectors always contains a subset S' that is independent and has the same linear span as S.

(g) If a linearly independent set S contains at least two vectors and we remove one vector, the remaining set S' is independent but $\mathbb{K}\text{-span}(S') \subsetneqq \mathbb{K}\text{-span}(S)$.

(h) For $n > 2$, the set of rank-2 matrices in $\mathrm{M}(n, \mathbb{K})$ is a vector subspace of $\mathrm{M}(n, \mathbb{K})$.

2. Let $E = \mathbb{R}\text{-span}\{M_1, M_2, , M_3, M_4\}$ be the linear span of the 2×2 matrices in $\mathrm{M}(2, \mathbb{R})$ listed below.

$$M_1 = \begin{pmatrix} 1 & 2 \\ -3 & 4 \end{pmatrix}, \quad M_2 = \begin{pmatrix} 1 & 0 \\ -1 & 0 \end{pmatrix}, \quad M_3 = \begin{pmatrix} 0 & 1 \\ 0 & 1 \end{pmatrix}, \quad M_4 = \begin{pmatrix} 1 & 1 \\ 0 & 0 \end{pmatrix}$$

Use row operations to do the following:

(a) Verify that the vectors $\{M_1, \dots, M_4\}$ are not linearly independent.

(b) Find a set of matrices $A_1, \dots, A_r$ in $\mathrm{M}(2, \mathbb{R})$ that are a basis for E. What is $\dim(E)$?

(c) Find a proper subset $S \subseteq \{M_1, \dots, M_4\}$ that is independent and spans E, and hence is a basis for E.

HINT: You can identify 2×2 matrices with elements of $\mathbb{R}^4$ by listing the matrix entries as 4-tuples $(a_{11}, a_{12}, a_{21}, a_{22})$. $\mathrm{M}(2, \mathbb{R})$ and $\mathbb{R}^4$ are isomorphic as vector spaces.

3. Consider the following vectors in $\mathbb{K}^3$

$$\mathbf{u}_1 = (1, 2, 3), \quad \mathbf{u}_2 = (2, 3, 4), \quad \mathbf{u}_3 = (2, 3, 4)$$

(a) Are these vectors independent? Do they span $\mathbb{K}^3$?

(b) Describe their linear span W by identifying a set of basis vectors.

(c) Does the vector $\mathbf{b} = (2, -1, -1)$ lie in the subspace W?

4. In matrix space $\mathrm{M}(n, \mathbb{K})$ the "matrix units" E_{ij} $(1 \leq i, j \leq n)$ are defined by

$$E_{ij} = \text{ the matrix with entries } (E_{ij})_{rs} = \begin{cases} 1 & \text{if } r = i \text{ and } s = j \\ 0 & \text{for all other } (r, s) \end{cases}$$

so the only nonzero entry in E_{ij} is a "1" in the (i, j) spot.

(a) Write down the four matrix units E_{ij} in $\mathrm{M}(2, \mathbb{K})$.

(b) Explain why the matrix units $\{E_{ij} : 1 \leq i, j \leq n\}$ are a basis for $\mathrm{M}(n, \mathbb{K})$.

(c) Prove that products of these units are given by the simple law

$$E_{ij} \cdot E_{k\ell} = \delta_{i,k} \cdot \delta_{j,\ell} \cdot E_{i\ell}$$

where δ_{pq} is the *Kronecker delta symbol*, which is equal to 1 when $p = q$ and to 0 otherwise.

NOTE: Property (c) is what makes the basis E_{ij} so handy in matrix calculations. Once it has been verified, (c) eliminates the need to carry out tedious row/column multiplications in computing matrix products.

5. If nonzero polynomials $f_1, \ldots, f_n$ in $\mathbb{K}[x]$ have different degrees explain why this set of polynomials is linearly independent.

6. Show that the matrices

$$M_1 = \begin{pmatrix} 1 & 0 \\ 0 & 0 \end{pmatrix}, \quad M_2 = \begin{pmatrix} 0 & 0 \\ 0 & 1 \end{pmatrix}, \quad M_2 = \begin{pmatrix} 0 & 1 \\ 1 & 0 \end{pmatrix}$$

are a basis for the subspace $\mathcal{S}_2$ of symmetric matrices in $\mathrm{M}(2, \mathbb{K})$.

7. What is the dimension of the space of symmetric matrices $\mathcal{S}_n$ that are in $\mathrm{M}(n, \mathbb{C})$?

8. The *trace* of a matrix $A \in \mathrm{M}(n, \mathbb{K})$ is $\mathrm{Tr}(A) = \sum_{i=1}^{n} A_{ii}$ (sum of the diagonal entries). Verify that $E = \{A \in \mathrm{M}(n, \mathbb{K}) : \mathrm{Tr}(A) = 0\}$ is a vector subspace and find a basis. What is $\dim(E)$?

9. Explain why the polynomials $\mathbb{1}, x, x^2, x^3, \ldots, x^n, \ldots$ are a basis for the infinite-dimensional vector space $\mathbb{K}[x]$.

10. Let E be the linear span of the following vectors in $\mathbb{R}^4$

$$\mathbf{v}_1 = (1, 0, -1, 2), \quad \mathbf{v}_2 = (4, 5, 1, -9), \quad \mathbf{v}_3 = (2, 1, 1, -3),$$
$$\mathbf{v}_4 = (-2, 1, 0, -2), \quad \mathbf{v}_5 = (-1, 1, -1, -1)$$

(a) Find a basis for E. What is $\dim(E)$?

(b) A basis can also be found by deleting certain vectors from the original list. Exhibit such a basis.

11. If $S_1 \subseteq S_2$ are nonempty sets in a vector space V,

(a) Explain why $\mathbb{K}\text{-span}(S_1) \subseteq \mathbb{K}\text{-span}(S_2)$.

(b) Give an example of sets in $\mathbb{K}^n$ such that

$$S_1 \subseteq S_2, \ S_1 \neq S_2, \text{ and yet } \mathbb{K}\text{-span}(S_1) = \mathbb{K}\text{-span}(S_2).$$

12. Verify that $W = \{\mathbf{z} = (z_1, \dots, z_n) \in \mathbb{C}^n : z_1 + \cdots + z_n = 0\}$ is a subspace of the complex vector space $V = \mathbb{C}^n$. Find a $\mathbb{C}$-basis and the dimension of E as a vector space over $\mathbb{C}$.

13. Let E be the linear span in $\mathbb{K}^4$ of the vectors

$$\mathbf{v}_1 = (1,2,3,4), \quad \mathbf{v}_2 = (2,0,4,5), \quad \mathbf{v}_3 = (-1,1,-3,-4), \quad \mathbf{v}_4 = (2,3,4,5)$$

(a) Give a parametric description of E by finding a basis. What is $\dim(E)$?

(b) Give an implicit description: find a system of equations $A\mathbf{x} = \mathbf{0}$ whose solution set is equal to E.

14. Suppose a finite-dimensional vector space V contains subspaces E_1, E_2 and that $\dim V = 10$, $\dim E_1 = 7$, $\dim E_2 = 5$. Let $E_1 + E_2$ be the vector subspace spanned by E_1 and E_2.

(a) What is the *maximum* possible dimension for $E_1 + E_2$? Give an example of subspaces in $\mathbb{R}^{10}$ for which maximum dimension is achieved.

(b) What is the *minimum* possible dimension for $E_1 + E_2$? Give an example of subspaces in $\mathbb{R}^{10}$ for which the minimum is achieved.

(c) Can V be a direct sum $V = E_1 \oplus E_2$ of these subspaces?

(d) What is the maximum possible dimension for $E_1 \cap E_2$?

(e) If $E_1, E_2 \subseteq \mathbb{R}^{10}$, find the minimum possible dimension of $E_1 \cap E_2$. What if they lie in $\mathbb{R}^{15}$?

15. In $V = \mathbb{R}^4$ consider the subspaces $E_1 = \mathbb{R}\text{-span}\{\mathbf{v}_1, \mathbf{v}_2\}$ and $E_2 = \mathbb{R}\text{-span}\{\mathbf{u}_1, \mathbf{u}_2\}$, where

$$\begin{aligned} \mathbf{v}_1 &= (1,0,-1,2), & \mathbf{u}_1 &= (3,2,4,12), \\ \mathbf{v}_2 &= (2,1,1,8), & \mathbf{u}_2 &= (1,1,3,4), \end{aligned}$$

which we may regard as the rows of a 4×4 matrix

$$A = \begin{pmatrix} 1 & 0 & -1 & 2 \\ 2 & 1 & 1 & 8 \\ 3 & 2 & 4 & 12 \\ 1 & 1 & 3 & 4 \end{pmatrix}.$$

Suppose elementary row operations on A and its transpose A^{T} yield the following echelon forms:

$$A \xrightarrow[\text{ops}]{\text{row}} \begin{pmatrix} 1 & 0 & -1 & 2 \\ 0 & 1 & 3 & 4 \\ 0 & 0 & 1 & -2 \\ 0 & 0 & 0 & 0 \end{pmatrix}, \qquad A^{\mathrm{T}} \xrightarrow[\text{ops}]{\text{row}} \begin{pmatrix} 1 & 2 & 3 & 1 \\ 0 & -1 & 2 & 1 \\ 0 & 0 & 1 & 1 \\ 0 & 0 & 0 & 0 \end{pmatrix}.$$

Use this information to answer the following questions.

(a) What are the column ranks of A and A^{T}?

(b) Find a basis in $\mathbb{R}^4$ for the linear span $E_1 + E_2$. What is the dimension of $E_1 + E_2$?

(c) What is $\dim(E_1 \cap E_2)$?

HINT: For (c) recall Exercise 1.67. Explain your answers.

16. For the spaces E_1, E_2 in Additional Exercise 15:
(a) Find the dimension of the intersection $E_1 \cap E_2$.
(b) Find a basis for $E_1 \cap E_2$.

NOTE: Part (a) can be answered from the data given in Additional Exercise 15 using the dimension formula of Exercise 1.67. The usual efforts with row operations do not yield an answer in part (b); a solution requires direct calculations involving the basis vectors $\mathbf{u}_1, \mathbf{u}_2$ and $\mathbf{v}_1, \mathbf{v}_2$.

HINTS: Find implicit descriptions of the E_i as solutions of matrix equations $A_i\mathbf{x} = \mathbf{0}$ for $\mathbf{x} \in \mathbb{R}^4$. To be in $E_1 \cap E_2$ a vector $\mathbf{x}$ must be a solution for each system. Combining these systems yields a system of 4 equations in 4 unknowns whose solution set is precisely $E_1 \cap E_2$. Finding a basis for the solution set of the combined system is then a routine matter.

17. In Example 1.37 we determined the linear span E of particular vectors $\mathbf{u}_1, \ldots, \mathbf{u}_5$ in $\mathbb{K}^3$. Regarding them as column vectors we formed the 3×5 matrix $A = \text{col}(\mathbf{u}_1; \ldots; \mathbf{u}_5)$, and then found E using the fact that the span is precisely the set of vectors $\mathbf{b} \in \mathbb{K}^3$ for which the system $A\mathbf{x} = \mathbf{b}$ has solutions.

In Example 1.40 we showed that E could be calculated more directly using row operations on the 5×3 transpose matrix $B = A^{\mathrm{T}}$ whose rows are the $\mathbf{u}_i$ regarded as row vectors, so $R_i(B) = \mathbf{u}_i$. Carry out this approach, using row operations to find basis vectors for E.

NOTE: Both approaches must yield sets of vectors that span E, but a vector space can have many different bases.

Section 1.5. Quotient Spaces.

1. TRUE/FALSE QUESTIONS: ("True" if statement is *always* true.)
(a) If W is a one-dimensional subspace in $V = \mathbb{R}^3$ (a line through the origin), the quotient space V/W can be interpreted as the set of all lines in $\mathbb{R}^3$ that are parallel to W.
(b) If $V \supseteq W$ and both vector spaces are infinite dimensional, the quotient space V/W must also be infinite dimensional.
(c) If $E \supseteq V \supseteq W$ are subspaces in a finite-dimensional vector space, then $\dim E/W = \dim E/V + \dim V/W$.

2. How are the quotient space V/W and the (linear) quotient map $\pi : V \to V/W$ with $\pi(x) = x + W$ in V/W to be interpreted when:
(a) $W = \{0\}$ (the trivial subspace in V)? (b) When $W = V$?

3. Find the dimension $\dim(V/W)$ and an explicit basis for this quotient space when $V = \mathbb{R}^4$ and

$$W = \{\mathbf{x} \in \mathbb{R}^4 : x_1 + x_2 - x_3 - x_4 = 0 \quad\text{and}\quad 4x_1 - 3x_2 - x_4 = 0\}$$

It will suffice to give coset representatives $\mathbf{v} \in \mathbb{R}^4$ for the basis vectors $\mathbf{v} + W$ in V/W.

4. Let W be the linear span of the following vectors in $V = \mathbb{C}^5$:

$$\mathbf{v}_1 = (1, 1, 2, 0, 3), \quad \mathbf{v}_2 = (3, 2, 1, 5, -1), \quad \mathbf{v}_3 = (2, 1, 2, 0, -1)$$

(a) Find vectors $\mathbf{e}_{i_1}, \mathbf{e}_{i_2}$ selected from the standard basis $\mathbf{e}_1, \dots, \mathbf{e}_5$ in $V = \mathbb{C}^5$ such that $\{\mathbf{v}_1, \mathbf{v}_2, \mathbf{v}_3, \mathbf{e}_{i_1}, \mathbf{e}_{i_2}\}$ is a basis for V.

(b) Prove that the cosets $\overline{\mathbf{e}}_{i_1}, \overline{\mathbf{e}}_{i_2}$ are a basis for the quotient space V/W.

5. What is $\dim(E_1 + E_2)/(E_1 \cap E_2)$ in Additional Exercise 14 for Section 1.4?

6. If E_1, E_2 are subspaces of V, find formulas for calculating the dimensions of the following quotient spaces in terms of $|V|, |E_1|, |E_2|$, and $|E_1 \cap E_2|$:

(a) $(E_1 + E_2)/E_1$

(b) $(E_1 + E_2)/(E_1 \cap E_2)$

Appendix: The Degree Formula for $\mathbb{K}[\mathbf{x}_1, \dots, \mathbf{x}_N]$

Let $\mathbb{K}[\mathbf{x}] = \mathbb{K}[x_1, \dots, x_N]$ be the algebra of polynomials with coefficients in a field $\mathbb{K}$ and let $\mathbb{1}$ (constant polynomial $\equiv 1$) be its multiplicative identity element. Using the multi-index notation introduced in Section 1.1 we can write any such polynomial as a finite sum (finitely many nonzero coefficients)

$$f(\mathbf{x}) = \sum_{\alpha \in \mathbb{Z}_+^N} a_\alpha x^\alpha \quad (x^\alpha = x_1^{\alpha_1} \cdot \dots \cdot x_N^{\alpha_N},\ c_\alpha \in \mathbb{K}) \tag{1}$$

The degree of a monomial x^α is $|\alpha| = \alpha_1 + \dots + \alpha_N$ and if $f \in \mathbb{K}[\mathbf{x}]$ is not the zero polynomial (all $a_\alpha = 0$), its degree is defined to be

$$m = \deg(f) = \max\{|\alpha| : c_\alpha \neq 0\} >$$

When $N > 1$ several different monomials x^α of the same maximum total degree $|\alpha| = m$ may have nonzero coefficients.

Let $f, g \neq 0$ in $\mathbb{K}[\mathbf{x}]$ with degrees $m = \deg(f), n = \deg(g)$. Their product is

$$\begin{aligned}(f \cdot g)(\mathbf{x}) = \Big(\sum_\alpha a_\alpha x^\alpha\Big) \cdot \Big(\sum_\beta b_\beta x^\beta\Big) &= \sum_{\alpha,\beta} a_\alpha b_\beta x^{\alpha+\beta} \\ &= \sum_\gamma \Big(\sum_{\alpha+\beta=\gamma} a_\alpha b_\beta\Big) x^\gamma = \sum_\gamma c_\gamma x^\gamma\end{aligned}$$

where $\alpha + \beta = (\alpha_1 + \beta_1, \dots, \alpha_N + \beta_n)$. If $a_\alpha b_\beta x^{\alpha+\beta} \neq 0$ in (1) we must have $|\alpha| \le m$ and $|\beta| \le n$, so that $|\alpha + \beta| \le m + n$; consequently, $\deg(f \cdot g) \le \deg(f) + \deg(g)$.

Let us split off the monomials of maximum degree, writing

$$f(\mathbf{x}) = \sum_{|\alpha|=m} a_\alpha x^\alpha + \cdots,$$
$$g(\mathbf{x}) = \sum_{|\beta|=n} b_\beta x^\beta + \cdots,$$
$$(f \cdot g)(\mathbf{x}) = \sum_{|\gamma|=m+n} c_\gamma x^\gamma + \cdots,$$

where $(\cdots)$ are terms of lower degree. To prove the degree formula,

(2) DEGREE FORMULA: $\deg(f \cdot g) = \deg(f) + \deg(g)$

for $f, g \neq 0$ in $\mathbb{K}[\mathbf{x}]$, it suffices to show

(3) There is at least one monomial x^{γ_0} having the maximal degree $m + n$ whose coefficient $c_{\gamma_0} = \sum_{\alpha+\beta=\gamma_0} a_\alpha b_\beta$ is nonzero.

This is trivial for $N = 1$, but problematic when $N \geq 2$ because this sum of products can be zero if there is more than one term, even if the individual terms are nonzero. On the other hand, the degree formula (2) follows immediately if we can prove the following:

(4) There exists some monomial x^γ of maximal degree $m + n$ for which the sum (3) consists of a single nonzero term.

The key to proving (4) is to introduce a ranking of the monomials x^γ, γ in $\mathbb{Z}_+^N$, more refined than ranking by total degree $\deg(x^\gamma) = |\gamma|$, which cannot distinguish between the various monomials of the same degree. The tool for doing this is *lexicographic* or *dictionary* ordering of the indices in $\mathbb{Z}_+^N$, an idea that has proved useful in many parts of mathematics.

DEFINITION A.1 (Lexicographic order). For multi-indices $\alpha = (\alpha_1, \dots, \alpha_N)$, $\beta = (\beta_1, \dots, \beta_N)$ in $\mathbb{Z}_+^N$, we define the relation $\alpha \succ \beta$ to mean

$$\alpha_i > \beta_i \text{ at the first index } i = 1, 2, \dots, N \text{ at which } \alpha_i \text{ differs from } \beta_i$$

Thus

$$\alpha_1 = \beta_1, \dots, \alpha_{i-1} = \beta_{i-1} \text{ and } \alpha_i > \beta_i \quad \text{(other entries in } \alpha, \beta \text{ are irrelevant).}$$

This is a *linear ordering* of multi-indices: given α, β exactly one of the possibilities

$$\alpha \succ \beta \qquad \alpha = \beta \qquad \beta \succ \alpha$$

holds. We write $\alpha \succeq \beta$ when the possibility $\alpha = \beta$ is allowed.

Obviously $\alpha = (0, \dots, 0)$ is the lowest multi-index in lexicographic order, and any finite set of multi-indices has a unique highest element. Note carefully that $\alpha \succ \beta$ does not imply that $|\alpha| \geq |\beta|$. For instance we have

$$\alpha = (1, 0, 0) \succ \beta = (0, 2, 2) \text{ in lexicographic order, but } |\beta| = 4 > |\alpha| = 1\,.$$

Other elementary properties of lexicographic order are easily verified once you understand the definitions.

EXERCISE A.2. For lexicographic order in $\mathbb{Z}_+^N$ verify that we have:

(a) LINEAR ORDERING. For any pair α, β we have exactly one of the possibilities $\alpha \succ \beta$, $\alpha = \beta$, $\beta \succ \alpha$.
(b) TRANSITIVITY OF ORDER. If $\alpha \succ \beta$ and $\beta \succ \gamma$ then $\alpha \succ \gamma$.
(c) If $\alpha \succ \alpha'$, then $\alpha + \beta \succ \alpha' + \beta$ for all indices β.
(d) If $\alpha \succ \alpha'$ and $\beta \succ \beta'$, then $\alpha + \beta \succ \alpha' + \beta'$.

HINT: It might help to make diagrams showing how the various N-tuples are related. You will have to do some "casework" in (c).

We now outline a proof of the crucial fact (4), leaving the final details as an exercise for the reader. If $f \neq 0$ with $m = \deg(f)$, so $f = \sum_{|\alpha| \leq m} a_\alpha x^\alpha$, there may be several monomials having maximal degree m with $a_\alpha \neq 0$, but just one of them is maximal with respect to lexicographic order, namely,

$$\alpha_0 = \max_{\succ}\{\alpha : |\alpha| = m \text{ and } a_\alpha \neq 0.\}$$

Likewise, there is a unique index β_0 for $h = \sum_{|\beta| \leq n} b_\beta x^\beta$ such that $|\beta| = \deg(h) = n$ and

$$\beta_0 = \max_{\succ}\{\beta : |\beta| = n \text{ and } b_\beta \neq 0\}.$$

The multi-index $\gamma_0 = \alpha_0 + \beta_0$ has $|\gamma_0| = m + n$ and is a likely candidate for the solution to (4); note that $a_{\alpha_0} b_{\beta_0} \neq 0$ *by definition*. We leave the reader to verify a few simple properties of this particular multi-index.

EXERCISE A.3. Explain why $\alpha_0 = \max_{\succ}\{\alpha : |\alpha| = m \text{ and } a_\alpha \neq 0\}$ might not be the same as $\alpha_0' = \max_{\succ}\{\alpha : a_\alpha \neq 0\}$. Is there any reason to expect α_0' to have maximal degree $|\alpha_0'| = m$?

EXERCISE A.4. In $\gamma_0 = \alpha_0 + \beta_0$ we have $|\alpha_0| = m$, $|\beta_0| = n$, and $a_{\alpha_0} b_{\beta_0} \neq 0$, by definition. If α, β are any indices such that

$$|\alpha + \beta| = |\alpha_0 + \beta_0| = m + n \quad \text{and} \quad a_\alpha b_\beta \neq 0$$

prove that we must have $|\alpha| = |\alpha_0| = m$ and $|\beta| = |\beta_0| = n$.

Defining $\alpha_0, \beta_0, \gamma_0 = \alpha_0 + \beta_0$ as above, we make the following claim:

CLAIM: If $\alpha + \beta = \alpha_0 + \beta_0$ and $a_\alpha b_\beta \neq 0$, then $\alpha = \alpha_0$ and $\beta = \beta_0$. Hence the sum

$$c_{\gamma_0} = \sum_{\alpha + \beta = \gamma_0} a_\alpha b_\beta \tag{5}$$

reduces to the single nonzero term $a_{\gamma_0} b_{\beta_0}$.

EXERCISE A.5. Prove the claim made in (5) using the facts assembled in the preceding discussion.

That will complete the proof of the degree formula.

CHAPTER 2

Linear Operators $T : V \to W$

2.1. Definitions and General Facts

DEFINITION 2.1. A map $T : V \to W$ between vector spaces is a *linear operator* if for any $v, v_1, v_2 \in V$ and $\lambda \in \mathbb{K}$:

1. Scaling operations are preserved: $T(\lambda \cdot v) = \lambda \cdot T(v)$
2. Sums are preserved: $T(v_1 + v_2) = T(v_1) + T(v_2)$

This is equivalent to saying

$$T\Big(\sum_i \lambda_i v_i\Big) = \sum_i \lambda_i T(v_i) \quad \text{in } W$$

for all finite linear combinations of vectors in V. We refer to V as the *initial space* and W as the *target space* of T. The *range* $R(T) = T(V)$ of the operator is a subspace of W, and we often refer to its dimension $\operatorname{rank}(T) = \dim R(T)$ as the *rank* of T.

A trivial example is the *zero operator* $T(v) = 0_W$ for every $v \in V$. If $W = V$ the *identity operator*, $\mathrm{id}_V : V \to V$, is given by $\mathrm{id}(v) = v$ for all vectors. Some basic properties of any linear operator $T : V \to W$ are:

1. $T(0_V) = 0_W$. [PROOF: $T(0_V) = T(0 \cdot 0_V) = 0 \cdot T(0_V) = 0_W$.]
2. $T(-v) = -T(v)$. [PROOF: $T(-v) = T((-1) \cdot v) = (-1) \cdot T(v) = -T(v)$.]
3. A linear operator is determined by its action on any set S of vectors that span V. If $T_1, T_2 : V \to W$ are linear operators and

$$T_1(s) = T_2(s) \quad \text{for all } s \in S,$$

then $T_1 = T_2$ everywhere on V. [PROOF: Any $v \in V$ is a finite linear combination $v = \sum_i c_i s_i$ with $s_i \in V$; then $T_1(v) = \sum_i c_i T_1(s_i) = \sum_i c_i T_2(s_i) = T_2(v)$.]

EXERCISE 2.2. If $S \subseteq V$ and $T : V \to W$ is a linear operator, prove that

$$T(\mathbb{K}\text{-span}\{S\}) \quad \text{is equal to} \quad \mathbb{K}\text{-span}\{T(S)\}.$$

DEFINITION 2.3. We write $\mathrm{Hom}_{\mathbb{K}}(V, W)$ for the space of linear operators $T : V \to W$. This becomes a vector space over $\mathbb{K}$ if for any $v \in V$, $\lambda \in \mathbb{K}$ we define:

1. $(T_1 + T_2)(v) = T_1(v) + T_2(v)$.
2. $(\lambda \cdot T)(v) = \lambda \cdot (T(v))$.

The vector space axioms are easily verified. The zero element in $\mathrm{Hom}(V, W)$ is the zero operator $\mathbf{0}(v) = 0_W$ for $v \in V$. The additive inverse $-T$ is the operator $-T(v) = (-1) \cdot T(v) = T(-v)$. It is a scalar multiple of T: $-T = (-1) \cdot T$.

If $V = W$ we can also define the *composition product* $S \circ T$ of two operators,

$$(S \circ T)(v) = S(T(v)) \quad \text{for all } v \in V.$$

This makes $\mathrm{Hom}_{\mathbb{K}}(V) = \mathrm{Hom}_{\mathbb{K}}(V, V)$ a noncommutative associative algebra with identity $I = \mathrm{id}_V$.

A linear operator $T : V \to W$ over $\mathbb{K}$ determines two important vector subspaces, the *kernel* $K(T) = \ker(T)$ in the initial space V and the *range* $R(T) = \mathrm{range}(T)$ in the target space W.

1. $K(T) = \ker(T) = \{v \in V : T(v) = 0_W\}$. The dimension of this space is sometimes referred to as $\mathrm{Nullity}(T)$.
2. $R(T) = \mathrm{range}(T)$ is the image set $T(V) = \{T(v) : v \in V\}$. The *rank* of T is defined to be the dimension of $R(T)$,

$$\mathrm{rank}(T) = \dim_{\mathbb{K}}(R(T)),$$

which is sometimes abbreviated as $\mathrm{rk}(T)$. It is easily seen that $\ker(T)$ and $\mathrm{range}(T)$ are vector subspaces of V and W, respectively.

DEFINITION 2.4. We say that a linear map $T : V \to W$ is:

1. *One-to-one* if $x \neq y$ in $V \Rightarrow T(x) \neq T(y)$ in W, so distinct points in V cannot have the same image in W.
2. *Surjective*, or an "onto" map, if every point $y \in W$ is the image of at least one $x \in V$.
3. *Bijective* if T is both one-to-one and onto.

When T is bijective we say that T is *invertible*: obviously a set-theoretic inverse map $T^{-1} : W \to V$ exists that undoes the action of T, and we will see in Section 2.2 that T^{-1} is in fact a linear map.

We often have to decide whether a linear map is one-to-one, onto, a bijection, or "none of the above." For surjectivity we must compute the range $R(T)$; determining whether T is one-to-one is easier, and amounts to computing the kernel $K(T)$. The diagram in Figure 2.1 illustrates the general behavior of any linear operator. Each coset $v + K(T)$ of the kernel gets mapped to a single point in W because

$$T(v + K(T)) = T(v) + T(K(T)) = T(v) + 0_W = T(v)$$

and distinct cosets go to different points in W. All points in V get mapped into the range $R(T) \subseteq W$.

LEMMA 2.5. *A linear operator $T : V \to W$ is one-to-one if and only if* $\ker(T) = 0$.

PROOF. ($\Leftarrow$) If $T(v_1) = T(v_2)$ for $v_1 \neq v_2$, then $0 = T(v_1) - T(v_2) = T(v_1 - v_2)$, so $v_2 - v_1 \neq 0$ so $\ker(T)$ is nontrivial. Writing "$\neg P$" for "not (P)", we

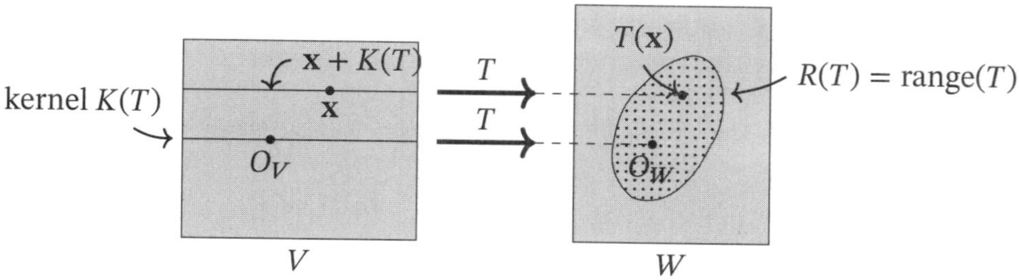

FIGURE 2.1. A linear map $T : V \to W$ sends all points in a coset $x + K(T)$ of the kernel to a single point $T(x)$ in W. Different cosets map to different points, and all images land within the range $R(T)$. The zero coset $0_V + K(T)$ collapses to the origin 0_W in W.

have just proved the contrapositive $\neg(T \text{ is one-to-one}) \Rightarrow \neg(K(T) = \{0\})$ of the statement ($\Leftarrow$) we want, but the two are logically equivalent. □

MORAL: If you want to prove ($P \Rightarrow Q$) it is sometimes easier to prove the equivalent *contrapositive statement* ($\neg Q \Rightarrow \neg P$), as was the case here.

PROOF. ($\Rightarrow$) If T is one-to-one and $v \neq 0_V$, then $Tv \neq T(0_V) = 0_W$ so $v \notin K(T)$ and $K(T)$ reduces to $\{0\}$. □

Dimension Theorems for Kernel $K(T)$ and Range $R(T)$. The following important result is closely related to Theorem 1.79 for quotient spaces.

THEOREM 2.6 (Dimension Theorem). *If $T : V \to W$ is a linear operator and V is finite dimensional, the range $R(T)$ is finite dimensional and is related to the kernel $K(T)$ via*

$$\dim(R(T)) + \dim(K(T)) = \dim(V). \tag{2.1}$$

In words, "rank + nullity = dimension of the initial space V." This can also be expressed concisely by writing $|R(T)| + |K(T)| = |V|$, adopting our convention of writing $\dim(V) = |V|$ *whenever convenient.*

PROOF. The kernel is finite dimensional since $K(T) \subseteq V \Rightarrow \dim(K(T)) \leq \dim(V) < \infty$. The range $R(T)$ is also finite dimensional. In fact, if $\{v_1, \dots, v_n\}$ is a basis for V every $w \in R(T)$ has the form $w = T(v) = \sum_i c_i T(v_i)$, so the vectors $T(v_1), \dots, T(v_n)$ span $R(T)$. Therefore $\dim(R(T)) \leq n = \dim(V)$.

Let $\{w_1, \dots, w_m\}$ be a basis for $K(T)$. By adjoining additional vectors from V we can obtain a basis $\{w_1, \dots, w_m, v_{m+1}, \dots, v_{m+k}\}$ for V. Obviously, $m = \dim(K(T))$ and $m+k = \dim(V)$, so to prove $k = \dim(R(T))$ we show the vectors $T(v_{m+1}), \dots, T(v_{m+k})$ are a basis for $R(T)$. They certainly span $R(T)$ because $w \in R(T) \Rightarrow w = T(v)$ for some $v \in V$, which can be written

$$v = c_1 w_1 + \cdots + c_m w_m + c_{m+1} v_{m+1} + \cdots + c_{m+k} v_{m+k} \quad (c_j \in \mathbb{K}).$$

Since $w_j \in K(T) \Rightarrow T(w_j) = 0_W$ we see that

$$w = T(v) = 0_W + \cdots + 0_W + \sum_{j=1}^{k} c_{m+j} T(v_{m+j}),$$

so $w \in \mathbb{K}\text{-span}\{T(v_{m+1}), \dots, T(v_{m+k})\}$. These vectors are also independent, for if

$$0_W = \sum_{i=1}^{k} c_{m+i} T(v_{m+i}) = T\Big(\sum_{i=1}^{k} c_{m+i} v_{m+i}\Big)$$

that means $\sum_{i=1}^{k} c_{m+i} v_{m+i} \in K(T)$ and there are coefficients $c_1, \dots, c_m$ such that $\sum_{i=1}^{m} c_i w_i = \sum_{j=1}^{k} c_{m+j} v_{m+j}$, or

$$0_V = -c_1 w_1 - \cdots - c_m w_m + c_{m+1} v_{m+1} + \cdots + c_{m+k} v_{m+k}$$

Because $\{w_1, \dots, v_{m+k}\}$ is a basis for V, we must have $c_1 = \cdots = c_{m+k} = 0$, proving independence of $T(v_{m+1}), \dots, T(v_{m+k})$. Thus $\dim(R(T)) = k$ as claimed. $\square$

COROLLARY 2.7. *Let $T : V \to W$ be a linear operator between finite-dimensional vector spaces such that* $\dim(V) = \dim(W)$, *which certainly holds if $V = W$. Then the following assertions are equivalent*:

(i) *T is one-to-one.*
(ii) *T is surjective.*
(iii) *T is bijective.*

PROOF. By the dimension theorem we have $|K(T)| + |R(T)| = |V|$. If T is one-to-one then $K(T) = (0)$, so by (2.1) we get $|R(T)| = |V| = |W|$. Since $R(T) \subseteq W$, this only happens if $R(T) = W$; i.e., T is surjective. Finally, if T is surjective, then $|R(T)| = |W| = |V|$ by hypothesis. Invoking (2.1) we see that $|K(T)| = 0$, so the kernel is trivial and T is one-to-one.

We just proved that T is one-to-one if and only if T is surjective, so either condition implies T is bijective. $\square$

EXERCISE 2.8. Explain why a spanning set $\mathfrak{X} = \{v_1, \dots, v_n\}$ is a basis for a finite-dimensional space if and only the vectors in $\mathfrak{X}$ are independent (so $\sum_{i=1}^{n} c_i v_i = 0_V \Rightarrow c_1 = \cdots = c_n = 0$).

PROPOSITION 2.9. *Let V be a finite-dimensional vector space and $\{v_1, \dots, v_n\}$ a basis. Select any n vectors $w_1, \dots, w_n$ in some other vector space W. Then there is a unique linear operator $T : V \to W$ such that $T(v_i) = w_i$ for $1 \le i \le n$.*

PROOF. Uniqueness of T (if it exists) was proved in our initial comments about linear operators. To construct such a T, define $T(\sum_{i=1}^{n} \lambda_i v_i) = \sum_{i=1}^{n} \lambda_i w_i$ for all choices of $\lambda_1, \dots, \lambda_n \in \mathbb{K}$. This is obviously well-defined since $\{v_i\}$ is a basis; T is easily seen to be a linear operator from $V \to W$. $\square$

Computing $K(T)$ and $R(T)$. One prolific source of linear operators is the correspondence between $n \times m$ matrices A with entries in $\mathbb{K}$ and the linear operators $L_A : \mathbb{K}^m \to \mathbb{K}^n$ determined by matrix multiplication

$$A(\mathbf{v}) = A \cdot \mathbf{v} \quad (\text{matrix product } (n \times m) \cdot (m \times 1) = (n \times 1))$$

where we regard vectors $\mathbf{v} \in \mathbb{K}^m$ as $m \times 1$ column matrices.

EXAMPLE 2.10. Let $L_A : \mathbb{R}^3 \to \mathbb{R}^4$ (or $\mathbb{C}^3 \to \mathbb{C}^4$, same discussion) be the linear operator associated with the 4×3 matrix

$$A = \begin{pmatrix} 1 & 2 & 3 \\ 1 & 0 & 2 \\ 2 & 1 & 1 \\ 1 & 1 & 1 \end{pmatrix}.$$

Describe $\ker(L_A)$ and $\text{range}(L_A)$ and find explicit bases in each of these spaces.

SOLUTION: The range $R(L_A)$ of L_A is determined by finding all $\mathbf{y}$ for which there is an $\mathbf{x} \in \mathbb{K}^3$ such that $A\mathbf{x} = \mathbf{y}$. Symbolic row reduction of the augmented matrix $[A : \mathbf{y}]$ yields

$$\left(\begin{array}{ccc|c} 1 & 2 & 3 & y_1 \\ 1 & 0 & 2 & y_2 \\ 2 & 1 & 1 & y_3 \\ 1 & 1 & 1 & y_4 \end{array}\right) \to \left(\begin{array}{ccc|c} 1 & 2 & 3 & y_1 \\ 0 & -2 & -1 & y_2 - y_1 \\ 0 & -3 & -5 & y_3 - 2y_1 \\ 0 & -1 & -2 & y_4 - y_1 \end{array}\right) \to \left(\begin{array}{ccc|c} 1 & 2 & 3 & y_1 \\ 0 & 1 & 2 & y_1 - y_4 \\ 0 & 2 & 1 & y_1 - y_2 \\ 0 & 3 & 5 & 2y_1 - y_3 \end{array}\right)$$

$$\to \left(\begin{array}{ccc|c} 1 & 2 & 3 & y_1 \\ 0 & 1 & 2 & y_1 - y_4 \\ 0 & 0 & -3 & y_1 - y_2 - 2(y_1 - y_4) \\ 0 & 0 & -1 & 2y_1 - y_3 - 3(y_1 - y_4) \end{array}\right)$$

$$\to \left(\begin{array}{ccc|c} \boxed{1} & 2 & 3 & y_1 \\ 0 & \boxed{1} & 2 & y_1 - y_4 \\ 0 & 0 & \boxed{1} & y_1 + y_3 - 3y_4 \\ 0 & 0 & 0 & 2y_1 - y_2 + 3y_3 - 7y_4 \end{array}\right).$$

There are no solutions $\mathbf{x} \in \mathbb{K}^3$ unless $\mathbf{y} \in \mathbb{K}^4$ lies in the 3-dimensional solution set of the condition

$$2y_1 - y_2 + 3y_3 - 7y_4 = 0.$$

When this constraint is satisfied, backsolving yields exactly one solution for each such $\mathbf{y}$, and there are no free variables.

Thus $R(L_A)$ is the solution set of equation

$$2y_1 - y_2 + 3y_3 - 7y_4 = 0.$$

When written as a matrix equation $C\mathbf{y} = \mathbf{0}$ ($C = 1 \times 4$ matrix $\text{col}(2, -1, 3, -7)$), y_2, y_3, y_4 are free variables and then $y_1 = \frac{1}{2}(y_2 - 3y_3 + 7y_4)$, so a typical vector in $R(L_A)$ has the form

$$\mathbf{y} = \begin{pmatrix} \frac{1}{2}(y_2 - 3y_3 + 7y_4) \\ y_2 \\ y_3 \\ y_4 \end{pmatrix} = y_2 \cdot \begin{pmatrix} \frac{1}{2} \\ 1 \\ 0 \\ 0 \end{pmatrix} + y_3 \cdot \begin{pmatrix} -\frac{3}{2} \\ 0 \\ 1 \\ 0 \end{pmatrix} + y_4 \cdot \begin{pmatrix} \frac{7}{2} \\ 0 \\ 0 \\ 1 \end{pmatrix}$$

with $y_1, y_2, y_3 \in \mathbb{K}$. The rescaled column vectors $\mathbf{u}_2 = \text{col}(1,2,0,0)$, $\mathbf{u}_3 = \text{col}(-3,0,2,0)$, $\mathbf{u}_4 = \text{col}(7,0,0,2)$ obviously span $R(L_A)$ and are easily seen to be linearly independent, so they are a basis for the range, which has dimension $|R(L_A)| = 3$.

THE KERNEL: Now we want to find all solutions $\mathbf{x} \in \mathbb{K}^3$ of the homogeneous equation $A\mathbf{x} = \mathbf{0}$. The same row operations used above transform $[A : \mathbf{0}]$ to

$$\left(\begin{array}{ccc|c} \boxed{1} & 2 & 3 & 0 \\ 0 & \boxed{1} & 2 & 0 \\ 0 & 0 & \boxed{1} & 0 \\ 0 & 0 & 0 & 0 \end{array}\right).$$

Then $\mathbf{x} = \text{col}(x_1, x_2, x_3)$ has entries $x_1 = x_2 = x_3 = 0$. Therefore

$$K(L_A) = \{\mathbf{x} \in \mathbb{K}^3 : L_A(\mathbf{x}) = A\mathbf{x} = \mathbf{0}\} \text{ is the trivial subspace } \{\mathbf{0}\}$$

and there is no basis to be found.

Note that $|R(L_A)| + |K(L_A)| = 3 + 0 =$ dimension of the *initial space* $\mathbb{K}^3$, while the target space $W = \mathbb{K}^4$ has dimension $= 4$.

EXAMPLE 2.11. Let $L_A : \mathbb{R}^4 \to \mathbb{R}^4$ be the linear operator associated with the 4×4 matrix

$$A = \begin{pmatrix} 1 & 2 & 3 & 0 \\ 1 & 0 & 2 & -1 \\ 2 & 1 & 1 & 2 \\ 1 & 1 & 1 & 1 \end{pmatrix}.$$

Describe the kernel $K(L_A)$ and range $R(L_A)$ by finding basis vectors.

SOLUTION: The range of L_A is determined by finding all $\mathbf{y}$ such that $A\mathbf{x} = \mathbf{y}$ for some $\mathbf{x} \in \mathbb{R}^4$. Row reduction of $[A : \mathbf{y}]$ yields

$$\left(\begin{array}{cccc|c} 1 & 2 & 3 & 0 & y_1 \\ 1 & 0 & 2 & -1 & y_2 \\ 2 & 1 & 1 & 2 & y_3 \\ 1 & 1 & 1 & 1 & y_4 \end{array}\right) \to \left(\begin{array}{cccc|c} 1 & 2 & 3 & 0 & y_1 \\ 0 & -2 & -1 & -1 & y_2 - y_1 \\ 0 & -3 & -5 & 2 & y_3 - 2y_1 \\ 0 & -1 & -2 & 1 & y_4 - y_1 \end{array}\right)$$

$$\to \left(\begin{array}{cccc|c} 1 & 2 & 3 & 0 & y_1 \\ 0 & 1 & 2 & -1 & y_1 - y_4 \\ 0 & 2 & 1 & 1 & y_1 - y_2 \\ 0 & 3 & 5 & -2 & 2y_1 - y_3 \end{array}\right)$$

$$\to \left(\begin{array}{cccc|c} 1 & 2 & 3 & 0 & y_1 \\ 0 & 1 & 2 & -1 & y_1 - y_4 \\ 0 & 0 & -3 & 3 & y_1 - y_2 - 2(y_1 - y_4) \\ 0 & 0 & -1 & 1 & 2y_1 - y_3 - 3(y_1 - y_4) \end{array}\right)$$

$$\to \left(\begin{array}{cccc|c} \boxed{1} & 2 & 3 & 0 & y_1 \\ 0 & \boxed{1} & 2 & -1 & y_1 - y_4 \\ 0 & 0 & \boxed{1} & -1 & y_1 + y_3 - 3y_4 \\ \hline 0 & 0 & 0 & 0 & 2y_1 - y_2 + 3y_3 - 7y_4 \end{array}\right).$$

There are *no* solutions $\mathbf{x}$ in $\mathbb{R}^4$ unless $\mathbf{y}$ lies in the 3-dimensional solution set of the linear constraint equation

$$2y_1 - y_2 + 3y_3 - 7y_4 = 0$$

i.e., $\mathbf{y}$ is a solution of the matrix equation

$$C\mathbf{y} = \mathbf{0} \quad \text{where} \quad C = [2, -1, 3, -7]_{1\times 4}$$

There are multiple solutions of $C(\mathbf{y}) = \mathbf{0}$, so $R(L_A) = \{\mathbf{y} \in \mathbb{R}^4 : C\mathbf{y} = \mathbf{0}\}$ is nontrivial. Multiplying C by $\frac{1}{2}$ puts it in echelon form, so y_2, y_3, y_4 are free variables in solving $C\mathbf{y} = \mathbf{0}$ and $y_1 = \frac{1}{2}(y_2 - 3y_3 + 7y_4)$. Thus a vector $\mathbf{y} \in \mathbb{R}^4$ is in the range $R(L_A) \Leftrightarrow$

$$\mathbf{y} = \begin{pmatrix} \frac{1}{2}(y_2 - 3y_3 + 7y_4) \\ y_2 \\ y_3 \\ y_4 \end{pmatrix} = y_2 \cdot \begin{pmatrix} \frac{1}{2} \\ 1 \\ 0 \\ 0 \end{pmatrix} + y_3 \cdot \begin{pmatrix} -\frac{3}{2} \\ 0 \\ 1 \\ 0 \end{pmatrix} + y_4 \cdot \begin{pmatrix} \frac{7}{2} \\ 0 \\ 0 \\ 1 \end{pmatrix}$$

with $y_1, y_2, y_3 \in \mathbb{R}^3$. The column vectors

$$\mathbf{u}_2 = \text{col}(1, 2, 0, 0), \quad \mathbf{u}_3 = \text{col}(-3, 0, 2, 0), \quad \mathbf{u}_4 = \text{col}(7, 0, 0, 2),$$

obviously span $R(L_A)$ and are easily seen to be linearly independent, so they are a basis for the range and $\dim(R(L_A)) = 3$. (Hence also $|K(L_A)| = |V| - |R(L_A)| = 1$ by the dimension theorem.)

THE KERNEL: $K(L_A)$ can be found by setting $\mathbf{y} = 0$ in the preceding echelon form of $[A : \mathbf{y}]$, which becomes

$$\left(\begin{array}{cccc|c} \boxed{1} & 2 & 3 & 0 & 0 \\ 0 & \boxed{1} & 2 & -1 & 0 \\ 0 & 0 & \boxed{1} & -1 & 0 \\ 0 & 0 & 0 & 0 & 0 \end{array}\right)$$

Now x_4 is the free variable and backsolving yields

$$\begin{aligned} x_3 &= x_4 \\ x_2 &= -2x_3 + x_4 = -x_4 \\ x_1 &= -2x_2 - 3x_3 = -x_4 \end{aligned}$$

hence

$$K(L_A) = \mathbb{K} \cdot \begin{pmatrix} -1 \\ -1 \\ 1 \\ 1 \end{pmatrix}$$

is one-dimensional as expected.

Given a vector $\mathbf{y}$ in the range $R(L_A)$ we can find a particular solution $\mathbf{x}_0$ of $A\mathbf{x} = \mathbf{y}$ by setting the free variable $x_4 = 0$. Then

$$\begin{aligned} x_3 &= x_4 + y_1 + y_3 - 3y_4 = y_1 + y_3 - 3y_4, \\ x_2 &= -2x_3 + x_4 + y_1 - y_4 = -y_1 - 2y_3 + 5y_4, \\ x_1 &= -2x_2 - 3x_3 + y_1 = y_1 + (2y_1 + 4y_3 - 10y_4) + (-3y_1 - 3y_3 + 9y_4) \\ &= y_3 - y_4 \end{aligned}$$

and

$$\mathbf{x}_0 = \begin{pmatrix} y_3 - y_4 \\ -y_1 - 2y_3 + 5y_4 \\ y_1 + y_3 - 3y_4 \\ 0 \end{pmatrix}$$

is a particular solution for $A\mathbf{x} = \mathbf{y}$. The full set of solutions is the additive coset $\mathbf{x}_0 + K(L_A)$ of the kernel of L_A.

EXERCISE 2.12. If $A \in \mathrm{M}(n \times m, \mathbb{K})$ and $\mathbf{e}_1, \dots, \mathbf{e}_m$ are the standard basis vectors in $\mathbb{K}^m$, viewed as $m \times 1$ column vectors:

(a) Explain why the image $L_A(\mathbf{e}_i) = A \cdot \mathbf{e}_i$ in W is just the ith column of the matrix A.

(b) Prove that the range $R(L_A)$ is precisely the *column space* of A, the subspace of $\mathbb{K}^n$ spanned by the columns of A.

Thus in Example 2.11 we have $\dim R(L_A) = \dim(\mathrm{ColSpace}(A)) = 3$.

2.2. Isomorphisms and Invariant Subspaces

If $T : V \to V$ is a linear operator, a subspace $W \subseteq V$ is T-*invariant* if $T(W) \subseteq W$. As we shall see, invariant subspaces are important in determining the structure of T.

Eigenvalues, Eigenspaces, and the Characteristic Polynomial $p_T(x)$. The subspaces $\{\mathbf{0}\}$, $R(T) = \mathrm{range}(T) = T(V)$, $K(T) = \ker(T)$, and V are all T-invariant. Structural analysis of T proceeds initially by searching for *eigenvectors* nonzero vectors $v \in V$ such that $T(v)$ is a scalar multiple $T(v) = \lambda \cdot v$ for some $\lambda \in \mathbb{K}$. These are precisely the vectors in $\ker(T - \lambda I)$ where $I : V \to V$ is the identity operator on V. Eigenvectors may or may not exist; when they do they have a story to tell.

DEFINITION 2.13. Fix a linear map $T : V \to V$ and scalar $\lambda \in \mathbb{K}$. The *λ-eigenspace* is

$$E_\lambda(T) = \{v \in V : T(v) = \lambda \cdot v\} = \{v \in V : (T - \lambda I)v = 0\} = \ker(T - \lambda I)$$

A scalar $\lambda \in \mathbb{K}$ is an *eigenvalue* for T if the eigenspace is nontrivial, $E_\lambda(T) \neq (0)$.

If V is finite dimensional, we will eventually see that the number of eigenvalues is $\leq n = \dim(V)$ (and could be zero) because it is the set of roots in $\mathbb{K}$ of the *characteristic polynomial* p_T

$$p_T(x) = \det(T - xI) \in \mathbb{K}[x],$$

which has degree $n = \dim_{\mathbb{K}}(V)$. The set of *distinct* eigenvalues in $\mathbb{K}$ is called the *spectrum* of T and is denoted

$$\begin{aligned} \mathrm{sp}_{\mathbb{K}}(T) &= \{\lambda \in \mathbb{K} : \text{such that } T(v) = \lambda \cdot v \text{ for some } v \neq 0\} \\ &= \{\lambda \in \mathbb{K} : E_\lambda(T) \neq (0)\}. \end{aligned}$$

Depending on the nature of the ground field, $\mathrm{sp}_{\mathbb{K}}(T)$ may be the empty set; it is always nonempty if $\mathbb{K} = \mathbb{C}$, because every nonconstant polynomial has at least one root in $\mathbb{C}$ (Fundamental Theorem of Algebra). Furthermore, all eigenspaces E_λ are T-invariant subspaces because T and $(T - \lambda I)$ "commute," hence

$$(T - \lambda I)(Tv) = T((T - \lambda I)v) = T(0) = 0 \quad \text{if } v \in E_\lambda.$$

The E_λ are also "essentially disjoint" from each other in the sense that $E_\mu \cap E_\lambda = (0)$ if $\mu \neq \lambda$. (You can't have $\lambda \cdot v = \mu \cdot v$ (or $(\lambda - \mu) \cdot v = 0$) for nonzero v if $\mu \neq \lambda$.) Note that $\ker(T)$ is the eigenspace corresponding to $\lambda = 0$ since

$$E_{\lambda=0} = \{v : (T - 0 \cdot I)(v) = T(v) = 0\} = \ker(T),$$

so $\lambda = 0$ is an eigenvalue in $\mathrm{sp}_{\mathbb{K}}(T) \Leftrightarrow$ this kernel is nontrivial $\Leftrightarrow T$ is not one-to-one. When $\lambda = 1$, $E_{\lambda=1}$ is just the set of fixed points under the action of T,

$$E_{\lambda=1} = \mathrm{Fix}(T) = \{v : T(v) = v\} \quad \text{(fixed points in } V\text{)}.$$

Decomposition of Operators. If $W \subseteq V$ is an *invariant subspace*, so $T(W) \subseteq W$, then T induces linear operators in W and in the quotient space V/W:

1. RESTRICTION: $T|_W : W \to W$ is the restriction of T to W. Consequently, $(T|_W)(w) = T(w)$ for all $w \in W$.
2. QUOTIENT OPERATOR: The operator $\widetilde{T} : V/W \to V/W$, sometimes denoted $T_{V/W}$, is induced by the action of T on additive cosets:

$$T_{V/W}(x + W) = T(x) + W \quad \text{for all cosets in } V/W \tag{2.2}$$

The outcome is determined using a representative x for the coset, but different representatives yield the same result, so $T_{V/W}$ is well defined.

THEOREM 2.14. *Given a linear operator $T : V \to V$ and an invariant subspace W, there is a unique linear operator $\widetilde{T} : V/W \to V/W$ that makes the following diagram "commute" in the sense that $\pi \circ T = \widetilde{T} \circ \pi$, where $\pi : V \to V/W$ is the quotient map.*

$$\begin{array}{ccc} V & \xrightarrow{\ T\ } & V \\ {\scriptstyle \pi}\downarrow & & \downarrow{\scriptstyle \pi} \\ V/W & \xrightarrow{\ \widetilde{T}\ } & V/W \end{array}$$

NOTE: *We have already shown that the quotient map $\pi : V \to V/W$ is a linear operator between vector spaces.*

PROOF OF EXISTENCE. As for the induced map $\widetilde{T}$ on V/W, the fact that $T(W) \subseteq W$ implies

$$T(x + W) = T(x) + T(W) \subseteq T(x) + W \quad \text{for } x \in V$$

This suggests that (2.2) is the right definition. It certainly insures that $\widetilde{T} \circ \pi = \pi \circ T$; the problem is to show the outcome $\widetilde{T}(x+W) = T(x)+W$ is independent of the choice of coset representative $x \in V$. But if $x' + W = x + W$, we have $x' = x' + 0 = x + w_0$ for some $w_0 \in W$ and

$$T(x') + W = T(x + w_0) + W = T(x) + (T(w_0) + W)$$

Since W is invariant $T(w_0) \in W$ and $T(w_0) + W = W$ by Exercise 1.74 of Chapter 1. Hence $T(v') + W = T(v) + W$ and the induced operator in (2.2) is well-defined. Linearity of $\widetilde{T}$ is easily checked: once we know $\widetilde{T}$ is well-defined we get

$$\begin{aligned} &\widetilde{T}((v_1 + W) \oplus (v_2 + W)) \\ &\quad = \widetilde{T}(v_1 + v_2 + W) \quad (\text{definition of } (\oplus) \text{ in } V/W) \\ &\quad = T(v_1 + v_2) + W = T(v_1) + T(v_2) + W \\ &\quad = (T(v_1) + W) + (T(v_2) + W) \quad (\text{since } W + W = W) \\ &\quad = \widetilde{T}(v_1 + W) \oplus \widetilde{T}(v_2 + W), \end{aligned}$$

and similarly

$$\widetilde{T}(\lambda \odot (v + W)) = \widetilde{T}(\lambda \cdot v + W) = \lambda \odot \widetilde{T}(v + W).$$

PROOF OF UNIQUENESS. If $\widetilde{T}_1, \widetilde{T}_2$ both satisfy the commutation relation $\widetilde{T}_i \circ \pi = \pi \circ T_i$, then

$$\widetilde{T}_1(v + W) = \widetilde{T}_1 \circ \pi(v) = \pi(T(v)) = \widetilde{T}_2(\pi(v)) = \widetilde{T}_2(v + W),$$

so $\widetilde{T}_1 = \widetilde{T}_2$ on V/W.

A LOOK AHEAD. If $T : V \to V$ is a linear operator on a finite-dimensional space, we will explain in Section 2.4 how a matrix $[T]_{\mathfrak{X}\mathfrak{X}}$ is associated with T once a basis $\mathfrak{X}$ in V has been specified. If W is a T-invariant subspace we will see that much of the structural information about T resides in the induced operators $T|_W$ and $T_{V/W}$, and that in some sense (to be made precise) T is assembled by "joining together" these smaller pieces. This is a big help in trying to understand the action of T on V, but it does depend on being able to find invariant subspaces—the more the better! To illustrate: if a basis $\mathfrak{X} = \{w_1, \dots, w_m\}$ in W is augmented to get a basis for V,

$$\mathfrak{Z} = \{w_1, \dots, w_m, v_{m+1}, \dots, v_{m+k}\} \quad (m + k = n = \dim(V))$$

we have seen that the image vectors $\overline{v}_{m+i} = \pi(v_{m+i})$ are a basis $\mathfrak{Y}$ in the quotient space V/W. In Section 2.4 we will show that the matrix $[T]_{\mathfrak{Z}\mathfrak{Z}}$ associated with the particular basis $\mathfrak{Z}$ assumes a special "block-upper triangular form":

$$[T]_{\mathfrak{Z}\mathfrak{Z}} = \begin{pmatrix} \boxed{A}_{m\times m} & \boxed{*}_{m\times k} \\ \mathbf{0}_{k\times m} & \boxed{B}_{k\times k} \end{pmatrix}$$

where $A = [T|_W]_{\mathfrak{X}\mathfrak{X}}$ and $B = [T_{V/W}]_{\mathfrak{Y}\mathfrak{Y}}$. Clearly, much of the information about T is encoded in the two diagonal blocks A, B; but some information is lost in passing from T to $T|_W$ and $T_{V/W}$—the "cross-terms" in the upper right block $\boxed{*}$ cannot be determined if we only know the two induced operators. We need additional information to piece them together to recover T, and there may be more than one operator T yielding a particular pair of induced operators $(T|_W, T_{V/W})$.

Isomorphisms of Vector Spaces. A linear map $T : V \to W$ is an *isomorphism* between vector spaces if it is a bijection, and then there is a well-defined set-theoretic inverse map in the opposite direction $T^{-1} : W \to V$

$$T^{-1}(w) = (\textit{the unique } v \in V \textit{ such that } T(v) = w)$$

for any $w \in W$. In general, it might not be easy to describe the inverse of a bijection $f : X \to Y$ between two point sets in closed form

$$f^{-1}(y) = (\textit{some explicit formula involving } y).$$

(Try finding $x = f^{-1}(y)$ where $f : \mathbb{R} \to \mathbb{R}$ is $y = f(x) = x^3 + x + 1$, which is a bijection because $df/dx > 0$ for all x.) But the inverse of a linear map is automatically linear, if it exists. We write $V \cong W$ if there is an isomorphism between the spaces V and W.

EXERCISE 2.15. Suppose $T : V \to W$ is linear and a bijection. Prove that the set-theoretic inverse map $T^{-1} : W \to V$ must also be linear, so

$$T^{-1}(\lambda w) = \lambda \cdot T^{-1}(w) \quad \text{and} \quad T^{-1}(w_1 + w_2) = T^{-1}(w_1) + T^{-1}(w_2).$$

Thus T and T^{-1} are both *isomorphisms* between V and W.
HINT: T^{-1} reverses the action of T and vice versa, so $T \circ T^{-1} = \mathrm{id}_W$ and $T^{-1} \circ T = \mathrm{id}_V$.

EXERCISE 2.16. Show that the relation $V \cong W$ between vector spaces is an *RST equivalence relation*, so it is

(a) REFLEXIVE: $V \cong V$

(b) SYMMETRIC: $V \cong W \Rightarrow W \cong V$
(c) TRANSITIVE: $V \cong W$ and $W \cong Z \Rightarrow V \cong Z$

NOTE: This means the family of all vector spaces is partitioned into disjoint "equivalence classes" that gather together all isomorphic vector spaces in a single class. In particular, for finite-dimensional spaces, all members of a class have the same dimension.

We now observe that an isomorphism between vector spaces V and W identifies important features of V with those in W: it maps

$$\left\{\begin{array}{l}\text{independent sets}\\ \text{spanning sets}\\ \text{bases}\end{array}\right. \text{in } V \quad \longrightarrow \quad \left\{\begin{array}{l}\text{independent sets}\\ \text{spanning sets}\\ \text{bases}\end{array}\right. \text{in } W$$

To illustrate, if $\{v_1, \dots, v_n\}$ are independent in V, then

$$0_W = \sum_{i=1}^{n} c_i T(v_i) = T\Big(\sum_{i=1}^{n} c_i v_i\Big) \implies 0_V = \sum_{i=1}^{n} c_i v_i \text{ in } V \Rightarrow c_1 = \cdots = c_n = 0$$

because $T(0_V) = 0_W$ and T is one-to-one. Thus $T(v_1), \dots, T(v_n)$ are independent in W. Similar arguments yield the other two assertions.

EXERCISE 2.17. If $T : V \to W$ is an isomorphism of vector spaces, verify that

(a) $T(\mathbb{K}\text{-span}\{v_i\}) = \mathbb{K}\text{-span}\{T(v_i)\}$
(b) If $\mathfrak{X} = \{v_i\}$ is a basis in V then $\mathfrak{Y} = \{T(v_i)\}$ is a basis in W.

In particular, isomorphic vector spaces V, W are either both infinite dimensional or both finite-dimensional with the same dimension.

More on Quotient Spaces. Quotient spaces are abstract objects and are not so easy to visualize. The following results allow us to transfer questions about quotient spaces (and linear maps between them) to questions about linear operators $T : V \to V$. In effect, we shift from working with cosets $v + W$ in V/W to computations involving the coset representatives v, which are ordinary vectors. This yields the following result relating operators, quotient spaces, and isomorphisms. It will be cited often in analyzing the structure of linear operators and is even valid for infinite-dimensional spaces.

First and Second Isomorphism Theorems for Quotient Spaces.

THEOREM 2.18 (First Isomorphism Theorem). *Let $T : V \to W$ be a linear map with range $R(T) \subseteq W$. If $K(T) = \ker(T)$, T induces a unique bijective linear map $\widetilde{T} : V/K(T) \to R(T)$ that makes the following diagram "commute" so that $\widetilde{T} \circ \pi = T$, where $\pi : V \to V/W$ is the quotient map. In particular, we have $R(\widetilde{T}) = R(T)$.*

If V is finite dimensional and we let $|E|$ be the dimension of a vector space E, we have

$$|V/K(T)| = |R(\widetilde{T})| = |R(T)| = |V| - |K(T)|$$

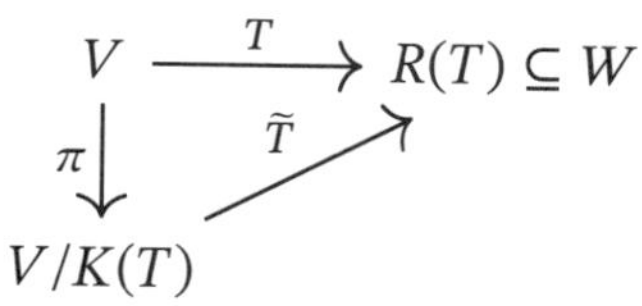

PROOF. On $V/K(T)$ define $\widetilde{T}(v + K(T)) = T(v)$. The outcome is independent of the particular coset representative v, for if $v' + K(T) = v + K(T)$ we get $v' = v+k$ for some $k \in K(T)$, hence $T(v') = T(v)+T(k) = T(v)+0 = T(v)$, and $\widetilde{T} : V/K(T) \to W$ is a well defined map with range $R(T)$. Linearity of $\widetilde{T}$ follows directly from the definition of vector space operations in quotient spaces (recall Section 1.5 of Chapter 1). In fact, since $(v+K(T))+(v'+K(T)) = (v+v')+K(T)$, we get

$$\begin{aligned}\widetilde{T}((v + K(T)) + (v' + K(T))) &= \widetilde{T}((v + v') + K(T)) \\ &= T(v + v') = T(v) + T(v') \quad \text{(definition of } \widetilde{T}\text{)} \\ &= \widetilde{T}(v + K(T)) + \widetilde{T}(v' + K(T)),\end{aligned}$$

and similarly $\widetilde{T}(\lambda \cdot (v + K(T))) = \lambda \cdot \widetilde{T}(v + K(T))$.

The commutation property $\widetilde{T} \circ \pi = T$ is built into the definition of $\widetilde{T}$. As for uniqueness of the map $\widetilde{T}$ that completes the diagram, suppose $S : V/K(T) \to W$ is another linear map such that $S \circ \pi = T$. Then $S(v + K(T)) = S(\pi(v)) = T(v)$, so $S = \widetilde{T}$ as maps on the quotient space $V/K(T)$.

Obviously $R(\widetilde{T}) = R(T)$ because $\widetilde{T}(v + K(T)) = T(v)$ for all $v \in V$. Furthermore, $\widetilde{T} : V/K(T) \to W$ is one-to-one because

$$\widetilde{T}(v + K(T)) = \widetilde{T}(v' + K(T)) \Rightarrow T(v) = T(v');$$

hence $T(v - v') = 0_W$. Then $k = v' - v \in K(T)$ and $v' + K(T) = v + k + K(T) = v + K(T)$ as required.

Since T is an isomorphism between $V/K(T)$ and $R(T)$, their dimensions are the same, so $|R(\widetilde{T})| = |R(T)|$. The Dimension Theorem 2.6 says $|K(T)|+|R(T)| = |V|$, so we conclude that

$$|V/K(T)| = |R(\widetilde{T})| = |R(T)| = |V| - |K(T)| \qquad \square$$

EXERCISE 2.19. If V is a finite-dimensional space and W any subspace, explain why

$$\text{(a) } \dim(W) < \infty \quad \text{and} \quad \text{(b) } \dim(V/W) < \infty$$

EXERCISE 2.20 (Second Isomorphism Theorem). Given finite-dimensional vector spaces $U \supseteq V \supseteq W$ we can form various quotient spaces U/W, U/V, and V/W. Prove that

$$\dim(U/W) = \dim(U/V) + \dim(V/W)\,.$$

HINT: Consider a basis for W, then add vectors to get a basis for v, and continue adding vectors until you have a basis for U.

2.3. Direct Sums of Vector Spaces

A vector space V is an *(internal) direct sum* of subspaces $V_1,\ldots,V_n$, indicated by writing $V = V_1 \oplus \cdots \oplus V_n$, if:

1. The linear span $\sum_{i=1}^n V_i = \{\sum_{i=1}^n v_i : v_i \in V\}$ is all of V.
2. Every $v \in V$ has a *unique* representation as a sum $v = \sum_{i=1}^n v_i$ with $v_i \in V_i$.

Once we know that the V_i span V, condition (2.) is equivalent to saying

$$2^*. \quad \sum_{i=1}^{n} v_i = 0 \quad \text{with } v_i \in V_i \;\Rightarrow\; v_i = 0 \text{ for all } i.$$

In fact, if a vector v has two different representations $v = \sum_i v_i = \sum v_i'$, then $0 = \sum_{i=1}^n v_i''$ with $v_i'' = (v_i' - v_i) \in V_i$, so (2*.) implies $v_i'' = 0$ and $v_i' = v_i$ for all i. Conversely, if we can write $0 = \sum w_i$ with $w_i \in V_i$ not all zero, then the representation of a vector as $v = \sum_i v_i$ $(v_i \in V_i)$ cannot be unique, for if

$$0 = \sum_{i=1}^{n} w_i \quad \text{with } w_i \neq 0 \text{ for some } i,$$

then $v = v + 0 = \sum_i (v_i + w_i)$ in which $v_i + w_i \in V_i$ is $\neq v_i$. Thus condition (2.) is equivalent to (2*).

EXAMPLE 2.21. Here are some examples of direct sum decompositions.

1. $\mathbb{K}^n = V \oplus W$ where

$$V = \{(x_1, x_2, 0, \ldots, 0) : x_1, x_2 \in \mathbb{K}\} \text{ and } W = \{(0, 0, x_3, \ldots, x_n) : x_k \in \mathbb{K}\}$$

   More generally, in an obvious sense we have $\mathbb{K}^{m+n} \cong \mathbb{K}^m \oplus \mathbb{K}^n$.

2. The space of polynomials $\mathbb{K}[x] = V \oplus W$ is a direct sum of the subspaces
   - EVEN POLYNOMIALS: $V = \{\sum_{i=0}^{\infty} a_i x^i : a_i = 0 \text{ for odd indices}\}$
   - ODD POLYNOMIALS: $W = \{\sum_{i=0}^{\infty} a_i x^i : a_i = 0 \text{ for even indices}\}$
3. For $\mathbb{K} = \mathbb{Q}, \mathbb{R}, \mathbb{C}$, matrix space $\mathrm{M}(n, \mathbb{K})$ is a direct sum $\mathcal{A} \oplus \mathcal{S}$ of
   - ANTISYMMETRIC MATRICES: $\mathcal{A} = \{A : A^{\mathsf{T}} = -A\}$
   - SYMMETRIC MATRICES: $\mathcal{S} = \{A : A^{\mathsf{T}} = A\}$.

   In fact, since $(A^{\mathsf{T}})^{\mathsf{T}} = A$ we can write any matrix as

$$A = \frac{1}{2}(A^{\mathsf{T}} + A) + \frac{1}{2}(A^{\mathsf{T}} - A).$$

   Since the first term is symmetric and the second is antisymmetric, $\mathrm{M}(n, \mathbb{K}) = \mathcal{A} + \mathcal{S}$ (linear span).

   If $B \in \mathcal{A} \cap \mathcal{S}$ then $B^{\mathsf{T}} = -B$ and also $B^{\mathsf{T}} = B$, hence $B = -B$ and $B = \mathbf{0}$ (the zero matrix). Thus $\mathcal{A} \cap \mathcal{S} = (0)$ and Exercise 2.22 (below) implies that $\mathrm{M}(n, \mathbb{K}) = \mathcal{A} \oplus \mathcal{S}$.

   NOTE: This actually works for any field $\mathbb{K}$ in which $2 = 1 + 1 \neq 0$ because the "$\frac{1}{2}$" in the formulas involves division by 2. In particular, it works for the finite fields $\mathbb{Z}_p = \mathbb{Z}/p\mathbb{Z}$ except for $\mathbb{Z}_2 = \mathbb{Z}/2\mathbb{Z}$, in which $[1] \oplus [1] = [1+1] = [2] = [0]$

If subspaces $V_1, \dots, V_n$ span V and $n \geq 3$, it can be tricky to verify that V is a direct sum, but if there are just two summands V_1 and V_2 (the case most often encountered) there is a simple criterion for deciding whether $V = V_1 \oplus V_2$.

EXERCISE 2.22. If E, F are subspaces of V show that V is the direct sum $E \oplus F$ if and only if

(a) They span V: $E + F = \{a + b : a \in E, b \in F\}$ is all of V.
(b) Their intersection is trivial: $E \cap F = \{0\}$.

This is NOT true when $n \geq 3$. If $\sum_{i=1}^n V_i = V$ and the spaces are only "pairwise disjoint," so that

$$V_i \cap V_j = (0) \quad \text{for } i \neq j;$$

this is not enough to insure that V is a direct sum of the given subspaces (see the following exercise).

EXERCISE 2.23. Find three distinct 1-dimensional subspaces V_i in the 2-dimensional space $\mathbb{R}^2$ such that

(a) $V_i \cap V_j = (0)$ for $i \neq j$
(b) $\sum_{i=1}^3 V_i = \mathbb{R}^2$

Explain why $\mathbb{R}^2$ is *not* a direct sum $V_1 \oplus V_2 \oplus V_3$ of these subspaces.

EXERCISE 2.24. If $V = V_1 \oplus \cdots \oplus V_n$ and V is finite dimensional, we have seen that each V_i must be finite-dimensional with $\dim(V_i) \leq \dim(V)$.

(a) Given bases $\mathfrak{X}_1 \subseteq V_1, \dots, \mathfrak{X}_n \subseteq V_n$, explain how to create a basis for all of V.
(b) Prove $\dim_{\mathbb{K}}(V_1 \oplus \cdots \oplus V_n) = \sum_{i=1}^n \dim_{\mathbb{K}}(V_i)$.

Direct sum decompositions play a large role in unraveling the structure of linear operators. Suppose $T : V \to V$ and $V = V_1 \oplus V_2$, and that *both* subspaces are T-invariant. We get restricted operators $T_1 = T|_{V_1} : V_1 \to V_1$ and $T_2 = T|_{V_2} : V_2 \to V_2$, but because both subspaces are T-invariant we can fully reconstruct the original operator T in V from its "components" T_1 and T_2. In fact, every $v \in V$ has a unique decomposition $v = v_1 + v_2$ ($v_i \in V_i$) and then

$$T(v) = T(v_1) + T(v_2) = T_1(v_1) + T_2(v_2).$$

We often indicate such decompositions by writing $T = T_1 \oplus T_2$.

This does not work if only one subspace is invariant. But when both are invariant and we take bases $\mathfrak{X}_1 = \{v_1, \dots, v_m\}$, $\mathfrak{X}_2 = \{v_{m+1}, \dots, v_{m+k}\}$ for V_1, V_2, we will soon see that the combined basis $\mathfrak{Y} = \{v_1, \dots, v_m, v_{m+1}, \dots, v_{m+k}\}$ for all of V yields a matrix of particularly simple *block-diagonal* form

$$[T]_{\mathfrak{Y}} = \begin{pmatrix} \boxed{A}_{m\times m} & \mathbf{0}_{m\times k} \\ \mathbf{0}_{k\times m} & \boxed{B}_{k\times k} \end{pmatrix}$$

where A, B are the matrices of T_1, T_2 with respect to the bases $\mathfrak{X}_1 \subseteq V_1$ and $\mathfrak{X}_2 \subseteq V_2$.

Projection Operators and Direct Sums. If $V = V_1 \oplus \cdots \oplus V_n$, then for each i there is a natural *projection operator* $P_i : V \to V_i \subseteq V$, the "*projection of V onto V_i along the complementary subspace* $\bigoplus_{j \neq i} V_j$." By definition we have

$$(2.3) \qquad P_i(v) = v_i \quad \text{if } v = \sum_{j=1}^{n} v_j \text{ is the unique decomposition with } v_j \in V_j.$$

Note that $\ker(P_i) \supseteq V_j$ for all $j \neq i$, so $K(P_i) \supseteq \bigoplus_{j \neq i} V_j$. The following basic properties of these projection operators are easily verified.

EXERCISE 2.25. Show that the projections P_i associated with a direct sum decomposition $V = V_1 \oplus \cdots \oplus V_n$ have the following properties:

(i) LINEARITY: Each $P_i : V \to V$ is a linear operator.
(ii) IDEMPOTENT PROPERTY: $P_i^2 = P_i \circ P_i = P_i$ for all i.
(iii) $P_i \circ P_j = 0$ if $i \neq j$.
(iv) $\text{Range}(P_i) = V_i$ and $\ker(P_i)$ is the linear span $\sum_{j \neq i} V_j$.
(v) $P_1 + \cdots + P_n = I$ (identity operator on V).

If we represent vectors $v \in V$ as ordered n-tuples $(v_1, \ldots, v_n)$ in the Cartesian product set $V_1 \times \ldots \times V_n$, the ith projection takes the form

$$P_i(v_1, \ldots, v_n) = (0, \ldots, 0, v_i, 0, \ldots, 0) \in V_i \subseteq V.$$

Don't be misled by this notation into thinking that we are speaking of orthogonal projections (onto orthogonal subspaces in $\mathbb{R}^n$). The following example and exercises illustrate what's really happening. A vector space must be equipped with additional structure such as an *inner product* if we want to speak of "orthogonality of vectors" or their "lengths." Such notions are meaningless in an unadorned vector space. Nevertheless, inner product spaces are important and will be fully discussed in Chapter 6.

EXAMPLE 2.26. The plane $\mathbb{R}^2$ is a direct sum of the subspaces $V_1 = \mathbb{R}\mathbf{e}_1$ and $V_2 = \mathbb{R}(\mathbf{e}_1 + \mathbf{e}_2)$, where $\mathfrak{X} = \{\mathbf{e}_1, \mathbf{e}_2\}$ are the standard basis vectors $\mathbf{e}_1 = (1, 0)$ and $\mathbf{e}_2 = (0, 1)$ in $\mathbb{R}^2$. The maps

- P_1 projecting V onto V_1 along V_2
- P_2 projecting V onto V_2 along V_1

are *oblique projections*; they are not the familiar *orthogonal projections* sending $\mathbf{x} = (x_1, x_2)$ to $(x_1, 0)$ and to $(0, x_2)$, respectively; see Figure 2.2. Find an explicit formula for $P_i(v_1, v_2)$ in $\mathbb{R}^2$ for arbitrary pairs (v_1, v_2) in $\mathbb{R}^2$.

DISCUSSION: To calculate these projections we must write an arbitrary vector $\mathbf{v} = (v_1, v_2) = v_1\mathbf{e}_1 + v_2\mathbf{e}_2$ in the form $a + b \in V_1 \oplus V_2$ where $a \in V_1 = \mathbb{R}\mathbf{e}_1$ and $b \in V_2 = \mathbb{R}(\mathbf{e}_1 + \mathbf{e}_2)$. The vectors $\mathfrak{Y} = \{\mathbf{f}_1, \mathbf{f}_2\}$,

$$(2.4) \qquad \mathbf{f}_1 = \mathbf{e}_1 \quad \text{and} \quad \mathbf{f}_2 = \mathbf{e}_1 + \mathbf{e}_2,$$

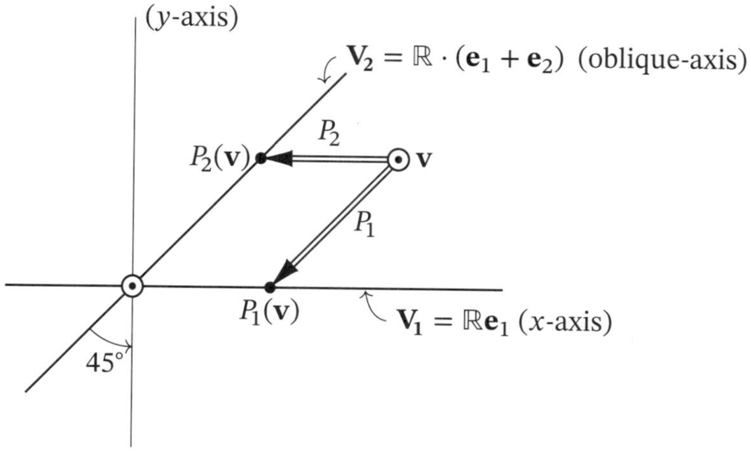

FIGURE 2.2. Projections P_1, P_2 determined by a direct sum decomposition $V = V_1 \oplus V_2$. Here $V = \mathbb{R}^2$, $V_1 = \mathbb{R}\mathbf{e}_1$, $V_2 = \mathbb{R}(\mathbf{e}_1 + \mathbf{e}_2)$; P_1 projects vectors $v \in V$ onto V_1 along V_2, and likewise for P_2.

that determine the 1-dimensional spaces V_1, V_2 are obviously a new basis for $\mathbb{R}^2$. If $v = c_1\mathbf{f}_1 + c_2\mathbf{f}_2$ in the new basis, the action of the projections P_1, P_2 can be written immediately based on the definitions

$$\begin{aligned} P_1(c_1\mathbf{f}_1 + c_2\mathbf{f}_2) &= c_1\mathbf{f}_1 \quad (= c_1\mathbf{e}_1), \\ P_2(c_1\mathbf{f}_1 + c_2\mathbf{f}_2) &= c_2\mathbf{f}_2 \quad (= c_2(\mathbf{e}_1 + \mathbf{e}_2)). \end{aligned} \tag{2.5}$$

In terms of the standard basis $\mathfrak{X}$ in $\mathbb{R}^2$ we have $v = (v_1, v_2) = v_1\mathbf{e}_1 + v_2\mathbf{e}_2$, and to answer our question we must describe v and the images $P_1(v), P_2(v)$ in terms of the same basis $\mathfrak{X}$.

First, observe that the action of P_2 is known as soon as we know the action of P_1 because $P_1 + P_2 = I$, hence

$$P_2(v) = (I - P_1)(v) = v - P_1(v) \quad \text{for all } v \in \mathbb{R}^2 .$$

Second, the projections P_i are linear, so their action is known once we determine the images $P_i(\mathbf{e}_k)$ of the basis vectors $\mathbf{e}_k$ because

$$P_i(v) = P_i(v_1, v_2) = P_i(v_1\mathbf{e}_1 + v_2\mathbf{e}_2) = v_1 \cdot P_i(\mathbf{e}_1) + v_2 \cdot P_i(\mathbf{e}_2) \quad (v_1, v_2 \in \mathbb{R})$$

The last step is to use the vector identities (2.4) to write the standard basis vectors $\{\mathbf{e}_1, \mathbf{e}_2\}$ in terms of the new basis $\{\mathbf{f}_1, \mathbf{f}_2\}$; then the action of P_1 in the standard basis is easily evaluated by applying (2.5). From (2.4) we get

$$\mathbf{e}_1 = \mathbf{f}_1 \quad \text{and} \quad \mathbf{e}_2 = \mathbf{f}_2 - \mathbf{f}_1.$$

Then (2.5) implies

$$\begin{aligned} P_1(\mathbf{e}_1) &= P_1(\mathbf{f}_1) = \mathbf{f}_1 = \mathbf{e}_1, \\ P_1(\mathbf{e}_2) &= P_1(\mathbf{f}_2 - \mathbf{f}_1) = -\mathbf{f}_1 = -\mathbf{e}_1, \end{aligned}$$

and similarly

$$P_2(\mathbf{e}_1) = P_2(\mathbf{f}_1) = \mathbf{0}$$
$$P_2(\mathbf{e}_2) = P_2(\mathbf{f}_2 - \mathbf{f}_1) = \mathbf{f}_2 = \mathbf{e}_1 + \mathbf{e}_2.$$

The projections P_i can now be rewritten in terms of the Cartesian coordinates in $\mathbb{R}^2$,

$$\begin{aligned} P_1(v_1, v_2) &= v_1 P_1(\mathbf{e}_1) + v_2 P_1(\mathbf{e}_2) \\ &= v_1 \cdot \mathbf{e}_1 - v_2 \cdot \mathbf{e}_1 = (v_1 - v_2, 0), \end{aligned}$$

$$\begin{aligned} P_2(v_1, v_2) &= v_1 P_2(\mathbf{e}_1) + v_2 P_2(\mathbf{e}_2) \\ &= v_1 \cdot \mathbf{0} + v_2 \cdot (\mathbf{e}_1 + \mathbf{e}_2) = (v_2, v_2). \end{aligned}$$

It is interesting to calculate P_1^2, P_2^2, and $P_1 \circ P_2$ using the preceding formulas to verify the properties listed in Exercise 2.25.

The "idempotent property" $P^2 = P$ for a linear operator is characteristic of projections associated with a direct sum decomposition $V = V_1 \oplus V_2$. We have already seen that if $P, Q = (I - P)$ are the projections associated with such a decomposition, then

(i) $P^2 = P$ and $Q^2 = Q$,
(ii) $PQ = QP = 0$,
(iii) $P + Q = I$ (identity operator).

But the converse is also true.

PROPOSITION 2.27. *If $P : V \to V$ is any linear operator such that $P^2 = P$, then V is a direct sum $V = R(P) \oplus K(P)$ and P is the projection of V onto the range $R(P)$ along the kernel $K(P)$. The operator $Q = I - P$ is idempotent, with*

$$R(Q) = R(I - P) = K(P) \quad \textit{and} \quad K(Q) = K(I - P) = R(P), \tag{2.6}$$

and projects V onto $R(Q) = K(P)$ along $K(Q) = R(P)$

PROOF. First observe that $Q = (I - P)$ is idempotent since

$$(I - P)^2 = I - 2P + P^2 = I - 2P + P = (I - P).$$

Next, note that[1]

$$v \in K(Q) \Leftrightarrow Q(v) = (I - P)v = 0 \Leftrightarrow P(v) = v \Leftrightarrow v \in R(P).$$

Thus:

1. $K(Q) = K(I - P)$ is equal to $R(P)$
2. $R(Q) = R(I - P)$ is equal to $K(P)$

which proves (2.6).

Obviously $P + Q = I$ because $v = P(v) + (I - P)(v)$ implies $P(v) \in R(P)$, while $(I - P)(v) \in R(Q) = K(P)$ by (2.6); thus the span $R(P) + K(P) = R(P) + R(Q)$ is all of V. Furthermore, $K(P) \cap R(P) = (0)$, for if v is in the intersection we have

[1]Implication ($\Rightarrow$) in the last step is obvious. Conversely, if $v \in R(P)$ then $v = P(w)$ for some w and then $P(v) = P^2(w) = P(w) = v$, proving ($\Leftarrow$).

$v \in K(P) \Rightarrow P(v) = 0$. But we also have $v \in R(P)$, so $v = P(w)$ for some w, and then

$$0 = P(v) = P^2(w) = P(w) = v.$$

By Exercise 2.22 we conclude that $V = R(P) \oplus K(P)$.

Finally, let $\widetilde{P}$ be the projection onto $R(P)$ along $K(P)$ associated with this decomposition; we claim that $P = \widetilde{P}$. By definition $\widetilde{P}$ maps $v = r + k \in R(P) \oplus K(P)$ to r; that, however, is exactly what our original operator P does:

$$P(r + k) = P(r) + P(k) = r + 0 = r.$$

Therefore $\widetilde{P} = P$ as operators on V. □

Direct Sums and Diagonalization. Let $T : V \to V$ be a linear operator on a finite-dimensional space V. As above, the spectrum of T is the set of *distinct* eigenvalues $\text{sp}(T) = \{\lambda \in \mathbb{K} : E_\lambda \neq 0\}$, where E_λ is the (nontrivial) λ-eigenspace of T,

$$E_\lambda = \{v \in V : T(v) = \lambda \cdot v\} = \ker(T - \lambda I) \quad (I = \text{id}_V).$$

DEFINITION 2.28. A linear operator $T : V \to V$ is *diagonalizable* if V is the direct sum of its nontrivial eigenspaces,

$$V = \bigoplus_{\lambda \in \text{sp}(T)} E_\lambda(T).$$

We will see below that this happens if and only if V has a basis $\mathbf{f}_1, \dots, \mathbf{f}_n$ of eigenvectors,

$$T(\mathbf{f}_i) = \mu_i \cdot \mathbf{f}_i \quad \text{for some } \mu_i \in \mathbb{K}$$

for $1 \le i \le n$.

Independence of the Eigenspaces. Our next result shows that T is diagonalizable if we only know that the eigenspaces *span* V, with

$$V = \sum_{\lambda \in \text{sp}(T)} E_\lambda(T) = \mathbb{K}\text{-span}\{E_\lambda(T) : \lambda \in \text{sp}_{\mathbb{K}}(T)\}.$$

This property is much easier to verify.

PROPOSITION 2.29. *If $T : V \to V$ is a linear operator on a finite-dimensional space, let W be the span $\sum_{\lambda \in \text{sp}_{\mathbb{K}}(T)} E_\lambda(T)$ of the eigenspaces. This subspace is T-invariant and is in fact a direct sum $W = \bigoplus_\lambda E_\lambda(T)$ of the eigenspaces.*

PROOF. Since each E_λ is invariant, their span W is also T-invariant. The E_λ span W by hypothesis, so each $w \in W$ has some decomposition $w = \sum_\lambda w_\lambda$ with $w_\lambda \in E_\lambda$. For uniqueness of this decomposition, it suffices to show that

$$0 = \sum_{\lambda \in \text{sp}(T)} w_\lambda \quad \text{with } w_\lambda \in E_\lambda \Rightarrow \text{ each } w_\lambda = 0.$$

The operators T, $(T - \lambda I)$, and $(T - \mu I)$ commute for all $\mu, \lambda \in \mathbb{K}$ since the identity element I and its scalar multiples commute with everybody. Let us fix an eigenvalue λ_0; we will show $w_{\lambda_0} = 0$. With this λ_0 in mind, we define the product

$$A = \prod_{\lambda \neq \lambda_0, \lambda \in \mathrm{sp}(T)} (T - \lambda I).$$

Then

$$0 = A(0) = A\Big(\sum_\lambda w_\lambda\Big) = \sum_\lambda A(w_\lambda) = A(w_{\lambda_0}) + \sum_{\lambda \neq \lambda_0} A(w_\lambda).$$

If $\lambda \neq \lambda_0$ we have

$$A(w_\lambda) = \Big(\prod_{\mu \neq \lambda_0} (T - \mu I)\Big) w_\lambda = \Big(\prod_{\mu \neq \lambda_0, \lambda} (T - \mu I)\Big) \cdot (T - \lambda I) w_\lambda = 0$$

because $w_\lambda \in E_\lambda$. On the other hand, by writing $(T - \lambda_0 I) + (\lambda_0 - \mu)I$, we find that

$$A(w_{\lambda_0}) = \Big(\prod_{\mu \neq \lambda_0} (T - \mu I)\Big) w_{\lambda_0} = \Big(\prod_{\mu \neq \lambda_0} (T - \lambda_0 I) + (\lambda_0 - \mu) I\Big) w_{\lambda_0}.$$

When we expand this product of sums, every term but one includes a factor $(T - \lambda_0 I)$ that kills w_{λ_0}:

$$(Term) = (\textit{product of operators}) \cdot (T - \lambda_0 I)\, w_{\lambda_0} = 0.$$

The one exception is the product

$$\Big(\prod_{\mu \neq \lambda_0} (\lambda_0 - \mu)\Big) \cdot w_{\lambda_0}.$$

The scalar out front cannot be zero because each $\mu \neq \lambda_0$, so

$$A(w_{\lambda_0}) = \prod_{\mu \neq \lambda_0} (\lambda_0 - \mu) \cdot w_{\lambda_0} \neq 0.$$

But we already observed that

$$0 = A\Big(\sum_\lambda w_\lambda\Big) = 0 + A(w_{\lambda_0}),$$

so we get a contradiction unless $w_{\lambda_0} = 0$. Thus each term in $\sum_\lambda w_\lambda$ is zero and W is the *direct* sum of the eigenspaces E_λ. □

If $W \subsetneqq V$ this result tells us nothing about the behavior of T off of the subspace W, but if we list the distinct eigenvalues as $\lambda_1, \dots, \lambda_r$ we can construct a basis for W that runs first through E_{λ_1}, then through E_{λ_2}, etc., to obtain a basis for W,

$$\mathfrak{X} = \{f_1^{(1)}, \dots, f_{d_1}^{(1)}; f_1^{(2)}, \dots, f_{d_2}^{(2)}; \dots; f_1^{(r)}, \dots, f_{d_r}^{(r)}\},$$

where $d_i = \dim(E_{\lambda_i})$ and $\sum_i d_i = m = \dim(W)$. The corresponding matrix describing $T|_W$ is *diagonal*, so $T|_W$ is a diagonalizable operator on W even if T

is not diagonalizable on all of V,

$$[T|_W]_{\mathfrak{X}\mathfrak{X}} = \begin{pmatrix} \lambda_1 & & & & & & 0 \\ & \ddots & & & & & \\ & & \lambda_1 & & & & \\ & & & \ddots & & & \\ & & & & \lambda_r & & \\ & & & & & \ddots & \\ 0 & & & & & & \lambda_r \end{pmatrix} \tag{2.7}$$

where $\text{sp}(T) = \{\lambda_1, \dots, \lambda_r\}$.

EXERCISE 2.30. If a linear operator $T : V \to V$ acts on a finite-dimensional space, prove that the following statements are equivalent.

(a) T is diagonalizable: $V = \bigoplus_{\lambda \in \text{sp}(T)} E_\lambda(T)$.

(b) There is a basis $f_1, \dots, f_n$ for V such that each f_i is an eigenvector, with $Tf_i = \mu_i f_i$ for some $\mu_i \in \mathbb{K}$.

2.4. Representing Linear Operators as Matrices

Let $T : V \to W$ be a linear operator between finite-dimensional vector spaces with $\dim(V) = m$, $\dim(W) = n$. An *ordered basis* in V is an ordered list $\mathfrak{X} = \{e_1, \dots, e_m\}$ of vectors that are independent and span V; let $\mathfrak{Y} = \{f_1, \dots, f_n\}$ be an ordered basis for the target space W.

The Associated Matrices $[T]_{\mathfrak{Y}\mathfrak{X}}$. The behavior of a linear map $T : V \to W$ is completely determined by what it does to the basis vectors in V because every $v = \sum_{i=1}^m c_i e_i$ (uniquely) and

$$T\Big(\sum_{i=1}^m c_i e_i\Big) = \sum_{i=1}^m c_i T(e_i).$$

Each image $T(e_i)$ can be expressed uniquely as a linear combination of vectors in the $\mathfrak{Y}$ basis,

$$T(e_i) = \sum_{j=1}^n t_{ji} f_j \quad \text{for } 1 \le i \le m,$$

yielding a system of $m = \dim(V)$ vector equations that tell us how to rewrite T-images of vectors in the $\mathfrak{X}$-basis in terms of vectors in the $\mathfrak{Y}$-basis

$$\begin{aligned} T(e_1) &= b_{11} f_1 + \cdots + b_{1n} f_n \\ &\ \ \vdots \\ T(e_m) &= b_{m1} f_1 + \cdots + b_{mn} f_n. \end{aligned} \tag{2.8}$$

We *define* the *matrix of T with respect to the bases $\mathfrak{X}$, $\mathfrak{Y}$* to be the $n \times m$ matrix

$$[T]_{\mathfrak{Y}\mathfrak{X}} = [t_{ij}] \quad \text{(the } \textit{transpose} \text{ of the array of coefficients } B = [b_{ij}] \text{ in (2.8)).}$$

Since $(B^{\mathrm{T}})_{k\ell} = B_{\ell k}$ that means $t_{ij} = b_{ji}$; to put it differently, the entries t_{ij} in $[T]_{\mathfrak{Y}\mathfrak{X}}$ satisfy the following identities derived from (2.8):

$$T(e_i) = \sum_{j=1}^{n} t_{ji} f_j \quad \text{for } 1 \le i \le m. \tag{2.9}$$

Note carefully: the basis vector f_j in (2.9) is paired with t_{ji} and not t_{ij}.

AN IMPORTANT SPECIAL CASE. When $V = W$, $T = I$ is the identity operator $\mathrm{id}(v) = v$ on V, and we take different bases $\mathfrak{X} = \{e_i\}$, $\mathfrak{Y} = \{f_j\}$ on V and W, then (2.9) takes the form

$$e_i = I(e_i) = \sum_{j=1}^{n} t_{ji} f_j \quad \text{for } 1 \le i \le \dim(V).$$

Taking $T = I$ in the system (2.8) we obtain the *change-of-basis formula* that allows us to rewrite the initial basis vectors e_i in terms of new basis vectors $f_j \in \mathfrak{Y}$. As before, the matrix $[\mathrm{id}]_{\mathfrak{Y}\mathfrak{X}} = [t_{ij}]$ is the transpose of the array of coefficients $[b_{ij}]$ in the system of vector identities (2.8).

The matrix description of $T : V \to W$ changes if we take different bases. Nevertheless, the same operator T (which has a coordinate-independent existence) underlies all of these descriptions. One objective in analyzing T is to find bases that yield simple matrix descriptions. When $V = W$ the best possible outcome is of course a basis that diagonalizes T as in (2.7), but alas, not all operators are diagonalizable.

Another issue worth considering is the following: If T is the identity operator $I = \mathrm{id}$ on a vector space V, and we compute $[\mathrm{id}]_{\mathfrak{X}\mathfrak{X}}$, the outcome is the same for all bases $\mathfrak{X}$,

$$[\mathrm{id}]_{\mathfrak{X}\mathfrak{X}} = I_{n\times n} \quad (\text{the } n \times n \text{ identity matrix}).$$

But there is no reason why we couldn't take different bases in the initial and final spaces (even if they are the same space), regarding $T = \mathrm{id}_V$ as a map from $(V, \mathfrak{X})$ to $(V, \mathfrak{Y})$. Then there are some surprises when you compute $[T]_{\mathfrak{X}\mathfrak{Y}}$.

EXERCISE 2.31. Let V be a 2-dimensional coordinate space $\mathbb{R}^2$. Let $I : V \to V$ be the identity map $I = \mathrm{id}_V$, but take different bases $\mathfrak{X} = \{\mathbf{e}_1, \mathbf{e}_2\}$ and $\mathfrak{Y} = \{\mathbf{f}_1, \mathbf{f}_2\}$ in the initial and final spaces. Letting $\mathbf{e}_1, \mathbf{e}_2$ be the standard basis vectors in $\mathbb{R}^2$ and $\mathbf{f}_1 = (1, 0)$, $\mathbf{f}_2 = (1, 2)$, compute the matrices

(a) $[I]_{\mathfrak{X}\mathfrak{X}}$ (b) $[I]_{\mathfrak{Y}\mathfrak{X}}$ (c) $[I]_{\mathfrak{Y}\mathfrak{Y}}$ (d) $[I]_{\mathfrak{X}\mathfrak{Y}}$

EXERCISE 2.32. If V, W are finite dimensional and $T : V \to W$ is linear, explain why there must exist bases $\mathfrak{X}, \mathfrak{Y}$ and $\mathfrak{X}', \mathfrak{Y}'$ in V, W such that

$$\text{(a) } [T]_{\mathfrak{Y}\mathfrak{X}} = \begin{pmatrix} 0 & 0 \\ 0 & I_{r\times r} \end{pmatrix} \qquad \text{(b) } [T]_{\mathfrak{Y}'\mathfrak{X}'} = \begin{pmatrix} I_{r\times r} & 0 \\ 0 & 0 \end{pmatrix}$$

where $r = \mathrm{rank}(T) = \dim(\mathrm{range}(T))$ and $I_{r\times r}$ is the $r \times r$ identity matrix.

HINT: Finding a basis that produces (a) is fairly easy; part (b) requires some thought about the order in which basis vectors are listed. Both matrices represent the same operator $T : V \to W$.

The matrix description of T could hardly be simpler than those in Exercise 2.32, but at the same time much information about T has been lost in allowing arbitrary unrelated bases in V and W. Most operators encode far more information than can be captured by the single number rank(T).

In addition to our description of a linear operator $T : V \to W$ as a matrix, we can also describe vectors $v \in V$, $w \in W$ as column matrices once bases $\mathfrak{X} = \{e_i\}$ and $\mathfrak{Y} = \{f_j\}$ are specified. The correspondence $\phi_{\mathfrak{X}} : V \to \mathbb{K}^m$ is a linear bijection (an isomorphism of vector spaces) defined by letting

$$\phi_{\mathfrak{X}}(v) = [v]_{\mathfrak{X}} = \begin{pmatrix} v_1 \\ \vdots \\ v_m \end{pmatrix} \quad \text{if } v = \sum_{i=1}^{m} v_i e_i \text{ (unique expansion).}$$

Similarly $\phi_{\mathfrak{Y}} : W \to \mathbb{K}^n$ is given by

$$\phi_{\mathfrak{Y}}(w) = [w]_{\mathfrak{Y}} = \begin{pmatrix} w_1 \\ \vdots \\ w_n \end{pmatrix} \quad \text{if } w = \sum_{j=1}^{n} w_j f_j \text{ (unique expansion).}$$

These coordinate descriptions of linear operators and vectors are closely related.

PROPOSITION 2.33. *If $T : V \to W$ is a linear operator and $\mathfrak{X}$, $\mathfrak{Y}$ are bases in V, W, then for all $v \in V$,*

$$\phi_{\mathfrak{Y}}(Tv) = [T]_{\mathfrak{Y}\mathfrak{X}} \cdot \phi_{\mathfrak{X}}(v),$$

or equivalently,

$$[T(v)]_{\mathfrak{Y}} = [T]_{\mathfrak{Y}\mathfrak{X}} \cdot [v]_{\mathfrak{X}} \quad \textit{(an } (n \times m) \cdot (m \times 1) \textit{ matrix product).}$$

Thus the ith component $(Tv)_i$ of $\phi_{\mathfrak{Y}}(Tv)$ is given by the familiar formula

$$(Tv)_i = \sum_{k=1}^{m} t_{ik} v_k \quad \textit{for } 1 \le i \le n$$

if $v = \sum_{k=1}^{m} v_k e_k$ and $[T]_{\mathfrak{Y}\mathfrak{X}} = [t_{ij}]$.

PROOF. If $T(e_i) = \sum_{j=1}^{n} t_{ji} f_j$ and $v = \sum_{k=1}^{m} v_k e_k$, then

$$T(v) = T\Big(\sum_{k=1}^{m} v_k e_k\Big) = \sum_{k=1}^{m} v_k T(e_k) = \sum_{k=1}^{m} v_k \Big(\sum_{j=1}^{n} t_{jk} f_j\Big) = \sum_{j}\Big(\sum_{k} t_{jk} v_k\Big) f_j.$$

So, the ith component $(Tv)_i$ of $\phi_{\mathfrak{Y}}(T(v))$ is $\sum_{k=1}^{m} t_{ik} v_k$, as claimed. □

In Lemma 2.34 we will show that $\widetilde{T}$ = the multiplication operator $L_A : \mathbb{K}^m \to \mathbb{K}^n$ if we take $A = [T]_{\mathfrak{Y}\mathfrak{X}}$.

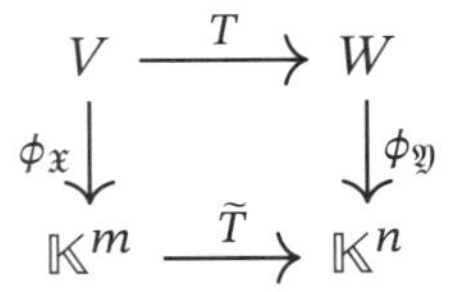

FIGURE 2.3. This diagram commutes, with $\phi_{\mathfrak{Y}} \circ T = \widetilde{T} \circ \phi_{\mathfrak{X}}$.

The Correspondence Between Matrices and Linear Operators. The natural linear maps $\phi_{\mathfrak{X}} : V \to \mathbb{K}^m$ and $\phi_{\mathfrak{Y}} : W \to \mathbb{K}^n$, with $m = \dim V$ and $n = \dim(W)$, are bijective isomorphisms. Therefore a unique linear map $\widetilde{T} = \phi_{\mathfrak{Y}} \circ T \circ \phi_{\mathfrak{X}}^{-1}$ is induced from $\mathbb{K}^m \to \mathbb{K}^n$ that makes the diagram in Figure 2.3 commute: It follows from Proposition 2.33 that the map $\widetilde{T}$ we get when $T : V \to W$ is transferred to a map between coordinate spaces is precisely the multiplication operator $L_A : \mathbb{K}^m \to \mathbb{K}^n$, where

$$A = [T]_{\mathfrak{Y}\mathfrak{X}}$$

is the coordinate matrix that describes T as in (2.8) and (2.9); see also Exercise 2.43 below.

Once bases $\mathfrak{X}, \mathfrak{Y}$ are specified there is also a natural linear isomorphism between the space of linear operators $\mathrm{Hom}_{\mathbb{K}}(V, W)$ and the space of matrices $\mathrm{M}(n \times m, \mathbb{K})$.

LEMMA 2.34. *If $\mathfrak{X}, \mathfrak{Y}$ are bases for finite-dimensional vector spaces V, W, the map ϕ from $\mathrm{Hom}_{\mathbb{K}}(V, W) \to \mathrm{M}(n \times m, \mathbb{K})$ given by*

$$\phi(T) = [T]_{\mathfrak{Y}\mathfrak{X}}$$

is a linear bijection, so these vector spaces are isomorphic and

$$\dim_{\mathbb{K}} (\mathrm{Hom}_{\mathbb{K}}(V, W)) = \dim_{\mathbb{K}} (\mathrm{M}(n \times m, \mathbb{K})) = m \cdot n.$$

PROOF. Linearity of ϕ follows because if $\mathfrak{X} = \{e_i\}$ and $\mathfrak{Y} = \{f_j\}$, we have

$$(\lambda \cdot T)(e_i) = \lambda \cdot (T(e_i)) = \lambda \cdot \Big(\sum_{j=1}^{n} t_{ji} f_j\Big) = \sum_{j=1}^{n} (\lambda t_{ji}) f_j$$

for $1 \leq i \leq m$, which means that $[\lambda T]_{ij} = \lambda \cdot [T]_{ij}$. Similarly, if we write $[T_k]_{\mathfrak{Y}\mathfrak{X}} = [t_{ij}^{(k)}]$ for $k = 1, 2$ we get

$$(T_1 + T_2)(e_i) = T_1(e_i) + T_2(e_i) = \sum_j t_{ji}^{(1)} f_j + \sum_j t_{ji}^{(2)} f_j = \sum_j (t_{ji}^{(1)} + t_{ji}^{(2)}) f_j.$$

So $[T_1 + T_2]_{ij} = [T_1]_{ij} + [T_2]_{ij}$, proving linearity of ϕ as a map from operators to matrices.

ONE-TO-ONE. To see that ϕ is one-to-one it suffices to show that $\ker(\phi) = (0)$; i.e., if $\phi(T) = [T]_{\mathfrak{Y}\mathfrak{X}} = [0]$ (the zero matrix), then T is the zero operator on V. This is clear: If $t_{ji} = 0$ for all i, j, then $T(e_i) = \sum_{j=1}^{m} t_{ji} e_j = 0$ for all i, and

$T(v) = 0$ for all v because $\{e_i\}$ is a basis.

SURJECTIVE. To prove ϕ surjective: given an $n \times m$ matrix $A = [a_{ij}]$, we must produce a linear operator $T : V \to W$ such that $[T]_{\mathfrak{Y}\mathfrak{X}} = [a_{ij}]$. This can done by working the definition of $[T]_{\mathfrak{Y}\mathfrak{X}}$ backward: we saw earlier that there is a unique linear operator $T : V \to W$ such that $T(e_i) = \sum_{j=1}^{n} a_{ji} f_j$, because $\{e_i\} = \mathfrak{X}$ is a basis in V. Then, by definition of $[T]_{\mathfrak{Y}\mathfrak{X}}$ as in (2.8) and (2.9) we have $t_{ij} = a_{ij}$. □

When $V = W$ and we take the same initial and final bases $\mathfrak{X} = \mathfrak{Y}$, composition of operators $S \circ T$ makes sense and the space of linear operators $\mathrm{Hom}_{\mathbb{K}}(V, V)$ becomes a (noncommutative) *associative algebra*, with the identity operator $I = \mathrm{id}_V$ as the multiplicative identity element. The set of matrices $\mathrm{M}(n, \mathbb{K})$ is also an associative algebra under matrix multiplication, its identity element being the $n \times n$ identity matrix $I_{n\times n} = \mathrm{diag}(1, \dots, 1)$. These systems are "isomorphic" as associative algebras as well as vector spaces, because the bijection $\phi : \mathrm{Hom}(V, V) \to \mathrm{M}(n, \mathbb{K})$ intertertwines the multiplication operations $(\circ)$ and $(\cdot)$.

PROPOSITION 2.35. *The bijective linear map $\phi : \mathrm{Hom}_{\mathbb{K}}(V, V) \to \mathrm{M}(n, \mathbb{K})$ intertwines the product operations in these algebras:*

$$\phi(S \circ T) = \phi(S) \cdot \phi(T) \quad (\textit{matrix product}) \tag{2.10}$$

for all $S, T \in \mathrm{Hom}(V, V)$. Given the correspondence between operators and their matrix representations, this is equivalent to saying that

$$[S \circ T]_{\mathfrak{X}\mathfrak{X}} = [S]_{\mathfrak{X}\mathfrak{X}} \cdot [T]_{\mathfrak{X}\mathfrak{X}}$$

for every basis $\mathfrak{X}$ in V, where we take the matrix product on the right.

This is a special case of a much more general result.

PROPOSITION 2.36. *Let $U \xrightarrow{T} V \xrightarrow{S} W$ be linear maps and let $\mathfrak{X} = \{u_i\}$, $\mathfrak{Y} = \{v_i\}$, $\mathfrak{Z} = \{w_i\}$ be bases in U, V, W. Then the correspondence between operators and their matrix realizations is covariant in the sense that*

$$[S \circ T]_{\mathfrak{Z}\mathfrak{X}} = [S]_{\mathfrak{Z}\mathfrak{Y}} \cdot [T]_{\mathfrak{Y}\mathfrak{X}}$$

(a matrix product of compatible nonsquare matrices).

PROOF. We have $S \circ T(u_i) = \sum_k [S \circ T]_{ki}\, w_k$ by definition, and also

$$\begin{aligned} S \circ T(u_i) &= S\,(T(u_i)) = S\Big(\sum_j t_{ji} v_j\Big) = \sum_j t_{ji} S(v_j) \\ &= \sum_j \Big(\sum_k s_{kj} w_k\Big) t_{ji} = \sum_k \Big(\sum_j s_{kj} t_{ji}\Big) w_k \\ &= \sum_k ([S][T])_{ki}\, w_k \quad (\text{definition of matrix product}) \end{aligned}$$

Thus $[S \circ T]_{ki} = ([S][T])_{ki}$ for all i, k. □

EXERCISE 2.37. E_{ij} with a "1" in the (i, j) spot and zeros elsewhere are a basis for matrix space $\mathrm{M}(n \times m, \mathbb{K})$ since $[a_{ij}] = \sum_{i,j} a_{ij}E_{ij}$. When $m = n$ the matrices E_{ij} have useful algebraic properties. Prove that

(a) These matrices satisfy the identities
$$E_{ij}E_{k\ell} = \delta_{jk} \cdot E_{i\ell}$$
where δ_{rs} is the *Kronecker delta symbol*, equal to 1 if $r = s$ and zero otherwise.

(b) The diagonal elements E_{ii} are projections, with $E_{ii}^2 = E_{ii}$.

(c) $E_{11} + \cdots + E_{nn} = I_{n\times n}$ (the identity matrix).

If $T : V \to W$ is an invertible (bijective) linear operator between finite-dimensional spaces, then $\dim(V) = \dim(W) = n$ and the inverse map $T^{-1} : W \to V$ is also linear (recall Exercise 2.15). From the definition of the inverse map T^{-1} we have
$$T^{-1} \circ T = \mathrm{id}_V \quad \text{and} \quad T \circ T^{-1} = \mathrm{id}_W,$$
so each operator undoes the action of the other. For any bases $\mathfrak{X}, \mathfrak{Y}$ in V, W the corresponding matrix realizations of T and T^{-1} are inverses of each other under matrix multiplication. To see why, first recall that
$$[\mathrm{id}_V]_{\mathfrak{X}\mathfrak{X}} = I_{n\times n} \quad \text{and} \quad [\mathrm{id}_W]_{\mathfrak{Y}\mathfrak{Y}} = I_{n\times n}.$$
Then by Proposition 2.36,
$$[T^{-1}]_{\mathfrak{X}\mathfrak{Y}} \cdot [T]_{\mathfrak{Y}\mathfrak{X}} = I_{n\times n} \quad \text{and} \quad [T]_{\mathfrak{Y}\mathfrak{X}} \cdot [T^{-1}]_{\mathfrak{X}\mathfrak{Y}} = I_{n\times n},$$
which means that $[T^{-1}]_{\mathfrak{X}\mathfrak{Y}}$ is the inverse $[T]_{\mathfrak{Y}\mathfrak{X}}^{-1}$ of the matrix of T. When $V = W$ and there is just one basis $\mathfrak{X}$, this reduces to the simpler statement $[T^{-1}] = [T]^{-1}$.

EXERCISE 2.38. Explain why *isomorphic* vector spaces must have the same dimension, even if one of them is infinite dimensional.

EXERCISE 2.39. If $T : V \to W$ is an invertible linear operator, prove that $(T^{-1})^{-1} = T$.

EXERCISE 2.40. If $U \xrightarrow{T} V \xrightarrow{S} W$ are invertible linear operators, explain why $S \circ T : U \to W$ is invertible, with $(S \circ T)^{-1} = T^{-1} \circ S^{-1}$. (Note the reversal of order.)

EXERCISE 2.41. Prove that the correspondence
$$L : \mathrm{M}(n, \mathbb{K}) \to \mathrm{Hom}_{\mathbb{K}}(\mathbb{K}^n, \mathbb{K}^n)$$
with $L_A(\mathbf{x}) = A\mathbf{x}$, has the following algebraic properties:

(a) $L_{A+B} = L_A + L_B$ and $L_{\lambda \cdot A} = \lambda \cdot L_A$ for all $\lambda \in \mathbb{K}$.

(b) $L_{AB} = L_A \circ L_B$.

(c) If $I = I_{n\times n}$ is the identity matrix, then $L_I = \mathrm{id}_{\mathbb{K}^n}$.

(d) L_A is an invertible linear operator if and only if the matrix inverse A^{-1} exists in $\mathrm{M}(n, \mathbb{K})$, and then we have $(L_A)^{-1} = (L_{A^{-1}})$.

EXERCISE 2.42. Explain why the map $L : \mathrm{M}(n,\mathbb{K}) \to \mathrm{Hom}_{\mathbb{K}}(\mathbb{K}^n,\mathbb{K}^n)$ is a linear bijection and an isomorphism between these associative algebras.

EXERCISE 2.43. If $A \in \mathrm{M}(n \times m,\mathbb{K})$ and $\mathfrak{X}$, $\mathfrak{Y}$ are the *standard bases* in coordinate spaces $\mathbb{K}^m$ and K^m, prove that the matrix $B = [L_A]_{\mathfrak{Y}\mathfrak{X}}$ that describes $L_A : \mathbb{K}^m \to \mathbb{K}^n$ for this particular choice of bases is just the original matrix A. Does this work for arbitrary bases in $\mathbb{K}^m$ and $\mathbb{K}^n$?

EXERCISE 2.44. Let $\mathcal{P} = \mathbb{K}[x]$ be the infinite-dimensional space of polynomials over $\mathbb{K}$. Consider the following linear operators:

(i) DERIVATIVE: $D(a_0 + a_1x + \cdots + a_nx^n) = a_1 + 2a_2x + \cdots + na_nx^{n-1}$

(ii) ANTIDERIVATIVE: $A(a_0 + a_1x + \cdots + a_nx^n) = a_0x + \frac{1}{2}a_1x^2 + \cdots + \frac{1}{n+1}a_nx^{n+1}$

Show that $D \circ A = \mathrm{id}_{\mathcal{P}}$ but that $A \circ D \neq \mathrm{id}_{\mathcal{P}}$. (What is $A \circ D$?) Show that D is surjective and A is one-to-one, but $\ker(D) \neq (0)$.

Such behavior is possible only in an infinite-dimensional space. We have already observed (recall Corollary 2.7) that if finite-dimensional spaces V, W *have the same dimension*, the following statements regarding a linear map $T : V \to W$ are equivalent:

T is one-to-one. $\qquad$ T is surjective. $\qquad$ T is bijective.

EXERCISE 2.45. Consider the operator $L = D^2 - cI$ on the $(n+1)$-dimensional space $\mathcal{P}_n \subseteq \mathbb{K}[x]$ of polynomials with degree $\leq n$, where $D = \dfrac{d}{dx}$, $I = \mathrm{id}$, and $c \in \mathbb{K}$.

(a) Find the matrix of $D^2 - cI$ with respect to the basis $\mathfrak{X} = \{1, x, x^2, \ldots, x^n\}$ in $\mathcal{P}_n$, and determine the values of c for which $(D^2 - cI) : \mathcal{P}_n \to \mathcal{P}_n$ fails to be invertible.

(b) For $n = 2$ and $c \neq 0$, determine the 3×3 inverse matrix $[D^2 - cI]^{-1}_{\mathfrak{X}\mathfrak{X}}$ and solve the equation $(D^2 - cI) \cdot f = 1 + 3x^2$ for $f \in \mathcal{P}_2$.

NOTE: Your answer in (b) will involve the undetermined constant c, and you might have to employ symbolic row operations to find the inverse matrix.

EXERCISE 2.46. If $A \in \mathrm{M}(n,\mathbb{K})$, define its *trace* to be

$$\mathrm{Tr}(A) = \sum_{i=1}^{n} a_{ii} \quad \text{(sum of the diagonal entries)}.$$

Show that, for any $A, B \in \mathrm{M}(n,\mathbb{K})$, the following hold:

(a) $\mathrm{Tr} : \mathrm{M}(n,\mathbb{K}) \to \mathbb{K}$ is a linear map.

(b) $\mathrm{Tr}(AB) = \mathrm{Tr}(BA)$.

(c) If $B = SAS^{-1}$ for some invertible $S \in \mathrm{M}(n,\mathbb{K})$, then $\mathrm{Tr}\left(SAS^{-1}\right) = \mathrm{Tr}(A)$.

Change of Basis and Similarity Transformations. If $T : V \to W$ is a linear map between finite-dimensional spaces and $\mathfrak{X}, \mathfrak{X}' \subseteq V$ and $\mathfrak{Y}, \mathfrak{Y}' \subseteq W$ are different bases, it is important to understand how the matrix models $[T]_{\mathfrak{Y}\mathfrak{X}}$ and $[T]_{\mathfrak{Y}'\mathfrak{X}'}$ are related as we seek particular bases yielding simple descriptions of T. For instance, if $T : V \to V$, we may ask if T is diagonalizable over $\mathbb{K}$: Is there a basis such that

$$[T]_{\mathfrak{X}\mathfrak{X}} = \begin{pmatrix} \lambda_1 & & & \\ & \lambda_2 & & 0 \\ & & \ddots & \\ & 0 & & \ddots \\ & & & & \lambda_n \end{pmatrix}$$

(repeats allowed among the λ_i)? Not all operators are so nice, and if T is not diagonalizable we will eventually work out a satisfactory but more complicated "Plan B" for dealing with such operators. All this requires a clear understanding of how matrix descriptions behave under a change ef basis.

THEOREM 2.47 (Change of Basis). *Let $T : V \to V$ be a linear operator on a finite-dimensional space, and let* id_V *be the identity operator on V. If $\mathfrak{X} = \{e_1, \dots, e_n\}$ and $\mathfrak{Y} = \{f_1, \dots, f_n\}$ are bases in V, then*

$$[T]_{\mathfrak{Y}\mathfrak{Y}} = [\mathrm{id}_V]_{\mathfrak{Y}\mathfrak{X}} \cdot [T]_{\mathfrak{X}\mathfrak{X}} \cdot [\mathrm{id}_V]_{\mathfrak{X}\mathfrak{Y}}. \tag{2.11}$$

Furthermore $[\mathrm{id}_V]_{\mathfrak{X}\mathfrak{Y}}$ *and* $[\mathrm{id}_V]_{\mathfrak{Y}\mathfrak{X}}$ *are inverses of each other.*

PROOF. Since $T = \mathrm{id}_V \circ T \circ \mathrm{id}_V : V \to V \to V$, repeated application of Proposition 2.34 yields 2.11, as in the following system of commutative diagrams:

$$\begin{array}{ccccccc}
\mathbb{K}^n & \xrightarrow{[\mathrm{id}_V]_{\mathfrak{X}\mathfrak{Y}}} & \mathbb{K}^n & \xrightarrow{L_A} & \mathbb{K}^n & \xrightarrow{[\mathrm{id}_V]_{\mathfrak{Y}\mathfrak{X}}} & \mathbb{K}^n \\
\uparrow & & \uparrow & & \uparrow & & \uparrow \\
(V, \mathfrak{Y}) & \xrightarrow{\mathrm{id}_V} & (V, \mathfrak{X}) & \xrightarrow{T} & (V, \mathfrak{X}) & \xrightarrow{\mathrm{id}_V} & (V, \mathfrak{Y})
\end{array}$$

where $A = [T]_{\mathfrak{X}\mathfrak{X}}$. Applying the same proposition to the maps $\mathrm{id}_V = \mathrm{id}_V \circ \mathrm{id}_V$ we get

$$I_{n\times n} = [\mathrm{id}_V]_{\mathfrak{Y}\mathfrak{Y}} = [\mathrm{id}_V]_{\mathfrak{Y}\mathfrak{X}} \cdot [\mathrm{id}_V]_{\mathfrak{X}\mathfrak{Y}},$$

which proves $[\mathrm{id}_V]_{\mathfrak{X}\mathfrak{Y}}$ and $[\mathrm{id}_V]_{\mathfrak{Y}\mathfrak{X}}$ are mutual inverses. □

To summarize: there is a unique, invertible "transition matrix" $S \in \mathrm{M}(n, \mathbb{K})$ such that

$$[T]_{\mathfrak{Y}\mathfrak{Y}} = S \cdot [T]_{\mathfrak{X}\mathfrak{X}} \cdot S^{-1} \quad \text{where } S = [\mathrm{id}_V]_{\mathfrak{Y}\mathfrak{X}} \text{ and } S^{-1} = [\mathrm{id}_V]_{\mathfrak{X}\mathfrak{Y}} \tag{2.12}$$

If we have explicit vector equations expressing the $\mathfrak{Y}$-basis vectors in terms of the $\mathfrak{X}$-basis vectors, the matrix $S^{-1} = [\mathrm{id}]_{\mathfrak{X}\mathfrak{Y}}$ can be written down by inspection of the vector identities

$$\mathbf{f}_i = \mathrm{id}(\mathbf{f}_i) = \sum_{j=1}^{n} t_{ji}\mathbf{e}_j,$$

from which we may compute $S = (S^{-1})^{-1}$.

DEFINITION 2.48 (Similarity transformations). Two matrices A and B in $\mathrm{M}(n, \mathbb{K})$ are *similar* if there is an invertible matrix $S \in \mathrm{GL}(n, \mathbb{K})$ such that $B = SAS^{-1}$. The mapping $\sigma_S : \mathrm{M}(n, \mathbb{K}) \to \mathrm{M}(n, \mathbb{K})$ given by $\sigma_S(A) = SAS^{-1}$ is referred to as a *similarity transformation* of A. It is also referred to by algebraists as "conjugation" of arbitrary matrices A by a particular invertible matrix S.

Each individual conjugation operator $\sigma_S(A) = SAS^{-1}$ is an *automorphism* of the associative matrix algebra—it is a bijection that respects all algebraic operations in $\mathrm{M}(n, \mathbb{K})$:

$$\begin{aligned}
\sigma_S(A \cdot B) &= \sigma_S(A) \cdot \sigma_S(B), \\
\sigma_S(A + B) &= \sigma_S(A) + \sigma_S(B), \\
\sigma_S(\lambda \cdot A) &= \lambda \cdot \sigma_S(A) \quad \text{for } \lambda \in \mathbb{K}, \\
\sigma_S\left(I_{n\times n}\right) &= I_{n\times n},
\end{aligned}$$

for all matrices A, B and all "conjugators" $S \in \mathrm{GL}(n, \mathbb{K})$. But there is even more to be said: the correspondence $\psi : S \to \sigma_S$ has important algebraic properties of its own,

$$\begin{aligned}
\sigma_{S_1S_2} &= \sigma_{S_1} \circ \sigma_{S_2} \quad \text{for all invertible matrices } S_1, S_2, \\
\sigma_{I_{n\times n}} &= (\text{the identity operator } \mathrm{id}_{\mathrm{M}} \text{ on matrix space } \mathrm{M} = \mathrm{M}(n, \mathbb{K}))\,,
\end{aligned}$$

from which we automatically conclude that

The operator $\sigma_{S^{-1}}$ is the inverse $(\sigma_S)^{-1}$ of conjugation by S.

Thus the conjugation operators $\{\sigma_S : S \in \mathrm{GL}\}$ form a *group* of automorphisms acting on the algebra of $n \times n$ matrices. When a linear operator $T : V \to V$ is described with respect to different bases in V, the resulting matrices must be similar as in (2.12). The converse is also true: if $A = [T]_{\mathfrak{X}\mathfrak{X}}$ and $B = SAS^{-1}$ for some invertible matrix S, there is a basis $\mathfrak{Y}$ such that $B = [T]_{\mathfrak{Y}\mathfrak{Y}}$. Thus, the different matrix models of T corresponding to bases $\mathfrak{Y}$ other than $\mathfrak{X}$ are precisely the transforms $\{S\,[T]_{\mathfrak{X}\mathfrak{X}}S^{-1} : S \text{ is invertible}\}$.

LEMMA 2.49. *If $T : V \to V$ is a linear operator on a finite-dimensional vector space V and if $A = [T]_{\mathfrak{X}\mathfrak{X}}$, then an $n \times n$ matrix B is equal to $[T]_{\mathfrak{Y}\mathfrak{Y}}$ for some basis $\mathfrak{Y}$ if and only if $B = SAS^{-1}$ for some invertible matrix S.*

PROOF. ($\Rightarrow$) follows from (2.12). For ($\Leftarrow$): since S is invertible it has a matrix inverse S^{-1}. (Later, in Chapter 4, we will discuss effective methods to compute matrix inverses such as S^{-1}.) According to Theorem 2.47, what we need is a basis $\mathfrak{Y} = \{f_1, \dots, f_n\}$ such that $[\mathrm{id}_V]_{\mathfrak{X}\mathfrak{Y}} = S^{-1}$; then $[\mathrm{id}_V]_{\mathfrak{Y}\mathfrak{X}} = (S^{-1})^{-1} = S$ and $B = S\,[T]_{\mathfrak{X}\mathfrak{X}}S^{-1}$. If we write $S^{-1} = [b_{ij}]$ and $S = [s_{ij}]$, the identity $S^{-1} = [\mathrm{id}_V]_{\mathfrak{X}\mathfrak{Y}}$ means that

$$f_i = \mathrm{id}_V(f_i) = \sum_j b_{ji} e_j \quad \text{for } 1 \le i \le n$$

where $\mathfrak{X} = \{e_1, \dots, e_n\}$. This is the desired new basis $\mathfrak{Y} = \{f_j\}$. To see it is a basis, we have $\{f_j\} \subseteq \mathbb{K}\text{-span}\{e_1, \dots, e_n\}$ by definition, but $\{e_i\}$ is in $\mathbb{K}\text{-span}\{f_j\}$ because the matrix $S^{-1} = [b_{ij}]$ is invertible; in fact, $S^{-1}S = I$ implies that $\sum_i b_{ji}s_{ik} = \delta_{jk}$ (Kronecker delta). Then

$$\sum_i s_{ik} f_i = \sum_i s_{ik}\Big(\sum_j b_{ji}e_j\Big) = \sum_j \Big(\sum_i b_{ji}s_{ik}\Big)e_j = \sum_j \delta_{jk}e_j = e_k,$$

for $1 \leq k \leq n$, so $\{e_k\} \subseteq \mathbb{K}\text{-span}\{f_j\}$ as claimed. Therefore $\{e_1, \dots, e_n\}$ and $\{f_1, \dots f_n\}$ both span V, and because $\{e_i\}$ is already a basis, $\{f_j\}$ must also be a basis. □

The next example shows that it can be difficult to tell by inspection whether an operator T is diagonalizable.

EXAMPLE 2.50. Let $T : \mathbb{C}^2 \to \mathbb{C}^2$ be the linear operator whose action on the standard basis $\mathfrak{X} = \{\mathbf{e}_1 = (1, 0), \mathbf{e}_2 = (0, 1)\}$ is

$$T(\mathbf{e}_1) = 4\mathbf{e}_1, \quad T(\mathbf{e}_2) = -\mathbf{e}_2.$$

Clearly T is diagonalized by the $\mathfrak{X}$-basis since

$$[T]_{\mathfrak{X}\mathfrak{X}} = \begin{pmatrix} 4 & 0 \\ 0 & -1 \end{pmatrix}.$$

Compute $[T]_{\mathfrak{Y}\mathfrak{Y}}$ for the basis

$$\mathbf{f}_1 = \frac{1}{\sqrt{2}}(\mathbf{e}_1 + \mathbf{e}_2), \quad \mathbf{f}_2 = \frac{1}{\sqrt{2}}(\mathbf{e}_1 - \mathbf{e}_2).$$

SOLUTION: We have $[T]_{\mathfrak{Y}\mathfrak{Y}} = S\,[T]_{\mathfrak{X}\mathfrak{X}}S^{-1}$ where $S = [\mathrm{id}_V]_{\mathfrak{Y}\mathfrak{X}}$ and $S^{-1} = [\mathrm{id}_V]_{\mathfrak{X}\mathfrak{Y}}$. This inverse can be computed easily from our definition of the vectors $\mathbf{f}_1, \mathbf{f}_2$. As in (2.9), the matrix $[\mathrm{id}_V]_{\mathfrak{X}\mathfrak{Y}}$ is the *transpose* of the array of coefficients in the system of vector identities

$$\mathbf{f}_1 = \mathrm{id}(\mathbf{f}_1) = \frac{1}{\sqrt{2}}\mathbf{e}_1 + \frac{1}{\sqrt{2}}\mathbf{e}_2, \quad \mathbf{f}_2 = \mathrm{id}(\mathbf{f}_2) = \frac{1}{\sqrt{2}}\mathbf{e}_1 - \frac{1}{\sqrt{2}}\mathbf{e}_2;$$

hence

$$S^{-1} = [\mathrm{id}]_{\mathfrak{X}\mathfrak{Y}} = \begin{pmatrix} \frac{1}{\sqrt{2}} & \frac{1}{\sqrt{2}} \\ \frac{1}{\sqrt{2}} & -\frac{1}{\sqrt{2}} \end{pmatrix} = \frac{1}{\sqrt{2}} \cdot \begin{pmatrix} 1 & 1 \\ 1 & -1 \end{pmatrix}.$$

The inverse of this matrix (found by standard matrix algebra methods or simply by solving (2.12) for $\mathbf{e}_1, \mathbf{e}_2$ in terms of $\mathbf{f}_1, \mathbf{f}_2$ is

$$[\mathrm{id}]_{\mathfrak{Y}\mathfrak{X}} = (S^{-1})^{-1} = S = -\frac{1}{\sqrt{2}} \cdot \begin{pmatrix} -1 & -1 \\ -1 & 1 \end{pmatrix} = \frac{1}{\sqrt{2}} \cdot \begin{pmatrix} 1 & 1 \\ 1 & -1 \end{pmatrix}.$$

(Notice that $S = S^{-1}$; this is not usually the case.) Then we get

$$[T]_{\mathfrak{Y}\mathfrak{Y}} = S[T]_{\mathfrak{X}\mathfrak{X}}S^{-1} = \frac{1}{\sqrt{2}} \cdot \begin{pmatrix} 1 & 1 \\ 1 & -1 \end{pmatrix} \begin{pmatrix} 4 & 0 \\ 0 & -1 \end{pmatrix} \begin{pmatrix} 1 & 1 \\ 1 & -1 \end{pmatrix} \cdot \frac{1}{\sqrt{2}}$$

$$= \frac{1}{2} \cdot \begin{pmatrix} 4 & -1 \\ 4 & 1 \end{pmatrix} \begin{pmatrix} 1 & 1 \\ 1 & -1 \end{pmatrix} \frac{1}{2} \cdot \begin{pmatrix} 3 & 5 \\ 5 & 3 \end{pmatrix} = \begin{pmatrix} \frac{3}{2} & \frac{5}{2} \\ \frac{5}{2} & \frac{3}{2} \end{pmatrix}.$$

Diagonalizability of T would not be at all apparent if we used the basis $\mathfrak{Y} = \{\mathbf{f}_1, \mathbf{f}_2\}$ to represent T.

EXERCISE 2.51. Compute the matrix $[T]_{\mathfrak{Y}\mathfrak{Y}}$ for the linear operator T from $\mathbb{C}^2 \to \mathbb{C}^2$ of the previous example for each of the following bases $\mathfrak{Y} = \{\mathbf{f}_1, \mathbf{f}_2\}$:

(a) $\begin{cases} \mathbf{f}_1 = \frac{1}{\sqrt{2}}(\mathbf{e}_1 + \mathbf{e}_2) \\ \mathbf{f}_2 = \frac{1}{\sqrt{2}}(-\mathbf{e}_1 + \mathbf{e}_2) \end{cases}$ (obtained by rotating the standard basis vectors by $\theta = +45°$)

(b) $\begin{cases} \mathbf{f}_1 = \frac{\sqrt{3}}{2}\mathbf{e}_1 + \frac{1}{2}\mathbf{e}_2 \\ \mathbf{f}_2 = -\frac{1}{2}\mathbf{e}_1 + \frac{\sqrt{3}}{2}\mathbf{e}_2 \end{cases}$ (the standard basis vectors rotated by $\theta = +30°$)

(c) $\begin{cases} \mathbf{f}_1 = \mathbf{e}_1 + i\mathbf{e}_2 \\ \mathbf{f}_2 = \mathbf{e}_1 - i\mathbf{e}_2 \end{cases}$ where $i = \sqrt{-1}$ in $\mathbb{C}$ and $V = \mathbb{C}^2$

EXERCISE 2.52. Let $D = d/dx : \mathcal{P}_n \to \mathcal{P}_n$ be the formal derivative of a polynomial where $\mathcal{P}_n$ = polynomials of degree $\leq n$. Compute $[D]_{\mathfrak{X}\mathfrak{X}}$ with respect to the basis $\mathfrak{X} = \{1, x, \dots, x^n\}$. Compute $[D^2]_{\mathfrak{X}\mathfrak{X}}$ and $[D^{n+1}]_{\mathfrak{X}\mathfrak{X}}$ too.

Similarity Classes in Matrix Space. An *RST equivalence relation* on a set X is a rule declaring certain points $x, y \in X$ to be "related" (and others not). Writing $x \underset{R}{\sim} y$ when the points are related, the phrase "RST" means the relation is

(1) REFLEXIVE: $x \underset{R}{\sim} x$ for all $x \in X$.
(2) SYMMETRIC: $x \underset{R}{\sim} y \Rightarrow y \underset{R}{\sim} x$.
(3) TRANSITIVE: $x \underset{R}{\sim} y$ and $y \underset{R}{\sim} z \Rightarrow x \underset{R}{\sim} z$.

For each $x \in X$ we can then define its *equivalence class*, the subset

$$[x]_R = \{y \in X : y \underset{R}{\sim} x\}$$

The RST property forces distinct equivalence classes to be disjoint, so the whole space X decomposes into a disjoint union of these classes.

One example of an RST equivalence is *congruence mod a fixed prime p* in the set $X = \mathbb{Z}$,

$$k \sim \ell \Leftrightarrow k \equiv \ell \pmod{p} \Leftrightarrow k \text{ and } \ell \text{ differ by a multiple of } p.$$

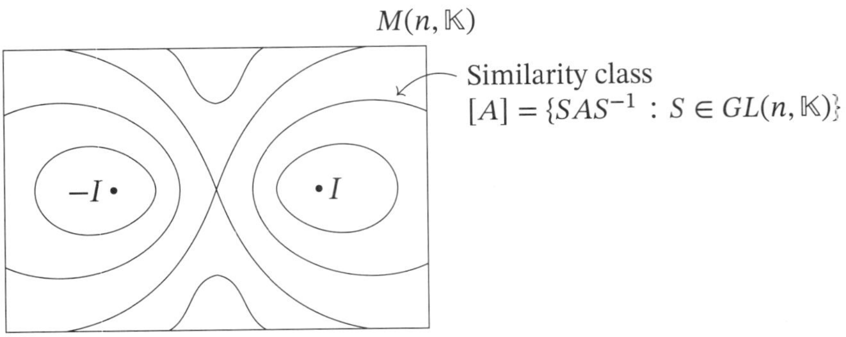

FIGURE 2.4. Similarity classes (= conjugacy classes) in $\mathrm{M}(n, \mathbb{K})$ partition matrix space into disjoint subsets $[A]$. Some classes are single points, for instance, $[-I]$, $[I]$, and $[0]$. Others are complicated hypersurfaces in $\mathbb{R}^{n^2} \cong \mathrm{M}(n, \mathbb{R})$. A similarity class could have several disconnected components.

It is easily verified that this is an RST relation and that the equivalence class of an integer m is its (mod p) congruence class

$$[m] = m + p\mathbb{Z} = \{k \in \mathbb{Z} : k \equiv m \,(\mathrm{mod}\, p)\}.$$

There are only finitely many distinct classes, namely $[0], [1], \dots, [p-1]$, which are disjoint and fill $\mathbb{Z}$. The finite field $\mathbb{Z}_p$ is precisely this set of equivalence classes equipped with suitable $\oplus$ and $\odot$ operations inherited from the system of integers $(\mathbb{Z}, +, \cdot)$.

Similarity and RST Equivalence Relations. The similarity of matrices $A \underset{R}{\sim} B \Leftrightarrow B = SAS^{-1}$ for some invertible matrix $S \in \mathrm{GL}(n, \mathbb{K})$ is an important example of an RST relation on matrix space $X = \mathrm{M}(n, \mathbb{K})$. The RST properties are easily verified.

EXERCISE 2.53. Prove that similarity of matrices (2.12) has the RST properties.

The equivalence classes partition $\mathrm{M}(n, \mathbb{K})$ into *similarity classes* (aka "conjugacy classes") that are disjoint; see Figure 2.4. The matrices $[T]_{\mathfrak{X}\mathfrak{X}}$ associated with a linear operator $T : V \to V$ by various bases constitute a single similarity class in matrix space—they are all the possible descriptions of T that arise from different choices of bases in V.

Similarity classes don't all look the same. Some are trivial, consisting of a single point; for instance, if

$$A = 0 \quad \text{or} \quad A = \lambda I_{n \times n} \text{ (a scalar multiple of the identity matrix),}$$

we have

$$SAS^{-1} = \lambda \cdot SIS^{-1} = \lambda \cdot SS^{-1} = \lambda I = A \quad \text{for all } S \in \mathrm{GL}(n, \mathbb{K}).$$

The similarity class $[A]$ consists of the single point A. In particular, $[0] = \{0\}$, $[I] = \{I\}$, and $[-I] = \{-I\}$. When $\mathbb{K} = \mathbb{R}$ and we identify $\mathrm{M}(n, \mathbb{R})$ with $\mathbb{R}^{n^2}$, other similarity classes can be large curvilinear surfaces in euclidean space. They can be quite a mess to classify and descibe.

EXERCISE 2.54. If $A \in \mathrm{M}(n, \mathbb{K})$, prove the following:

(a) A commutes with *all* $n \times n$ matrices $\Leftrightarrow A = \lambda \cdot I_{n\times n}$, a scalar multiple if the identity matrix for some $\lambda \in \mathbb{K}$.

(b) A commutes with all *invertible* matrices $\mathrm{GL}(n, \mathbb{K}) = \{A : \det(A) \neq 0\}$ $\Leftrightarrow A$ commutes with *all* $n \times n$ matrices.

HINT: Recall the matrices E_{ij} defined in Exercise 2.37, which are a basis for matrix space. To prove ($\Rightarrow$) in (b), if $i \neq j$ then $I + E_{ij}$ is invertible (verify that $(I - E_{ij})$ is the inverse) and commutes with A. Hence E_{ij} commutes with A; we leave you to figure out what to do when $i = j$. If A commutes with all basis vectors E_{ij}, it obviously commutes with all $n \times n$ matrices, and (a) can be applied.

This shows that a similarity class $[A]$ in $\mathrm{M}(n, \mathbb{K})$ consists of a single point $\Leftrightarrow A = \lambda I$ (a *scalar matrix*).

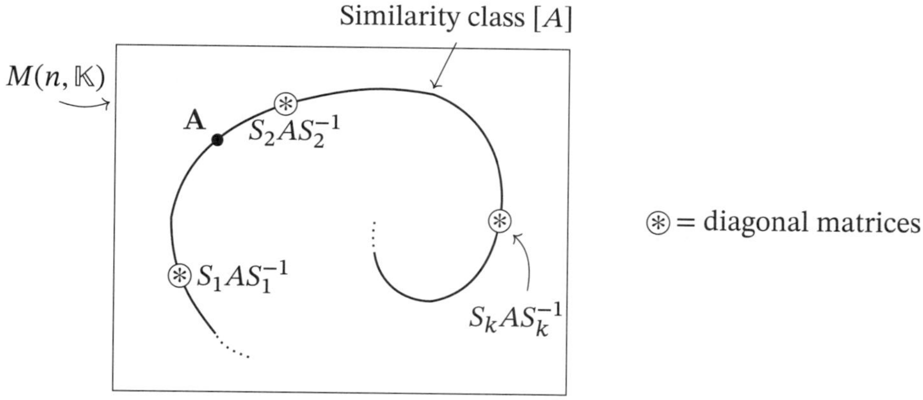

FIGURE 2.5. The diagonalization problem for $A \in \mathrm{M}(n, \mathbb{K})$ amounts to searching for one (or more) diagonal matrices lying in the similarity class $[A] = \{SAS^{-1} : S \in \mathrm{GL}(n, \mathbb{K})\}$.

EXERCISE 2.55. When we identify $\mathrm{M}(2, \mathbb{R}) \cong \mathbb{R}^4$ via the linear isomorphism

$$\mathbf{x} = \phi(A) = (a_{11}, a_{12}, a_{21}, a_{22}),$$

show that the similarity class $[A]$ of the matrix

$$A = \begin{pmatrix} 1 & 1 \\ 0 & 1 \end{pmatrix}$$

is a 2-dimensional surface in $\mathbb{R}^4$ that can be described in parametric form as the range of the polynomial map $\phi : \mathbb{R}^2 \to \mathrm{M}(2, \mathbb{R})$

$$\phi(s,t) = \begin{pmatrix} 1 - st & s^2 \\ -t^2 & 1 + st \end{pmatrix} \quad \text{where } s, t \in \mathbb{R} \text{ and } (s,t) \neq (0,0).$$

NOTE: As we will see in a later chapter, a 2×2 matrix $S = \left(\begin{smallmatrix} a & b \\ c & d \end{smallmatrix}\right)$ is invertible if the determinant $\det(S) = ad - bc$ is not 0, and then the inverse matrix is

$$S^{-1} = \frac{1}{\det(S)} \begin{pmatrix} d & -b \\ -c & a \end{pmatrix}.$$

We will have a lot more to say about change of basis, similarity classes, and the diagonalization problem in Chapter 5. Incidentally, not all matrices can be put into diagonal form by a similarity transformation. Our fondest hope is that in the equivalence class $[A]$ there will be at least one point SAS^{-1} that is diagonal (there may be several, as in Figure 2.5). If $A = [T]_{\mathfrak{X}\mathfrak{X}}$ for some linear operator $T : V \to V$, this is telling us which bases $\mathfrak{Y}$ make $[T]_{\mathfrak{Y}\mathfrak{Y}}$ diagonal, or whether there are any such bases at all.

EXERCISE 2.56. Suppose $T : \mathbb{R}^2 \to \mathbb{R}^2$ is the linear operator such that $T(\mathbf{e}_1) = 0$ and $T(\mathbf{e}_2) = \mathbf{e}_1$, so its matrix with respect to the standard basis $\mathfrak{X} = \{\mathbf{e}_1, \mathbf{e}_2\}$ is $[T]_{\mathfrak{X}\mathfrak{X}} = \left(\begin{smallmatrix} 0 & 1 \\ 0 & 0 \end{smallmatrix}\right)$. Prove that *no* basis $\mathfrak{Y} = \{f_1, f_2\}$ can make $[T]_{\mathfrak{Y}\mathfrak{Y}}$ diagonal.
HINT: $S = \left(\begin{smallmatrix} a_{11} & a_{12} \\ a_{21} & a_{22} \end{smallmatrix}\right)$ is invertible $\Leftrightarrow$ the determinant $\det(S) = a_{11}a_{22} - a_{12}a_{21}$ is nonzero.

We will eventually develop systematic methods to answer questions of this sort. For the moment, you will have to do it with "bare hands."

Additional Exercises

Section 2.1. Generalities.

1. TRUE/FALSE QUESTIONS: ("True" if the statement is *always* true.)
   (a) If $T : V \to W$ and $T(x + y) = T(x) + T(y)$ for all $x, y \in V$, then T is a linear operator.
   (b) A linear operator $T : V \to W$ is one-to-one if and only if the zero vector 0_V is the only element $v \in V$ such that $T(v) = 0_W$.
   (c) If $T : V \to W$ is linear, then $T(0_V) = 0_W$.
   (d) If $T : V \to W$ is linear, then $\text{nullity}(T) + \text{rank}(T) = \dim(W)$.
   (e) If $T : V \to W$ is linear, it sends linearly independent sets $S \subseteq V$ to linearly independent subsets in W.
   (f) If $S, T : V \to W$ are linear operators that agree on a *basis* for V, then $S = T$.
   (g) If linear operators $S, T : V \to W$ agree on a set that *spans* V, then $S = T$.
   (h) If $\{v_1, \dots, v_r\}$ and $\{w_1, \dots, w_r\}$ are subsets of vector spaces V, W, there is a linear operator $T : V \to W$ such that $T(v_i) = w_i$ for $1 \le i \le r$.
2. If V and W have equal dimensions, a linear operator $T : V \to W$ is

   (i) one-to-one $\Leftrightarrow$ (ii) surjective $\Leftrightarrow$ (iii) bijective.

   but if the spaces have different dimensions $|V| \neq |W|$, this is no longer true.

(a) Can T be one-to-one but not onto
(i) when $|V| > |W|$? (ii) when $|V| < |W|$?
(b) Can T be onto but not one-to-one
(i) when $|V| > |W|$? (ii) when $|V| < |W|$?
In each case, either provide an example or explain why this cannot happen.

3. In $V = \mathbb{R}^3$ consider the set $\mathfrak{X}$ of vectors

$$\mathbf{u}_1 = (1, 2, 3), \quad \mathbf{u}_2 = (2, 0, 2), \quad \mathbf{u}_3 = (-1, 2, 0)$$

(a) Explain why every $v \in V$ has a unique decomposition

$$V = \sum_{i=1}^{3} c_i \mathbf{u}_i (c_i \in \mathbb{R}).$$

(b) What are the coefficients in the decomposition $v = \sum_{i=1}^{3} c_i \mathbf{u}_i$ for the particular vector $v = (5, -2, -1)$?

4. For the linear operator $T = L_A : \mathbb{C}^4 \to \mathbb{C}^3$ determined by the 3×4 matrix

$$A = \begin{pmatrix} 2 & 4 & 8 & -2 \\ 0 & 1 & 2 & 4 \\ 1 & 3 & 6 & 3 \end{pmatrix},$$

find bases for $R(T) = \text{range}(T)$ and $K(T) = \ker(T)$. What are their dimensions?

5. Answer the following questions.
(a) Suppose $T : \mathbb{R}^2 \to \mathbb{R}^2$ is linear and $T(1, 0) = (1, 4)$, $T(1, 1) = (2, 5)$. What is $T(2, 3)$? Is T one-to-one?
(b) Prove that there exists a linear operator $T : \mathbb{R}^2 \to \mathbb{R}^3$ such that

$$T(1, 1) = (1, 0, 2) \quad \text{and} \quad T(2, 3) = (1, -1, 4).$$

What is $T(8, 11)$?
(c) Is there a linear map $T : \mathbb{R}^3 \to \mathbb{R}^2$ such that $T(1, 0, 3) = (1, 1)$ and $T(-2, 0, -6) = (2, 1)$?

6. Find a simple example of a linear operator $T : \mathbb{R}^5 \to \mathbb{R}^5$ such that $R(T) = K(T)$ so that $T^2 = 0$ (the zero operator).

HINT: How must $K(T)$ and $R(T)$ be related if $T^2 = 0$?

7. Exhibit linear operators $S, T : V \to V$ such that $K(S) = K(T)$ and $R(S) = R(T)$, but $S \neq T$.

Section 2.2. Invariant Subspaces and Invertibility.

1. TRUE/FALSE QUESTIONS: ("True" if the statement is *always* true.)
(a) If $T : V \to W$ and $S : W \to V$ are linear operators such that $S \circ T = \text{id}_V$, and if both spaces are finite dimensional, then $T \circ S = \text{id}_W$ and S is a two-sided inverse for T.
(b) If V is a vector space with $\dim_{\mathbb{K}} V \geq 2$, every linear operator $T : V \to V$ must have a *proper* invariant subspace $(0) \subsetneqq W \subsetneqq V$.

Suppose $S \circ T = \mathrm{id}_V$ for linear operators $T : V \to W$ and $S : W\ toV$ between finite-dimensional vector spaces. If $\dim(V) = \dim(W) < \infty$, prove that $T \circ S = \mathrm{id}_W$ so that S and T are bijective inverses of each other. NOTE: A similar result holds for matrices. If A, B are $n \times n$ matrices, we have $AB = I \Leftrightarrow BA = I \Leftrightarrow B = A^{-1}$.

EIGENVALUES AND EIGENVECTORS. If $T : V \to V$ is a linear operator over a field, we say that $\lambda \in \mathbb{K}$ is an *eigenvalue* of T if there is some *nonzero* $v \in V$ such that $Tv = \lambda v$. The eigenvalues are important in understanding the structure of any linear operator. In Chapter 4 on determinants, we will develop systematic methods for finding the eigenvalues of T.

3. Let $T : V \to V$ be a linear operator and $\lambda \in \mathbb{K}$. The *λ-eigenspace* is the set of vectors

$$E_\lambda(T) = \{v \in V : T(v) = \lambda \cdot v\} = \{v \in V : (T - \lambda I)v = 0\}$$

where I = the identity operator on V. The *generalized λ-eigenspace* consists of the vectors

$$M_\lambda(T) = \{v \in V : (T - \lambda I)^k v = 0 \text{ for some } k = k(v) \in \mathbb{N}\}$$
$$= \bigcup_{k=1}^{\infty} \ker\left((T - \lambda I)^k\right)$$

Since $M_\lambda \supseteq E_\lambda$ these spaces are nontrivial $\Leftrightarrow$ λ is an eigenvalue of T.

(a) Show that E_λ and M_λ are vector subspaces in V.

(b) Show that both spaces are invariant under the action of T.

4. *Sequence space* ℓ^∞ consists of all infinite sequences $\mathbf{z} = (a_1, a_2, a_3, \dots)$ with *bounded* complex entries. It is a vector space over $\mathbb{C}$ when it is equipped with the operations

$$\{a_i\} + \{b_i\} = \{a_i + b_i\} \quad \text{and} \quad \lambda \cdot \{a_i\} = \{\lambda \cdot a_i\} \text{ for } \lambda \in \mathbb{C}$$

The *left-shift operator* L and *right-shift operator* R have the following actions on elements $\mathbf{z} = (a_1, a_2, a_3, \dots)$ in ℓ^∞.

$$L(\mathbf{z}) = (a_2, a_3, a_4, \dots) \quad \text{and} \quad R(\mathbf{z}) = (0, a_1, a_2, a_3, \dots),$$

so $L(\mathbf{z})_i = a_{i+1}$, while $R(\mathbf{z})_i = a_{i-1}$ for $i \geq 2$ and $R(\mathbf{z})_1 = 0$.

(a) Prove ℓ^∞ is infinite dimensional.

(b) Determine all the complex eigenvalues for L and for R. Describe their eigenspaces $E_\lambda(L)$ and $E_\lambda(R)$.

HINTS: The relation $L(\mathbf{z}) = \lambda \cdot \mathbf{z}$ forces certain recursive relationships between the entries a_i, but an arbitrary sequence satisfying these relations might not be *bounded*, and hence not an element of ℓ^∞. The same considerations apply to the right-shift R. For (a) you must show ℓ^∞ contains arbitrarily large linearly independent sets.

NOTE: It is worth observing that $L \circ R = \mathrm{id}$ so L is a "left inverse" for R, but it is not a "right inverse" because $R \circ L$ has a nontrivial kernel and is not one-to-one.

Section 2.3. Direct Sums of Vector Spaces.

1. TRUE/FALSE QUESTIONS: ("True" if the statement is *always* true.)
   (a) We have $V = W_1 \oplus W_2 \oplus W_3$ if: (i) the W_i span V (so $W_1+W_2+W_3 = V$), and (ii) $W_i \cap W_j = (0)$ when $i \neq j$.
   (b) Coordinate space $\mathbb{K}^n$ is a direct sum $\mathbb{K} \oplus \cdots \oplus \mathbb{K}$ of n copies of the ground field $\mathbb{K}$ regarded as a 1-dimensional vector space.
   (c) If $T : V \rightarrow V$ is linear and $E_1, \ldots, E_r$ are T-invariant subspaces in V, their linear span $E = E_1 + \cdots + E_r$ is also T-invariant.
   (d) If W_1 is a nonzero subspace of a finite-dimensional space V, there always exists a subspace W_2 such that $V = W_1 \oplus W_2$.
2. If $V_1, \ldots, V_r$ are subspaces of a vector space V such that $V = V_1 \oplus \cdots \oplus V_r$, prove that the V_i must be pairwise essentially disjoint, so $V_i \cap V_j = (0)$ for $i \neq j$.

   NOTE: The converse is only true for direct sums $V_1 \oplus V_2$ of *two* subspaces. Recall Exercises 2.22 and 2.23.
3. In the preceding group of exercises we defined the λ-eigenspace E_λ of a linear operator $T : V \rightarrow V$ (see Additional Exercise 3 for Section 2.2).
   (a) If $\lambda \neq \mu$ in $\mathbb{K}$, show that the corresponding eigenspaces must be pairwise essentially disjoint: $E_\lambda \cap E_\mu = (0)$ when $\lambda \neq \mu$.
   (b) When $\lambda \neq \mu$, the subspace $M = E_\lambda + E_\mu$ that is spanned by the eigenspaces is in fact a *direct sum*: $M = E_\lambda \oplus E_\mu$ (recall Exercise 2.22).

   BEWARE: If $\lambda_1, \lambda_2, \lambda_3$ are distinct eigenvalues of $T : V \rightarrow V$, the eigenspaces are still "pairwise essentially disjoint" with $E_{\lambda_i} \cap E_{\lambda_j} = (0)$ for $i \neq j$. But this is no longer enough to insure that the subspace $M = E_{\lambda_1} + E_{\lambda_2} + E_{\lambda_3}$ spanned by the E_{λ_i} is a direct sum (as in the previous exercise, which involved just two summands). The difficulty is already apparent from the simple example in Exercise 2.23.
4. Prove that $\mathbb{R}^3 = W_1 \oplus W_2 \oplus W_3$ where

$$W_1 = \mathbb{R} \cdot \mathbf{e}_1, \quad W_2 = \mathbb{R} \cdot (\mathbf{e}_1 + \mathbf{e}_2), \quad W_3 = \mathbb{R} \cdot (\mathbf{e}_1 - 2\mathbf{e}_3),$$

   and $\mathbf{e}_1 = (1, 0, 0), \ldots, \mathbf{e}_3 = (0, 0, 1)$ are the standard basis vectors in $\mathbb{R}^3$.

Section 2.4. Representing Linear Operators as Matrices.

1. TRUE/FALSE QUESTIONS: ("True" if the statement is *always* true.) In this section A is an $n \times m$ matrix and $L_A : \mathbb{K}^m \rightarrow \mathbb{K}^n$ is the multiplication operator $L_A(\mathbf{x}) = A \cdot \mathbf{x}$.
   (a) If A is an $n \times m$ matrix and $\mathfrak{X}, \mathfrak{Y}$ are bases in $\mathbb{K}^m, \mathbb{K}^n$, the matrix $[L_A]_{\mathfrak{Y}\mathfrak{X}}$ of the linear operator $L_A(\mathbf{x}) = A \cdot \mathbf{x}$ is equal to the original matrix A.
   (b) If A is a square matrix we can only have $A^2 = I$ (the $n \times n$ identity matrix) when $A = I$ or $A = -I$.
   (c) If A is a square matrix, then $A^2 = 0_{n \times n} \Rightarrow A = 0_{n \times n}$.

(d) If $T : V \to W$ is a linear map between spaces of the same dimension n, we can find bases $\mathfrak{X} \subseteq V$, $\mathfrak{Y} \subseteq W$ such that $[T]_{\mathfrak{Y}\mathfrak{X}}$ is diagonal.

(e) If $T : V \to W$ is an *invertible* linear map between spaces of the same dimension n, we can find bases $\mathfrak{X} \subseteq V$, $\mathfrak{Y} \subseteq W$ such that $[T]_{\mathfrak{Y}\mathfrak{X}} = I_{n\times n}$ (identity matrix).

(f) Let $(V, \mathfrak{X})$ and $(V, \mathfrak{Y})$ be copies of the same finite-dimensional vector space, equipped with different bases $\mathfrak{X} = \{e_i\}$ and $\mathfrak{Y} = \{f_j\}$, and let $I : V \to V$ be the identity map. Then entries in the matrix $[I]_{\mathfrak{Y}\mathfrak{X}} = [t_{ij}]$ are the coefficients in the system of vector identities $f_i = \sum_{i=1}^{n} t_{ji} e_j$, $1 \le j \le n = \dim(V)$.

2. Let $T : \mathbb{C}^2 \to \mathbb{C}^3$ be defined by

$$(w_1, w_2, w_3) = T(z_1, z_2) = (z_1 - iz_2,\ (1+i)z_2,\ 2z_1 + iz_2).$$

Consider $\mathbb{C}$-basis $\mathfrak{Y} = \{\mathbf{f}_1 = (1,1,0), \mathbf{f}_2 = (0,i,1), \mathbf{f}_3 = (-2i, 2, -3)\}$ in $\mathbb{C}^3$ and the standard basis $\mathfrak{X} = \{\mathbf{e}_1 = (1,0), \mathbf{e}_2 = (0,1)\}$ in $\mathbb{C}^2$.

(a) Compute $[T]_{\mathfrak{Y}\mathfrak{X}}$.

(b) If $\mathfrak{Z} = \{(1,2),(2,3)\}$, compute $[T]_{\mathfrak{Y}\mathfrak{Z}}$.

3. Let $V = \mathrm{M}(2, \mathbb{R})$ and $W = \mathcal{P}_2$, the space of polynomials $f \in \mathbb{R}[x]$ with $\deg(f) \le 2$. If $T : V \to W$ is given by

$$T\begin{pmatrix} a & b \\ c & d \end{pmatrix} = (a+b) + (2d)x + cx^2,$$

find the matrix $[T]_{\mathfrak{Y}\mathfrak{X}}$ for the bases

$$\mathfrak{X} = \left\{ \begin{pmatrix} 1 & 0 \\ 0 & 0 \end{pmatrix}, \begin{pmatrix} 0 & 1 \\ 0 & 0 \end{pmatrix}, \begin{pmatrix} 0 & 0 \\ 1 & 0 \end{pmatrix}, \begin{pmatrix} 0 & 0 \\ 0 & 1 \end{pmatrix}, \right\},$$

$$\mathfrak{Y} = \{1, x, x^2\}.$$

4. Let $T : V \to W$ be a linear map on a vector space of dimension $|V| = m$, and let $\mathfrak{X} = \{e_1, \dots, e_n\}$ be a basis for $K(T) = \ker(T)$. Adjoin outside vectors to get a basis $\mathfrak{Y} = \{e_1, \dots, e_n, f_{n+1} \cdots f_m\}$ for V.

(a) Explain why the set of images $\mathfrak{Z} = \{u_1, \dots, u_{m-n}\}$ where $u_j = T(f_{n+j})$ for $1 \le j \le m - n$ is a basis for the range $R(T)$.

(b) Show that $[T]_{\mathfrak{Y}\mathfrak{Y}}$ has the general form $\begin{pmatrix} \boxed{\mathbf{0}} & \boxed{\mathbf{A}} \\ \boxed{\mathbf{0}} & \boxed{\mathbf{B}} \end{pmatrix}$

(c) What are the shapes for the rectangular blocks in (b)?

5. If $R_\theta : \mathbb{R}^2 \to \mathbb{R}^2$ is a rotation counterclockwise about the origin by θ radians and $\mathfrak{X} = \{\mathbf{e}_1, \mathbf{e}_2\}$ is the standard basis in $\mathbb{R}^2$,

(a) Verify that

$$[R_\theta]_{\mathfrak{X}\mathfrak{X}} = \begin{pmatrix} \cos\theta & -\sin\theta \\ \sin\theta & \cos\theta \end{pmatrix}.$$

(b) Explain why the kth power of this "rotation matrix" is

$$[R_\theta]^k_{\mathfrak{X}\mathfrak{X}} = \begin{pmatrix} \cos(k\theta) & -\sin(k\theta) \\ \sin(k\theta) & \cos(k\theta) \end{pmatrix}$$

for every exponent $k \in \mathbb{Z}$.

6. If ℓ is the line through the origin in $\mathbb{R}^2$ that makes a 30° angle with the $+x$-axis, and r_ℓ is the operation that reflects vectors across ℓ,

(a) Describe the action of r_ℓ in Cartesian coordinates.

(b) Compute $r_\ell(\mathbf{e}_1)$ and $r_\ell(\mathbf{e}_2)$ for the standard basis vectors in $\mathbb{R}^2$ and determine the matrix $A = [\,r_\ell]_{\mathfrak{X}\mathfrak{X}}$.

(c) Explain why $A^2 = I$ (the 2×2 identity matrix).

7. If $T : V \to W$ is a linear operator between vector spaces of dimensions m and n respectively, and $\mathfrak{X} = \{e_i\}$, $\mathfrak{Y} = \{f_j\}$ are bases, explain why each of the following equalities holds:

$$\text{rank}(T) = \text{rank}(L_A) = \text{ColRank}(A) = \dim\, R(T)\,,$$

where $A = [T]_{\mathfrak{Y}\mathfrak{X}}$ and $L_A(\mathbf{x}) = A \cdot \mathbf{x}$.

NOTE: ColRank(A) is the dimension of the column space of A.

HINTS: Recall the commutative diagram Figure 2.3 associated with Proposition 2.33 and Lemma 2.34.

8. The set $\mathfrak{X} = \{1, x, x^2, \dots, x^n\}$ is a basis for the subspace $\mathcal{P}_n$ of polynomials in $\mathbb{R}[x]$ of degree $\deg(f) \leq n$. Let $D : \mathcal{P}_n \to \mathcal{P}_n$ be the (linear) *differentiation operator* on polynomials

$$D(c_0 1 + c_1 x + c_2 x^2 + \cdots + c_n x^n) = c_1 1 + 2c_2 x + 3c_3 x^2 + \cdots + nc_n x^{n-1}.$$

(a) What are the kernel $K(D)$ and range $R(D)$ when we regard differentiation as a linear operator $D : \mathcal{P}_n \to \mathcal{P}_n$? Find bases for each of these subspaces. What are their dimensions?

(b) Determine the matrix $[D]_{\mathfrak{X}\mathfrak{X}}$ with respect to the basis $\mathfrak{X}$ in $\mathcal{P}_n$.

(c) What are $K(D)$ and $R(D)$ when we regard differentiation D as a linear operator on the infinite-dimensional space of all polynomials $\mathbb{R}[x]$?

9. Consider the linear differential operator $T = (D - xI)$ acting on the infinite-dimensional space of polynomials $\mathbb{R}[x]$, where

$$D = (\text{differentiation operator}) \quad \text{and} \quad xI \text{ maps } f(x) \text{ to } xf(x).$$

Is $(D - xI)$ linear? Is it one-to-one? Is it surjective?

HINTS: For $K(T)$ start by writing out long-form versions $c_0 + c_1x + c_2x^2 + \cdots$ for $f(x), Df(x), xf(x), (D-xI)f(x)$ and compare them. Certain recursion relations between the c_i will emerge if $f \in K(T)$. Keep in mind: these formal sums are supposed to represent polynomials, so only finitely many coefficients can be nonzero if $f \in \ker(D - xI)$. The same recursive calculations can also be used to answer the corresponding question regarding $R(T)$ by observing that if T is surjective $R(T)$ must contain every basis vector $1, x, x^2, \dots$. Does it?

10. If the linear operator $T : \mathbb{R}^3 \to \mathbb{R}^2$ has matrix

$$[T]_{\mathfrak{Y}\mathfrak{X}} = \begin{pmatrix} 3 & 1 & 2 \\ 1 & 0 & -1 \end{pmatrix}$$

with respect to the standard bases $\mathfrak{X} = \{\mathbf{e}_1, \mathbf{e}_2, \mathbf{e}_3\}$ and $\mathfrak{Y} = \{\mathbf{f}_1, \mathbf{f}_2\}$, find the matrix $[T]_{\mathfrak{A}\mathfrak{B}}$ with respect to new bases

$$\mathfrak{A} : \mathbf{u}_1 = (0, 1, -1), \quad \mathbf{u}_2 = (2, 2, -1), \quad \mathbf{u}_3 = (4, 0, 1)$$
$$\mathfrak{B} : \mathbf{v}_1 = (1, -1), \quad \mathbf{v}_2 = (0, -1).$$

11. The vectors

$$\mathbf{u}_1 = (1, 2, 3), \quad \mathbf{u}_2 = (2, 0, 2), \quad \mathbf{u}_3 = (-1, 2, 0),$$

are a basis $\mathfrak{X}$ for $V = \mathbb{R}^3$. Let $\mathfrak{Y} = \{\mathbf{e}_1, \mathbf{e}_2, \mathbf{e}_3\}$ be the standard basis, let $\text{id} : \mathbb{R}^3 \to \mathbb{R}^3$ be the identity map, and let $P : V \to V$ be the projection of V onto M along N corresponding to the direct sum decomposition

$$\mathbb{R}^3 = M \oplus N \quad \text{where } \mathrm{M} = \mathbb{R} \cdot \mathbf{u}_1 \text{ and } N = \mathbb{R}\text{-span}\{\mathbf{u}_1, \mathbf{u}_2\}.$$

Compute the following matrices:

(a) $[\text{id}]_{\mathfrak{X}\mathfrak{X}}$ (b) $[\text{id}]_{\mathfrak{Y}\mathfrak{Y}}$ (c) $[\text{id}]_{\mathfrak{Y}\mathfrak{X}}$ (d) $[P]_{\mathfrak{X}\mathfrak{X}}$ (e) $[P]_{\mathfrak{Y}\mathfrak{Y}}$

HINT: Some of these are easy.

12. The matrix

$$A = \begin{pmatrix} 1 & 0 & -1 & 3 \\ 2 & -1 & 0 & 7 \\ 1 & -2 & 3 & 5 \end{pmatrix} \quad \text{is row equivalent to} \quad \begin{pmatrix} 1 & 0 & -1 & 3 \\ 0 & 1 & -2 & -1 \\ 0 & 0 & 0 & 0 \end{pmatrix}$$

and its transpose A^{T} is row equivalent to

$$\begin{pmatrix} 1 & 2 & 1 \\ 0 & 1 & 2 \\ 0 & 0 & 0 \\ 0 & 0 & 0 \end{pmatrix}.$$

For the linear map $L_A : \mathbb{K}^4 \to \mathbb{K}^3$:

(a) Find bases for $K(L_A) \subseteq V$ and $R(L_A) \subseteq W$.

(b) Find explicit bases $\mathfrak{X}, \mathfrak{Y}$ for V, W such that $[L_A]_{\mathfrak{Y}\mathfrak{X}}$ takes the form

$$\begin{pmatrix} \boxed{I_{r\times r}} & \mathbf{0} \\ \mathbf{0} & \boxed{\mathbf{0}_{s\times s}} \end{pmatrix}$$

where $r = |R(L_A)|$ and $s = |W| - r$.

CHAPTER 3

Duality and the Dual Space V^*

3.1. Definitions and Examples

DEFINITION 3.1. The *linear functionals* (also known as *dual vectors*) on a vector space V over $\mathbb{K}$ are the linear maps $\ell : V \to \mathbb{K}$. We denote the *dual space* of functionals by V^*, or equivalently $\mathrm{Hom}_{\mathbb{K}}(V, \mathbb{K})$. It becomes a vector space when we impose the operations

1. ADDITION: $(\ell_1 + \ell_2)(v) = \ell_1(v) + \ell_2(v)$ for all $v \in V$,
2. SCALING: $(\lambda \cdot \ell)(v) = \lambda \cdot \ell(v)$ for $\lambda \in \mathbb{K}$, $v \in V$.

The zero element in V^* is the *zero functional* $\ell(v) = 0_{\mathbb{K}}$ for all $v \in V$, for which $\ker(\ell) = V$ and $\mathrm{range}(\ell) = \{0_{\mathbb{K}}\}$.

NOTATION: We will often employ "angle bracket" notation in discussing functionals, writing

$$\langle \ell, v \rangle \quad \text{instead of} \quad \ell(v).$$

This notation combines inputs ℓ, v to create a map $V^* \times V \to \mathbb{K}$ that is linear in each entry when the other entry is held fixed. In angle bracket notation both inputs play equal roles, and either one can be held fixed while the other varies. As we shall see this has many advantages.

We begin with an example that is central in understanding what dual vectors are and what they do.

EXAMPLE 3.2. Let V be a finite-dimensional space and $\mathfrak{X} = \{e_1, \dots, e_n\}$ an ordered basis. Every $v \in V$ has a unique expansion

$$v = \sum_{i=1}^{n} c_i e_i \quad (c_i \in \mathbb{K}).$$

For each $1 \le i \le n$ the map $e_i^* : V \to \mathbb{K}$ that reads off the ith coefficient

$$\langle e_i^*, v \rangle = c_i$$

is a linear functional in V^*. We will soon see that the set of functionals $\mathfrak{X}^* = \{e_1^*, \dots, e_n^*\}$ is a basis for the dual space V^*, called the *dual basis* determined by $\mathfrak{X}$, from which it will follow that the dual space is finite dimensional with $\dim(V^*) = \dim(V) = n$.

The following examples give some idea of the ubiquity of dual spaces and vectors in linear algebra.

EXAMPLE 3.3. If V is the space of polynomials $\mathbb{K}[x]$, an element $a \in \mathbb{K}$ determines an "evaluation functional" $\epsilon_a \in V^*$:

$$\langle \epsilon_a, f \rangle = \sum_{k=0}^{n} c_k a^k \quad \text{if } f = \sum_{k=0}^{n} c_k x^k .$$

These do not by themselves form a vector subspace of V^* because $\langle \epsilon_a - \epsilon_b, f \rangle = f(a) - f(b)$ cannot always be written as $f(c)$ for some $c \in \mathbb{K}$.

More generally, if $V = \mathcal{C}[a, b]$ is the space of *continuous* complex-valued functions on the interval $X = [a, b] \subseteq \mathbb{R}$, we can define evaluation functionals $\langle \epsilon_s, f \rangle = f(s)$ for $a \le s \le b$, but many elements in V^* are of a quite different nature. Two examples:

$$I(f) = \int_a^b f(t)\, dt \quad \text{(Riemann integral of } f\text{)}, \tag{3.1}$$

$$I^x(f) = \int_a^x f(t)\, dt \quad \text{(for any endpoint } a \le x \le b\text{)}. \tag{3.2}$$

For another example, consider the space $V = \mathcal{C}^{(1)}(a, b)$ of real-valued functions on an interval $(a, b) \subseteq \mathbb{R}$ that have continuous first derivatives $df/dx(s)$. We can define the usual evaluation functionals $\epsilon_s \in V^*$, but since differentiation is a linear operator on $\mathcal{C}^{(1)}(a, b)$, there are also functionals ℓ_s involving derivatives, such as

$$\ell_s : f \to \frac{df}{dx}(s) \quad \text{for } a < s < b,$$

or even linear combinations such as $\ell_s(f) = f(s) + \dfrac{df}{dx}(s)$.

EXAMPLE 3.4. Suppose V is finite dimensional and that $\ell \in V^*$ is not the zero functional. The kernel $E = \ker(\ell) = \{v \in V : \langle \ell, v \rangle = 0\}$ is a *hyperplane* in V—a vector subspace of dimension $n - 1$ where $n = \dim(V)$.

PROOF. By the dimension formula,

$$\dim_{\mathbb{K}}(V) = \dim_{\mathbb{K}}(\ker(\ell)) + \dim_{\mathbb{K}}(\text{range}(\ell))$$

But if $\ell \neq 0$, say $\langle \ell, v_0 \rangle \neq 0$, then $\langle \ell, \mathbb{K} v_0 \rangle = \mathbb{K}$, so $\text{range}(\ell) = \mathbb{K}$ has dimension 1. □

Lengths and Orthogonality of Vectors in Inner Product Spaces.

EXAMPLE 3.5. On $\mathbb{R}^n$ we have the standard euclidean inner product

$$(\mathbf{x}, \mathbf{y}) = \sum_{k=1}^{n} x_k y_k \quad \text{for } \mathbf{x}, \mathbf{y} \in \mathbb{R}^n,$$

familiar from calculus, but this is just a special case of the standard inner product on complex n-dimensional coordinate space $\mathbb{C}^n$,

$$(\mathbf{z},\mathbf{w}) = \sum_{k=1}^{n} z_k\overline{w}_k \quad \text{for complex } n\text{-tuples } \mathbf{z},\mathbf{w} \text{ in } \mathbb{C}^n, \tag{3.3}$$

where $\overline{z} = x - iy$ is the complex conjugate of $z = x + iy$. We will focus on the complex case, because everything said here applies verbatim to the real case if you interpret "conjugation" to mean $\overline{x} = x$ for real numbers.

In either case, imposing an inner product on coordinate space $V = \mathbb{K}^n$ allows us to construct $\mathbb{K}$-linear functionals $\ell_{\mathbf{y}} \in V^*$ associated with individual vectors $\mathbf{y} \in V = \mathbb{K}^n$, by defining

$$\langle \ell_{\mathbf{y}}, \mathbf{x}\rangle = (\mathbf{x},\mathbf{y}) \quad \text{for any } \mathbf{x} \in V.$$

In this setting the right-hand vector $\mathbf{y}$ is fixed and acts on the left-hand entry to produce a scalar in $\mathbb{K}$. (Think of $\mathbf{y}$ as the "actor" and $\mathbf{x}$ as the "actee"—the vector that gets acted upon.)

The functional $\ell_{\mathbf{y}}$ is $\mathbb{K}$-linear because the inner product is linear in its first entry when the second entry $\mathbf{y}$ is held fixed, hence $\ell_{\mathbf{y}}$ is a dual vector in V^*. Note carefully the placement of the "actee" on the left side of the inner product; the inner product on a vector space over $\mathbb{K} = \mathbb{C}$ is a *conjugate-linear* function of the right-hand entry:

$$(\mathbf{z}, \lambda\cdot\mathbf{w}) = \overline{\lambda}\cdot(\mathbf{z},\mathbf{w}) \quad \text{while} \quad (\lambda\cdot\mathbf{z},\mathbf{w}) = \lambda\cdot(\mathbf{z},\mathbf{w})$$

for $\lambda \in \mathbb{C}$. Placing the "actee" on the right would not produce a $\mathbb{C}$-*linear* operation on input vectors. (When $\mathbb{K} = \mathbb{R}$, complex conjugation doesn't do anything, and "conjugate-linear" is the same as "linear.")

The special case $\mathbb{K}^n = \mathbb{R}^n$ is of course important in geometry. The standard inner product on $\mathbb{R}^n$ and the functionals $\ell_{\mathbf{y}}$ then have explicit geometric interpretations:

$$\begin{aligned}(\mathbf{x},\mathbf{y}) = \langle \ell_{\mathbf{y}},\mathbf{x}\rangle &= \|\mathbf{x}\|\cdot\|\mathbf{y}\|\cdot\cos(\theta)\\ &= \|\mathbf{x}\|\cdot(\|\mathbf{y}\|\cdot\cos\theta)\\ &= \|\mathbf{x}\|\cdot\begin{pmatrix}\text{orthogonally projected length of } \mathbf{y}\\ \text{on the 1-dimensional subspace } \mathbb{R}\cdot\mathbf{x}\end{pmatrix},\end{aligned}$$

where

$$\|\mathbf{x}\| = (\mathbf{x},\mathbf{x})^{1/2} = \left(\sum_{k=1}^{n} |x_k|^2\right)^{1/2}$$

is the euclidean length of vector $\mathbf{x} \in \mathbb{R}^n$. The angle $\theta = \theta(\mathbf{x},\mathbf{y})$ is the angle in radians between $\mathbf{x}$ and $\mathbf{y}$, measured counterclockwise from $\mathbf{x}$ to $\mathbf{y}$ in the plane (two-dimensional subspace) spanned by $\mathbf{x}$ and $\mathbf{y}$ as shown in Figure 3.1. Notice that nonzero vectors $\mathbf{x}$ and $\mathbf{y}$ are perpendicular if $(\mathbf{x},\mathbf{y}) = 0$, so $\cos(\theta) = 0$. (The zero vector $\mathbf{0}$ is perpendicular to everybody, by definition.)

NOTE: While the real inner product is natural in geometry, the complex inner product is the notion of choice in physics (as well as electrical engineering,

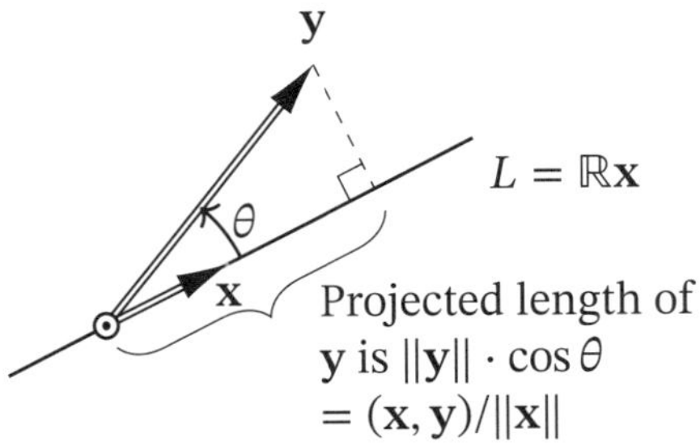

FIGURE 3.1. Geometric interpretation of the standard inner product $(\mathbf{x},\mathbf{y}) = \|\mathbf{x}\|\,\|\mathbf{y}\|\cdot\cos(\theta(\mathbf{x},\mathbf{y}))$ in $\mathbb{R}^n$. The projected length of a vector $\mathbf{y}$ onto the line $L = \mathbb{R}\mathbf{x}$ is $\|\mathbf{y}\|\cdot\cos(\theta)$. The angle $\theta(\mathbf{x},\mathbf{y})$ is measured within the two-dimensional subspace $M = \mathbb{R}\text{-span}\{\mathbf{x},\mathbf{y}\}$. Vectors are *orthogonal* when $(\mathbf{x},\mathbf{y}) = 0$, so $\cos\theta = 0$. The zero vector is orthogonal to everybody.

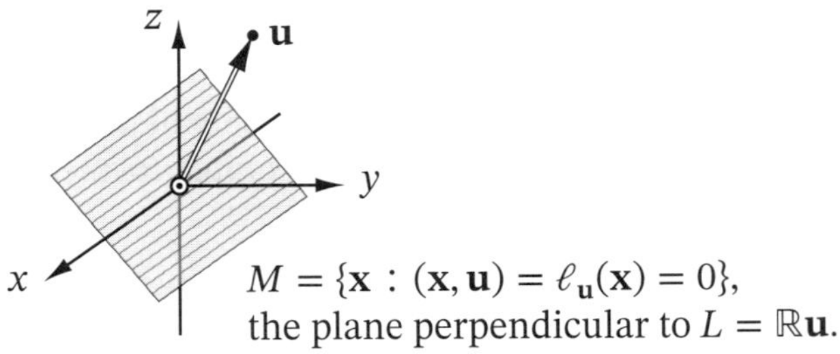

FIGURE 3.2. A vector u $\neq \mathbf{0}$ in $\mathbb{R}^n$ determines a *hyperplane* $M = (\mathbb{R}\mathbf{u})^{\perp} = \{\mathbf{x} : (\mathbf{x},\mathbf{y}) = 0\} = \ker(\ell_{\mathbf{u}})$. This $(n-1)$-dimensional subspace consists of vectors perpendicular to $\mathbf{u}$.

quantum mechanics, etc.). *But beware*: physicists often employ a convention opposite to ours, especially in quantum mechanics. For them an inner product is linear in the *right-hand* entry and conjugate-linear on the *left*. That can be confusing if you are not forewarned.

EXAMPLE 3.6. In $V = \mathbb{R}^3$ with the standard inner product $(\mathbf{x},\mathbf{y}) = \sum_i x_i y_i$, fix a vector $\mathbf{u} \neq \mathbf{0}$. The set of vectors $M = \{\mathbf{x} \in \mathbb{R}^3 : (\mathbf{x},\mathbf{u}) = 0\}$ is the hyperplane of vectors orthogonal to $\mathbf{u}$; see Figure 3.2. As an example, if $\mathbf{u} = (0,0,1) \in \mathbb{R}^3$ and $\ell_{\mathbf{u}}(x) = (\mathbf{x},\mathbf{u})$ as in Example 3.5, this orthogonal hyperplane coincides with the kernel of $\ell_{\mathbf{u}}$:

$$M = \ker(\ell_{\mathbf{u}}) = \{(x_1,x_2,0) : x_1,x_2 \in \mathbb{R}\} = \mathbb{R}\text{-span}\{\mathbf{e}_1,\mathbf{e}_2\}$$

EXERCISE 3.7. If $\mathbf{u} \neq \mathbf{0}$ in an inner product space of dimension n, explain why the orthogonal complement

$$M = (\mathbb{R}\cdot\mathbf{u})^{\perp} = \{\mathbf{x} : (\mathbf{x},\mathbf{u}) = 0\}$$

is a subspace of dimension $n-1$.

HINT: Reread Example 3.4.

Duality and the Fourier Transform.

EXAMPLE 3.8. Let $V = \mathcal{C}[0,1]$ be the ∞-dimensional space of all continuous complex-valued functions $f : [0,1] \to \mathbb{C}$. The standard inner product in $\mathcal{C}[0,1]$ is given by the integral $(f,h) = \int_0^1 f(t)\overline{h(t)}dt$. The *Fourier transform* of f is the function $f^\wedge : \mathbb{Z} \to \mathbb{C}$ defined by integrating $f(t)$ against the *complex trigonometric functions trigononetric functions* $E_n(t) = e^{2\pi int}$

$$E_n(t) = e^{2\pi int} = \cos(2\pi nt) + i\sin(2\pi nt) \quad (n \in \mathbb{Z})$$

on the real line. The nth *Fourier coefficient* of $f(t)$ is the integral

$$\begin{aligned} f^\wedge(n) = (f, e^{2\pi int}) &= \int_0^1 f(t)e^{-2\pi int}\,dt \\ &= \int_0^1 f(t)\cos(2\pi t)dt - i\int_0^1 f(t)\sin(2\pi t)dt \quad (n \in \mathbb{Z}). \end{aligned}$$

Note that each E_n is periodic with period $\Delta t = 1$ because $e^{2\pi in} = 1$ for all $n \in \mathbb{Z}$; then by the exponent law $e^{z+w} = e^z \cdot e^w$ (valid for complex as well as real exponents) we get

$$E_n(t+\Delta t) = e^{2\pi in(t+\Delta t)} = e^{2\pi int} \cdot e^{2\pi in\cdot\Delta t} = e^{2\pi in} \cdot E_n(t) = E_n(t)$$

when $\Delta t = 1$, for all $t \in \mathbb{R}$.

The preceding integral is taken over the period interval $0 \leq t \leq 1$ common to all E_n. Any function $f(t)$ that is smooth and periodic with $f(t+1) = f(t)$ for all t can be synthesized as a superposition of the periodic complex trigonometric functions $E_n(t)$, with complex-valued weights given by its Fourier coefficients:

$$f(t) = \sum_{n=-\infty}^{+\infty} f^\wedge(n) \cdot e^{2\pi int} = \sum_{n=-\infty}^{+\infty} f^\wedge(n) \cdot E_n(t).$$

The series converges pointwise on $\mathbb{R}$ if f is periodic and once continuously differentiable.

For each $n \in \mathbb{Z}$ the map

$$f \in \mathcal{C}[0,1] \xrightarrow{\phi_n} f^\wedge(n) \in \mathbb{C}$$

is a linear functional in V^*. It is actually another example of a functional determined via an inner product as in Example 3.5 because the standard inner product on $\mathcal{C}[0,1]$ is $(f,h) = \int_0^1 f(t)\overline{h(t)}\,dt$, and we have

$$\phi_n(f) = f^\wedge(n) = \int_0^1 f(t)\overline{E_n(t)}dt = (f, E_n)$$

for all $n \in \mathbb{Z}$, $f \in V$. So, ϕ_n is precisely the functional ℓ_{E_n} in Example 3.5.

3.2. Dual Bases in the Dual Space V^*

The dual space V^* of linear functionals can be viewed as the space of linear operators $\mathrm{Hom}_{\mathbb{K}}(V, \mathbb{K})$. For arbitrary vector spaces V, W of dimension m, n we saw earlier in Lemma 2.34 of Chapter 2 that $\mathrm{Hom}_{\mathbb{K}}(V, W)$ is isomorphic to the space $\mathrm{M}(n \times m, \mathbb{K})$ of $n \times m$ matrices, which obviously has dimension $m \cdot n$. In the special case when $W = \mathbb{K}$ we get $\dim(V^*) = m = \dim(V)$. This can also be seen by re-examining Example 3.2, which provides a natural way to construct a basis $\mathfrak{X}^* = \{e_1^*, \dots, e_n^*\}$ in V^* given an ordered basis $\mathfrak{X} = \{e_1, \dots, e_n\}$ in V. The functional e_i^* reads the ith coefficent in the unique expansion $v = \sum_i c_i e_i$ of a vector $v \in V$, so that

$$\left\langle e_i^*, \sum_{k=1}^{n} c_k e_k \right\rangle = c_i \quad \text{for } 1 \leq i \leq n. \tag{3.4}$$

As an immediate consequence, the linear functional $e_i^* : V \to \mathbb{K}$ is completely determined by the property

$$\langle e_i^*, e_j \rangle = \delta_{ij} \quad \text{(Kronecker delta symbol} = 1 \text{ if } i = j \text{ and } 0 \text{ otherwise).} \tag{3.5}$$

(3.5) follows because $e_j = 0 \cdot e_1 + \cdots + 1 \cdot e_j + \cdots + 0 \cdot e_n$; we recover (3.4) by observing that

$$\left\langle e_i^*, \sum_{k=1}^{n} c_k e_k \right\rangle = \sum_{k=1}^{n} c_k \langle e_i^*, e_k \rangle = \sum_{k=1}^{n} c_k \delta_{ik} = c_i \,.$$

We now show that the vectors $e_1^*, \dots, e_n^*$ form a basis in V^*, the *dual basis* to the original basis $\mathfrak{X}$ in V. This implies that $\dim(V^*) = \dim(V) = n$. Note, however, that to define the dual vectors e_i^* you must start with a basis in V; given a single vector "v" in V, there is no way to define a dual vector v^* in V^*.

THEOREM 3.9. *If V is finite dimensional and $\mathfrak{X}$ is a basis for V, the vectors $\mathfrak{X}^* = \{e_1^*, \dots, e_n^*\}$ are a basis for V^*.*

PROOF OF INDEPENDENCE. If $\ell = \sum_{j=1}^{n} c_j e_j^*$ is the zero vector in V^*, then $\langle \sum_j c_j e_j^*, v \rangle = 0$ for every $v \in V$, and in particular if $v = e_i$ we get

$$0 = \langle \ell, e_i \rangle = \sum_j c_j \langle e_j^*, e_i \rangle = \sum_j c_j \delta_{ji} = c_i$$

for $1 \leq i \leq n$, proving independence of the vectors e_i^*. □

PROOF OF SPANNING. If $\ell \in V^*$ and $c_i = \langle \ell, e_i \rangle$, we claim that ℓ is equal to

$$\ell' = \sum_{j=1}^{n} \langle \ell, e_j \rangle \cdot e_j^*.$$

It suffices to show that ℓ and ℓ' have the same values on the basis vectors $\{e_i\}$

in V, but that is obvious because

$$\langle \ell', e_i\rangle = \Big\langle \sum_j \langle \ell, e_j\rangle e_j^*\,,\, e_i\Big\rangle = \sum_j \langle \ell, e_j\rangle \cdot \langle e_j^*, e_i\rangle = \sum_j \langle \ell, e_j\rangle \cdot \delta_{ij} = \langle \ell, e_i\rangle$$

for $1 \leq i \leq n$ as claimed. □

The formula developed in this proof is often useful in computing dual bases.

COROLLARY 3.10. *If V is finite dimensional, $\mathfrak{X} = \{e_i\}$ a basis in V, and $\mathfrak{X}^* = \{e_i^*\}$ the dual basis in V^*, then any $\ell \in V^*$ has*

$$\ell = \sum_{i=1}^{n} \langle \ell, e_i\rangle \cdot e_i^*$$

as its expansion in the $\mathfrak{X}^*$-basis.

EXERCISE 3.11. If $v_1 \neq v_2$ in a finite-dimensional vector space V, prove that there is an $\ell \in V^*$ such that $\langle \ell, v_1\rangle \neq \langle \ell, v_2\rangle$. (Thus there are enough functionals in the dual V^* to distinguish vectors in V.)

HINT: It suffices to show $v_0 \neq 0 \Rightarrow \langle \ell, v_0\rangle \neq 0$ for some $\ell \in V^*$. (Why?) Think about bases in V that involve v_0 and their duals.

NOTE: This result is actually true for all infinite-dimensional spaces, but the proof is harder and requires "transcendental methods" involving the *Axiom of Choice*. These methods also show that every infinite-dimensional space has a basis $\mathfrak{X}$—an (infinite) set of independent vectors such that every $v \in V$ can be written uniquely as a *finite* $\mathbb{K}$-linear combination of vectors from $\mathfrak{X}$. As an example, the basic powers $\mathfrak{X} = \{1, x, x^2, \dots\}$ are a basis for the space of polynomials $\mathbb{K}[x]$ in this sense. A more challenging problem is to produce a basis for $V = \mathbb{R}$ when $\mathbb{R}$ is regarded as a vector space over the field of rationals $\mathbb{Q}$. Any such *Hamel basis* for $\mathbb{R}$ is necessarily uncountable.

EXAMPLE 3.12. Consider the basis $\mathbf{u}_1 = (1,0,1)$, $\mathbf{u}_2 = (1,-1,0)$, $\mathbf{u}_3 = (2,0,-1)$ in $\mathbb{R}^3$. We shall determine the dual basis vectors $\mathbf{u}_i^*$ by computing their action as functionals acting on an arbitrary vector $v = (x_1, x_2, x_3)$ in $\mathbb{R}^3$.

SOLUTION: Note that $(x_1, x_2, x_3) = \sum_{k=1}^{3} x_k \mathbf{e}_k$ where $\{\mathbf{e}_k\}$ is the standard basis in $\mathbb{R}^3$. The basis $\{\mathbf{e}_k^*\}$ dual to the standard basis $\{\mathbf{e}_k\}$ has the following action:

$$\langle \mathbf{e}_k^*, (x_1, x_2, x_3)\,\rangle = \Big\langle \mathbf{e}_k^*, \sum_{i=1}^{3} x_i \mathbf{e}_i \Big\rangle = x_k$$

because $\mathbf{e}_k^*$ reads the kth coefficient in $v = \sum_i x_i \mathbf{e}_i$.

For a different basis such as $\mathfrak{Y} = \{\mathbf{u}_i\}$, the dual vector $\mathbf{u}_k^*$ reads the kth coefficient c_k when we expand a typical vector $v \in \mathbb{R}^3$ as $v = \sum_{j=1}^{3} c_j \mathbf{u}_j$, and to determine the coefficients c_k we must write $v = (x_1, x_2, x_3) = \sum_{j=1}^{3} x_i \mathbf{e}_j$ in terms of the new basis $\{\mathbf{u}_k\}$.

To find c_k when we write

$$\begin{pmatrix} x_1 \\ x_2 \\ x_3 \end{pmatrix} = \sum_i c_i \mathbf{u}_i = c_1 \begin{pmatrix} 1 \\ 0 \\ 1 \end{pmatrix} + c_2 \begin{pmatrix} 1 \\ -1 \\ 0 \end{pmatrix} + c_3 \begin{pmatrix} 2 \\ 0 \\ -1 \end{pmatrix},$$

we must solve for the 3×1 matrix $\mathbf{c}$ in the equation

$$A\mathbf{c} = \mathbf{x} = \begin{pmatrix} x_1 \\ x_2 \\ x_3 \end{pmatrix} \quad \text{where } A = \begin{pmatrix} 1 & 1 & 2 \\ 0 & -1 & 0 \\ 1 & 0 & -1 \end{pmatrix}.$$

Row operations on the augmented matrix for $A\mathbf{x} = \mathbf{b}$ for this system yield

$$[A : \mathbf{x}] = \left(\begin{array}{ccc|c} 1 & 1 & 2 & x_1 \\ 0 & -1 & 0 & x_2 \\ 1 & 0 & -1 & x_3 \end{array}\right) \to \left(\begin{array}{ccc|c} 1 & 1 & 2 & x_1 \\ 0 & 1 & 0 & -x_2 \\ 0 & -1 & -3 & x_3 - x_1 \end{array}\right)$$

$$\to \left(\begin{array}{ccc|c} \boxed{1} & 1 & 2 & x_1 \\ 0 & \boxed{1} & 0 & -x_2 \\ 0 & 0 & \boxed{1} & \frac{1}{3}(x_1 + x_2 - x_3) \end{array}\right).$$

There are no free variables; backsolving yields the unique solution $\mathbf{c}$:

$$\begin{aligned} c_1 &= x_1 - c_2 - 2c_3 = \frac{1}{3}x_1 + \frac{1}{3}x_2 + \frac{2}{3}x_3, \\ c_2 &= -x_2, \\ c_3 &= \frac{1}{3}(x_1 + x_2 - x_3). \end{aligned}$$

Thus,

$$v = \left(\frac{1}{3}x_1 + \frac{1}{3}x_2 + \frac{2}{3}x_3\right)\mathbf{u}_1 - x_2\mathbf{u}_2 + \left(\frac{1}{3}x_1 + \frac{1}{3}x_2 - \frac{1}{3}x_3\right)\mathbf{u}_3.$$

For $i = 1, 2, 3$ apply $\mathbf{u}_i^*$ to both sides of this identity. Since $\langle \mathbf{u}_i^*, \mathbf{u}_j \rangle = \delta_{ij}$ we get

$$\begin{aligned} \langle \mathbf{u}_i^*, v \rangle &= \langle \mathbf{u}_i^*, (x_1, x_2, x_3) \rangle = \Big\langle \mathbf{u}_i^*, \sum_i x_i \mathbf{e}_i \Big\rangle \\ &= \Big\langle \mathbf{u}_i^*, \sum_j c_j \mathbf{u}_j \Big\rangle = c_i = \begin{cases} \frac{1}{3}x_1 + \frac{1}{3}x_2 + \frac{2}{3}x_3, & i = 1, \\ -x_2, & i = 2, \\ \frac{1}{3}x_1 + \frac{1}{3}x_2 - \frac{1}{3}x_3, & i = 3. \end{cases} \end{aligned}$$

Since $\langle \mathbf{e}_k^*, v \rangle = \langle \mathbf{e}_k^*, (x_1, x_2, x_3) \rangle = x_k$ we obtain the system of vector identities

$$\begin{aligned} \mathbf{u}_1^* &= \frac{1}{3}\mathbf{e}_1^* + \frac{1}{3}\mathbf{e}_2^* + \frac{2}{3}\mathbf{e}_3, \\ \mathbf{u}_2^* &= -\mathbf{e}_2^*, \\ \mathbf{u}_3^* &= \frac{1}{3}\mathbf{e}_1^* + \frac{1}{3}\mathbf{e}_2^* - \frac{1}{3}\mathbf{e}_3^*. \end{aligned}$$

3.3. The Transpose $T^{\top} : W^* \to V^*$ of $T : V \to W$

There is a natural connection between linear operators $T : V \to W$ and operators in the opposite direction, from $W^* \to V^*$.

THEOREM 3.13. *The transpose $T^{\top} : W^* \to V^*$ of a linear operator $T : V \to W$ between finite-dimensional vector spaces is a linear operator that is uniquely determined in a coordinate-free manner by requiring that*

$$\langle T^{\top}(\ell), v\rangle = \langle \ell, T(v)\rangle \quad \textit{for all } \ell \in W^*, v \in V. \tag{3.6}$$

PROOF. The right side of (3.6) defines a map $\phi_\ell : V \to \mathbb{K}$ such that $\phi_\ell(v) = \langle \ell, T(v)\rangle$. Observe that ϕ_ℓ is a linear functional on V, so each $\ell \in W^*$ determines a well defined element of V^*. Now let $T^{\top} : W^* \to V^*$ be the map $T^{\top}(\ell) = \phi_\ell$. The property (3.6) holds by definition, but we must prove $T^{\top}$ is linear and uniquely determined by the property (3.6).

Uniqueness is easy: if $S : W^* \to V^*$ is another operator such that

$$\langle S(\ell), v\rangle = \langle T^{\top}(\ell), v\rangle = \langle \ell, T(v)\rangle \quad \text{for all } \ell \in W^* \text{ and } v \in V,$$

these identities imply $S(\ell) = T^{\top}(\ell)$ for all ℓ (by Exercise 3.11), which means $S = T^{\top}$ as maps on W^*.

The easiest proof that $T^{\top}$ is linear uses the scalar identities (3.5) and the following general observation.

EXERCISE 3.14. If V, W are finite-dimensional vector spaces, explain why the following statements regarding two linear operators $A, B : V \to W$ are equivalent.

(a) $A = B$ as operators.
(b) $Av = Bv$ for all $v \in V$.
(c) $\langle \ell, Av\rangle = \langle \ell, Bv\rangle$ for all $v \in V, \ell \in W^*$.

HINT: Use Exercise 3.11 to prove (c) ⇒ (b); implications (b) ⇒ (a) ⇒ (c) are trivial.

To prove $T^{\top}(\ell_1+\ell_2) = T^{\top}(\ell_1)+T^{\top}(\ell_2)$, just bracket these with an arbitrary $v \in V$ and compute:

$$\begin{aligned}\langle T^{\top}(l_1 + l_2), v\rangle &= \langle l_1 + l_2, T(v)\rangle \\ &= \langle l_1, T(v)\rangle + \langle l_2, T(v)\rangle \quad \text{(definition of } (+) \text{ in } W^*) \\ &= \langle T^{\top}(l_1) + T^{\top}(l_2), v\rangle \quad \text{(definition of } (+) \text{ in } V^*)\end{aligned}$$

for all $v \in V$. The other identity we need, $T^{\top}(\lambda \cdot \ell) = \lambda \cdot T^{\top}(\ell)$, is proved similarly. Thus $T^{\top} : W^* \to V^*$ is a well-defined linear operator that acts in the opposite direction from $T : V \to W$.

Basic properties of the correspondence $T \to T^{\top}$ are left as exercises. The proofs are easy using the scalar identities (3.6). □

EXERCISE 3.15. Verify that

(a) The transpose $\mathbf{0}^{\top}$ of the zero operator $\mathbf{0}(v) \equiv 0_W$ from $V \to W$ is the zero operator from $W^* \to V^*$, so $\mathbf{0}^{\top}(\ell) = 0_{V^*}$ for all $\ell \in W^*$.

(b) When $V = W$ the transpose of the identity map $\mathrm{id}_V : V \to V$, with $\mathrm{id}_V(v) \equiv v$, is the identity map $\mathrm{id}_{V^*} : V^* \to V^*$ so that $(\mathrm{id}_V)^\top = \mathrm{id}_{V^*}$.

(c) $(\lambda_1 T_1 + \lambda_2 T_2)^\top = \lambda_1 T_1^\top + \lambda_2 T_2^\top$ for any $\lambda_1, \lambda_2 \in \mathbb{K}$ and $T_1, T_2 : V \to W$.

EXERCISE 3.16. If $U \xrightarrow{T} V \xrightarrow{S} W$ are linear maps between finite-dimensional vector spaces, prove that

$$(S \circ T)^\top = T^\top \circ S^\top.$$

Note the reversal of order when we compute the transpose of a product.

EXERCISE 3.17. If V, W are finite dimensional and $T : V \to W$ is an invertible linear operator (a bijection), prove that $T^\top : W^* \to V^*$ is invertible too, and $(T^{-1})^\top = (T^\top)^{-1}$ as maps from $V^* \to W^*$.

Matrix Description of a Transpose $T^\top$. Now for some computational issues.

THEOREM 3.18. *Let $T : V \to W$ be a linear operator between finite-dimensional spaces, let $\mathfrak{X} = \{v_1, \dots, v_m\}$, $\mathfrak{Y} = \{w_1, \dots, w_n\}$ be bases in V, W, and let $\mathfrak{X}^* = \{v_i^*\}$, $\mathfrak{Y}^* = \{w_j^*\}$ be the dual bases in V^*, W^*. We have defined the transpose $A^\top$ of an $n \times m$ matrix to be the $m \times n$ matrix such that $(A^\top)_{ij} = A_{ji}$. Then for operators T we have "$[T^\top] = [T]^\top$" in the sense that*

$$[T^\top]_{\mathfrak{X}^*\mathfrak{Y}^*} = ([T]_{\mathfrak{Y}\mathfrak{X}})^\top.$$

IMPORTANT NOTE: *This only works for the dual bases $\mathfrak{X}^*$, $\mathfrak{Y}^*$ in V^*, W^*. If $\mathfrak{A}, \mathfrak{B}$ are arbitrary bases in V^*, W^* unrelated to the dual bases there is no reason to expect that*

$$[T^\top]_{\mathfrak{A}\mathfrak{B}} = \textit{the transpose of the matrix } [T]_{\mathfrak{Y}\mathfrak{X}}$$

PROOF. To determine $[T^\top]$ we must calculate the coefficients in the system of vector equations

$$T^\top(w_i^*) = \sum_{j=1}^{m} [T^\top]_{ji} v_j^*, \quad 1 \le i \le n.$$

These can be found by applying each of these identities to a basis vector v_k in V:

$$\langle T^\top(w_i^*), v_k \rangle = \sum_{j=1}^{m} [T^\top]_{ji} \cdot \langle v_j^*, v_k \rangle = \sum_{j=1}^{m} [T^\top]_{ji} \delta_{jk} = [T^\top]_{ki} \tag{3.7}$$

for any $1 \le i \le n$ and $1 \le k \le m$. Thus

$$[T^\top]_{ki} = \langle T^\top(w_i^*), v_k \rangle.$$

dual lariz By definition of $T^{\top}$ and the matrix $[T]_{\mathfrak{Y}\mathfrak{X}}$, we can also write (3.7) as

$$\begin{aligned}\langle T^{\top}(w_i^*), v_k\rangle = \langle w_i^*, T(v_k)\rangle &= \left\langle w_i^*, \sum_{j=1}^{n} [T]_{jk} w_j \right\rangle \\ &= \sum_{j=1}^{n} [T]_{jk}\langle w_i^*, w_j\rangle = \sum_{j=1}^{n} [T]_{jk}\delta_{ij} = [T]_{ik}\end{aligned}$$

for any $1 \le i \le n, 1 \le k \le m$.

Comparing this with (3.7), we conclude that $[T]_{ik} = ([T]^{\top})_{ki}$. Thus the matrix $[T^{\top}]_{\mathfrak{X}^*\mathfrak{Y}^*}$ is the transpose of the matrix $[T]_{\mathfrak{Y}\mathfrak{X}}$. □

EXERCISE 3.19 (Computing Matrix Entries). If $T : V \to W$ and bases $\mathfrak{X} = \{e_i\}$, $\mathfrak{Y} = \{f_i\}$ are given in V, W, let $\mathfrak{X}^*$, $\mathfrak{Y}^*$ be the dual bases. Prove that

$$[T]_{\mathfrak{Y}\mathfrak{X}} = [t_{ij}] \text{ has entries } t_{ij} = \langle f_j^*, T(e_i)\rangle.$$

Calculating the Transpose of a Projection Operator. The transpose of a projection $P : V \to V$ is a projection $P^{\top} : V^* \to V^*$ because $P^{\top} \circ P^{\top} = (P \circ P)^{\top} = P^{\top}$, so $P^{\top}$ maps V^* onto the range $R(P^{\top})$ along the nullspace $K(P^{\top}) = \ker(P^{\top})$ in the direct sum $V^* = R(P^{\top}) \oplus K(P^{\top})$. The following example shows how to calculate these geometric objects in terms of dual bases.

EXAMPLE 3.20. Let $V = \mathbb{R}^2$ with basis $\mathfrak{Y} = \{\mathbf{u}_1, \mathbf{u}_2\}$ where $\mathbf{u}_1 = (1,0)$, $\mathbf{u}_2 = (1,1)$, and let P = projection onto $W_1 = \mathbb{R}\mathbf{u}_1$ along $W_2 = \mathbb{R}\mathbf{u}_2$. The standard basis $\mathfrak{X} = \{\mathbf{e}_1, \mathbf{e}_2\}$ or the basis $\mathfrak{Y} = \{\mathbf{u}_1, \mathbf{u}_2\}$ can be used to describe P. The description of P with respect to these bases has already been worked out in Example 2.26 of Chapter 2. We will now:

1. Compute the dual bases $\mathfrak{X}^*$, $\mathfrak{Y}^*$ as functions $\ell : \mathbb{R}^2 \to \mathbb{R}$ and find the matrix descriptions of $P^{\top}$:
$$[P^{\top}]_{\mathfrak{X}^*\mathfrak{X}^*} \quad \text{and} \quad [P^{\top}]_{\mathfrak{Y}^*\mathfrak{Y}^*}.$$
2. Compute the kernel $K(P^{\top})$ and the range $R(P^{\top})$ in terms of the basis $\mathfrak{X}^*$ dual to the standard basis $\mathfrak{X}$.
3. Repeat (2.) for the basis $\mathfrak{Y}^*$.

SOLUTION: First observe that

$$\begin{cases} \mathbf{u}_1 = \mathbf{e}_1, \\ \mathbf{u}_2 = \mathbf{e}_1 + \mathbf{e}_2, \end{cases} \quad \Rightarrow \quad \begin{cases} \mathbf{e}_1 = \mathbf{u}_1, \\ \mathbf{e}_2 = \mathbf{u}_2 - \mathbf{u}_1. \end{cases}$$

By definition, $\mathfrak{Y} = \{\mathbf{u}_1, \mathbf{u}_2\}$ is a diagonalizing basis for P, with

$$\begin{cases} P(\mathbf{u}_1) = \mathbf{u}_1, \\ P(\mathbf{u}_2) = 0, \end{cases} \quad \Rightarrow \quad [P]_{\mathfrak{Y}\mathfrak{Y}} = \begin{pmatrix} 1 & 0 \\ 0 & 0 \end{pmatrix}.$$

We then have

$$\begin{cases} P(\mathbf{e}_1) = P(\mathbf{u}_1) = \mathbf{u}_1 = \mathbf{e}_1 \\ P(\mathbf{e}_2) = P(\mathbf{u}_2 - \mathbf{u}_1) = -\mathbf{u}_1 = -\mathbf{e}_1 \end{cases} \quad \Rightarrow \quad [P]_{\mathfrak{X}\mathfrak{X}} = \begin{pmatrix} 1 & -1 \\ 0 & 0 \end{pmatrix}$$

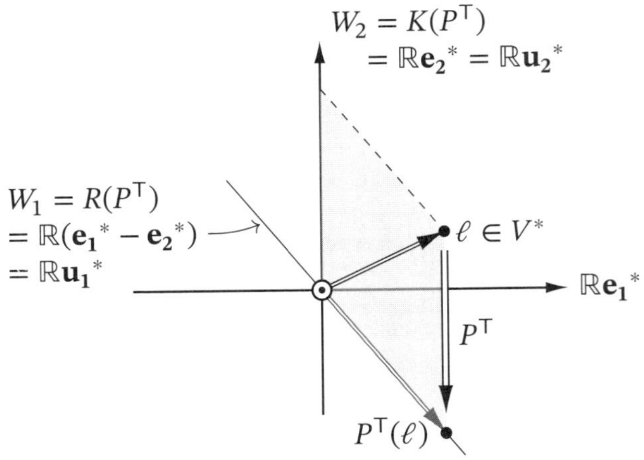

FIGURE 3.3. The decomposition $V = W_1 \oplus W_2$ determines the projection $P : \mathbb{R}^2 \to \mathbb{R}^2$ in Example 3.8 that maps V onto $W_1 = \mathbb{R}\mathbf{u}_1$ along $W_2 = \mathbb{R}\mathbf{u}_2$. We show that P^{T} projects V^* onto its range $R(P^{\mathsf{T}}) = \mathbb{R}\mathbf{u}_1^* = \mathbb{R}(\mathbf{e}_1^* - \mathbf{e}_2^*)$, along $K(P^{\mathsf{T}}) = \mathbb{R}\mathbf{u}_2^* = \mathbb{R}\mathbf{e}_2^*$. Horizontal axis in this picture is $\mathbb{R}\mathbf{e}_1^*$ and vertical axis is $\mathbb{R}\mathbf{e}_2^*$; a functional is then represented as $\ell = (\dot{x}_1\mathbf{e}_1^* + \dot{x}_2\mathbf{e}_2^*)$ with respect to the basis $\mathfrak{X}^* = \{\mathbf{e}_1^*, \mathbf{e}_2^*\}$ dual to the standard basis $\mathfrak{X} = \{\mathbf{e}_1, \mathbf{e}_2\}$.

Compute the dual basis vectors $\mathfrak{Y}^* = \{\mathbf{u}_1^*, \mathbf{u}_2^*\}$ as functions $\mathbb{R}^2 \to \mathbb{R}$ by observing that

$$\left.\begin{aligned} \mathbf{u}_1^*(v_1, v_2) &= \mathbf{u}_1^*(v_1\mathbf{e}_1 + v_2\mathbf{e}_2) \\ &= \mathbf{u}_1^*((v_1 - v_2)\mathbf{u}_1 + v_2\mathbf{u}_2) \\ &= v_1 - v_2 = (\mathbf{e}_1^* - \mathbf{e}_2^*)(v_1, v_2) \end{aligned}\right\} \Rightarrow \mathbf{u}_1^* = \mathbf{e}_1^* - \mathbf{e}_2^*$$

and

$$\left.\begin{aligned} \mathbf{u}_2^*(v_1, v_2) &= \mathbf{u}_2^*(v_1\mathbf{e}_1 + v_2\mathbf{e}_2) \\ &= \mathbf{u}_2^*((v_1 - v_2)\mathbf{u}_1 + v_2\mathbf{u}_2) \\ &= v_2 = \mathbf{e}_2^*(v_1, v_2) \end{aligned}\right\} \Rightarrow \mathbf{u}_2^* = \mathbf{e}_2^*.$$

No further calculations are needed to finish (1.); just apply Theorem 3.18 to get

$$[P^{\mathsf{T}}]_{\mathfrak{X}^*\mathfrak{X}^*} = ([P]_{\mathfrak{X}\mathfrak{X}})^{\mathsf{T}} = \begin{pmatrix} 1 & 0 \\ -1 & 0 \end{pmatrix}$$

Using the same ideas we see that

$$[P^{\mathsf{T}}]_{\mathfrak{Y}^*\mathfrak{Y}^*} = ([P]_{\mathfrak{Y}\mathfrak{Y}})^{\mathsf{T}} = \begin{pmatrix} 1 & 0 \\ 0 & 0 \end{pmatrix}^{\mathsf{T}} = \begin{pmatrix} 1 & 0 \\ 0 & 0 \end{pmatrix},$$

which resolves Question 1.

For (2.), a functional $\ell = \dot{x}_1\mathbf{e}_1^* + \dot{x}_2\mathbf{e}_2^*$ ($\dot{x}_i \in \mathbb{R}$) is in $K(P^{\mathsf{T}}) \Leftrightarrow$

$$[P^{\mathsf{T}}]_{\mathfrak{X}^*\mathfrak{X}^*}\begin{pmatrix} \dot{x}_1 \\ \dot{x}_2 \end{pmatrix} = \begin{pmatrix} 1 & 0 \\ -1 & 0 \end{pmatrix} \cdot \begin{pmatrix} \dot{x}_1 \\ \dot{x}_2 \end{pmatrix} = \begin{pmatrix} \dot{x}_1 \\ -\dot{x}_1 \end{pmatrix} \quad \text{is equal to} \quad \begin{pmatrix} 0 \\ 0 \end{pmatrix}.$$

That happens $\Leftrightarrow \dot{x}_1 = 0$, so $K(P^{\mathsf{T}}) = \mathbb{R}\mathbf{e}_2^*$ with respect to the $\mathfrak{X}^*$-basis. Since we know $\mathbf{e}_2^* = \mathbf{u}_2^*$, we get $K(P^{\mathsf{T}}) = \mathbb{R}\mathbf{u}_2^*$ with respect to the $\mathfrak{Y}^*$-basis.

As for $R(P^{\mathrm{T}})$, if $\ell = b_1\mathbf{e}_1^* + b_2\mathbf{e}_2^*$ in the $\mathfrak{X}^*$-basis and we let $\mathbf{b} = \mathrm{col}(b_1, b_2)$ and $\dot{\mathbf{x}} = \mathrm{col}(\dot{x}_1, \dot{x}_2)$, we must solve the matrix equation $A \cdot \dot{\mathbf{x}} = \mathbf{b}$, where $A = [P^{\mathrm{T}}]_{\mathfrak{X}^*\mathfrak{X}^*}$. Row operations on $[A : \mathbf{b}]$ yield

$$\left(\begin{array}{cc|c} 1 & 0 & b_1 \\ -1 & 0 & b_2 \end{array}\right) \to \left(\begin{array}{cc|c} 1 & 0 & b_1 \\ 0 & 0 & b_2 + b_1 \end{array}\right)$$

so $\mathbf{b} \in R(P^{\mathrm{T}}) \Leftrightarrow b_1 + b_2 = 0 \Leftrightarrow \ell \in \mathbb{R} \cdot (\mathbf{e}_1^* - \mathbf{e}_2^*)$. Thus $R(P^{\mathrm{T}}) = \mathbb{R} \cdot (\mathbf{e}_1^* - \mathbf{e}_2^*)$ in the $\mathfrak{X}^*$-basis, while in the $\mathfrak{Y}^*$-basis this becomes

$$R(P^{\mathrm{T}}) = \mathbb{R}(\mathbf{e}_1^* - \mathbf{e}_2^*) = \mathbb{R}((\mathbf{u}_2^* + \mathbf{u}_1^*) - \mathbf{u}_2^*) = \mathbb{R}\mathbf{u}_1^*.$$

The projection P^{T} and the corresponding decomposition $V^* = R(P^{\mathrm{T}}) \oplus K(P^{\mathrm{T}})$ both have coordinate-independent geometric meaning. But the components of the direct sum have different descriptions according to which dual basis we use to describe vectors in V^*:

$$V^* = R(P^{\mathrm{T}}) \oplus K(P^{\mathrm{T}}) = \begin{cases} \mathbb{R}(\mathbf{u}_1^*) \oplus \mathbb{R}(\mathbf{u}_2^*) & \text{for the } \mathfrak{Y}^*\text{-basis,} \\ \mathbb{R}(\mathbf{e}_1^* - \mathbf{e}_2^*) \oplus \mathbb{R}(\mathbf{e}_2^*) & \text{for the } \mathfrak{X}^*\text{-basis.} \end{cases}$$

EXERCISE 3.21. Round out the previous discussion by verifying the following statements:

(a) For the standard basis $\mathfrak{X}$ we have $R(P) = \mathbb{R}\mathbf{e}_1$ and $K(P) = \mathbb{R} \cdot (\mathbf{e}_1 + \mathbf{e}_2)$.
(b) For the $\mathfrak{Y}$-basis we have $R(P) = \mathbb{R}\mathbf{u}_1$ and $K(P) = \mathbb{R}\mathbf{u}_2$.

EXERCISE 3.22. Let $\mathfrak{X} = \{\mathbf{e}_1, \dots, \mathbf{e}_m\}$ and $\mathfrak{Y} = \{\mathbf{f}_1, \dots, \mathbf{f}_n\}$ be the standard bases in $V = \mathbb{K}^m, W = \mathbb{K}^n$. If S, T are linear maps such that

EMBEDDING MAP: $m \leq n$ and $S(x_1, \dots, x_m) = (x_1, \dots, x_m, 0, \dots, 0)$,
RESTRICTION MAP: $m \geq n$ and $T(x_1, \dots, x_n, \dots, x_m) = (x_1, \dots, x_n)$,

compute the transpose actions S^{T} and T^{T} with respect to the dual bases $\mathfrak{X}^* = \{\mathbf{e}_1^*, \dots, \mathbf{e}_m^*\}$ and $\mathfrak{Y}^* = \{\mathbf{f}_1^*, \dots, \mathbf{f}_m^*\}$.

Reflexivity of Finite-Dimensional Spaces. If V is finite dimensional, there is a natural "bracketing map"

$$\phi : V^* \times V \to \mathbb{K} \quad \text{given by} \quad \phi : (\ell, v) \mapsto \langle \ell, v \rangle.$$

The expression $\langle \ell, v \rangle$ is *bilinear*—linear in each variable when the other is held fixed. If ℓ is fixed, we get a linear functional $v \mapsto \ell(v)$ on V, but if we fix v the map $\ell \mapsto \langle \ell, v \rangle$ is a linear map from $V^* \to \mathbb{K}$, and hence is an element $j(v) \in V^{**} = (V^*)^*$, the "double dual" of V.

LEMMA 3.23. *If $\dim(V) < \infty$ the map $j : V \to V^{**}$ is linear and a bijection. It is a "natural" isomorphism (defined without reference to any coordinate system) that allows us to identify V^{**} with V.*

PROOF. For any $\ell \in V^*$ we have

$$\langle j(v_1 + v_2), \ell \rangle = \langle \ell, v_1 + v_2 \rangle = \langle \ell, v_1 \rangle + \langle \ell, v_2 \rangle = \langle j(v_1), \ell \rangle + \langle j(v_2), \ell \rangle$$

and similarly

$$\langle j(\lambda \cdot v), \ell \rangle = \langle \ell, \lambda v \rangle = \lambda \cdot \langle \ell, v \rangle = \langle \lambda \cdot j(v), \ell \rangle.$$

Since these relations are true for all $\ell \in V^*$ we see that $j(\lambda_1 v_1 + \lambda_2 v_2) = \lambda_1 j(v_1) + \lambda_2 j(v_2)$ in V^{**} and $j : V \to V^{**}$ is linear.

Finite dimensionality of V insures that $\dim(V^{**}) = \dim(V^*) = \dim(V)$, so j is a bijection $\Leftrightarrow$ j is onto $\Leftrightarrow$ j is one-to-one $\Leftrightarrow$ $\ker(j) = (0)$. But $j(v) = 0$ if and only if $0 = \langle j(v), \ell\rangle = \langle \ell, v\rangle$ for every $\ell \in V^*$. This forces $v = 0$ and hence $\ker(j) = (0)$, because if $v \neq 0$ there is a functional $\ell \in V^*$ such that $\langle \ell, v\rangle \neq 0$. In fact, we can extend $\{v\}$ to a basis $\{v, v_2, \dots, v_n\}$ of V. Then, if we form the dual basis $\{v^*, v_2^*, \dots, v_n^*\}$ in V^*, we have $\langle v^*, v\rangle = 1$. $\square$

There is, on the other hand, no *natural* (basis-independent) isomorphism from V to V^*. The spaces V and V^* are isomorphic because they have equal dimension, so there are many unnatural bijective linear maps between them. (We can create such a map given any basis $\{e_i\} \subseteq V$ and any basis $\{f_j\} \subseteq V^*$ by sending $e_i \to f_i$.)

If we identify $V = V^{**}$ via the natural map j, then the dual basis $(\mathfrak{X}^*)^*$ gets identified with the original basis $\mathfrak{X}$ in V. In fact, the vector e_i^{**} in the dual basis to $\mathfrak{X}^*$ coincides with the image vector $j(e_i)$ because

$$\langle j(e_i), e_j^*\rangle = \langle e_j^*, e_i\rangle = \delta_{ij},$$

which is the defining property of the vectors $(e_i^*)^*$ in $\mathfrak{X}^{**}$. Hence $j(e_i) = e_i^{**}$. By time-honored abuse of notation, mathematicians often write "$\mathfrak{X}^{**} = \mathfrak{X}$" even though this is not strictly true. Furthermore, when we identify $V^{**} \cong V$, the "double transpose" $T^{\mathrm{T}\mathrm{T}} = (T^{\mathrm{T}})^{\mathrm{T}}$ mapping $V^{**} \to V^{**}$ becomes the original operator T, allowing us to write

$$T^{\mathrm{T}\mathrm{T}} = T \quad \text{(again, by abuse of notation).}$$

The precise connection between T and $T^{\mathrm{T}\mathrm{T}}$ is shown in the following commutative diagram:

$$\begin{array}{ccc} V^{**} & \xrightarrow{(T^{\mathrm{T}})^{\mathrm{T}}} & V^{**} \\ {\scriptstyle j^{-1}}\downarrow & & \uparrow{\scriptstyle j} \\ V & \xrightarrow{\ T\ } & V \end{array} \qquad \text{(Diagram commutes: } T^{\mathrm{T}\mathrm{T}} = j \circ T \circ j^{-1}\text{)}$$

EXERCISE 3.24. Assume $|V| = \dim V < \infty$, $\mathfrak{X} = \{e_1, \dots, e_n\}$ is a basis, and $T : V \to V$ a linear operator.

(a) Fill in the details needed to show that $(T^{\mathrm{T}})^{\mathrm{T}} \circ j = j \circ T$, so the diagram above commutes.

(b) Prove the following useful fact relating matrix realizations of T and $T^{\mathrm{T}\mathrm{T}}$:

$$[T^{\mathrm{T}\mathrm{T}}]_{\mathfrak{X}^{**}\mathfrak{X}^{**}} = [T]_{\mathfrak{X}\mathfrak{X}}$$

for the bases $\mathfrak{X}$ and $\mathfrak{X}^{**} = j(\mathfrak{X})$.

For infinite-dimensional spaces there is still a natural linear embedding $j : V \to V^{**}$. Although j is again one-to-one, it is not necessarily onto and there is a chain of distinct dual spaces $V, V^*, V^{**}, V^{***}, \dots$. When $\dim(V) < \infty$, this process terminates with $V^{**} \cong V$. For this reason finite-dimensional vector spaces are said to be "*reflexive*." (Some infinite-dimensional spaces are reflexive too, but not many.)

The Annihilator M° of a Subspace M in V. Additional structure must be imposed on a vector space in order to speak of "lengths" or "orthogonality" of vectors, or the "orthogonal complement" $W^\perp$ of some subspace. When $\mathbb{K} = \mathbb{R}$ or $\mathbb{C}$, this is most often accomplished by imposing an "inner product" $B : V \times V \to \mathbb{K}$ on the space. However, in the absence of such extra structure there is still a natural notion of a "complementary subspace" to any subspace $W \subseteq V$, but this complement, the *annihilator* of W,

$$W^\circ = \{\ell \in V^* : \langle \ell, w \rangle = 0 \text{ for all } w \in W\},$$

lives in V^* rather than V. It is easily seen that W° is a vector subspace in V^*. Obviously $(0)^\circ = V^*$ and $V^\circ = (0)$ in V^*, and when W is a proper subspace in V, the annihilator W° is a proper subspace of V^*, with $(0) \subsetneqq W^\circ \subsetneqq V^*$.

LEMMA 3.25. *Let V be finite dimensional and $W \subsetneqq V$ a subspace where* $\dim(W) = r$. *If $v_0 \in V$ lies outside of W, there is a functional $\ell \in V^*$ such that* $\langle \ell, x_0 \rangle \neq 0$ *and* $\langle \ell, W \rangle = 0$ *so* $\ell \in W^\circ$.

PROOF. If $W \neq V$, so $r < n = \dim(V)$, the idea is to start with a *basis* $\{e_1, \dots, e_r\}$ for W. Given a vector $v_0 \notin W$, adjoin additional vectors $e_{r+1} = v_0$, $e_{r+2}, \dots, e_n$ to make a basis $\mathfrak{X}$ for V. The dual basis $\mathfrak{X}^*$ provides the answer. We leave the details as an exercise. □

A Dimension Formula for Annihilators M°. We list the basic properties of annihilators as a series of exercises, some of which are major theorems (hints provided). In proving one of these results you may use any prior exercise or theorem. In all cases we assume $\dim(V) < \infty$.

EXERCISE 3.26. Let W be a subspace and $\mathfrak{X} = \{e_1, \dots, e_r, \dots, e_n\}$ a basis for V such that $\{e_1, \dots, e_r\}$ is a basis in W. If $\mathfrak{X}^* = \{e_i^*\}$ is the dual basis, prove that $\{e_{r+1}^*, \dots, e_n^*\}$ is a basis for the annihilator $W^\circ \subseteq V^*$.

EXERCISE 3.27 (Dimension Theorem for Annihilators). If W is a subspace in a finite-dimensional vector space V, prove that

$$\dim_{\mathbb{K}}(W) + \dim_{\mathbb{K}}(W^\circ) = \dim_{\mathbb{K}}(V),$$

or in abbreviated form, $|W| + |W^\circ| = |V|$.

EXERCISE 3.28. Assume $T : V \to W$ is a linear operator between finite-dimensional vector spaces.

(a) Prove that in W^* we have

$$K(T^{\mathrm{T}}) = R(T)^\circ \quad \text{(annihilator in } W^* \text{ of the range } R(T) \subseteq W).$$

(b) Is it also true that $R(T^{\mathrm{T}}) = K(T)^\circ$? If not, what goes wrong?

EXERCISE 3.29. If V is finite dimensional, $T : V \to V$ is linear, and E is a subspace of V, prove that E is T-invariant if and only if its annihilator E° is invariant under the transpose T^T.

HINT: Implication ($\Rightarrow$) is easy; in the other direction think about dual bases.

EXERCISE 3.30. If $T : V \to W$ is a linear operator between finite-dimensional vector spaces, prove that $\text{rank}(T^\mathsf{T}) = \text{rank}(T)$.

Row Rank vs. Column Rank (Revisited). Recall that the rank of any linear operator $T : V \to W$ is the dimension $|R(T)| = \dim(R(T))$ of its range. If $A \in \text{M}(n\times m, \mathbb{K})$, we defined $L_A : \mathbb{K}^m \to \mathbb{K}^n$ via $L_A(v) = A\cdot v$ for $v \in \mathbb{K}^m$, and the rank of the matrix is $\text{rank}(A) = \dim(R(L_A))$. Furthermore, recall that the "column rank" of a matrix is the dimension $\text{ColRank}(A) = \dim(\text{ColSpace}(A))$ of its column space, and similarly $\text{RowRank}(A) = \dim(\text{RowSpace}(A))$. It is important to know that these numbers, computed in entirely different ways, are always equal—so that

$$\text{Row Rank} = \text{Column Rank} = \text{rank}(A)$$

for any matrix A, regardless of its shape. The following exercises address this issue.

Outline of a Proof That (RowRank) = (ColRank).

EXERCISE 3.31. Let $T : V \to W$ be a linear map between finite-dimensional spaces, with bases $\mathfrak{X} = \{e_1, \dots, e_m\}$ and $\mathfrak{Y} = \{f_1, \dots, f_n\}$. If $A = [T]_{\mathfrak{Y}\mathfrak{X}}$ and $L_A : \mathbb{K}^m \to \mathbb{K}^n$ is given by $L_A(\mathbf{x}) = A \cdot \mathbf{x}$, explain why:

(a) The range $R(L_A)$ is equal to column space $\text{ColSpace}(A)$.

(b) The identities

$$\text{rank}(A) = \text{rank}(L_A) = \dim(R(L_A)) = \dim(\text{ColSpace}(A)) = \text{ColRank}(A)$$

hold for any $n \times m$ matrix A.

(c) If $A = [T]_{\mathfrak{Y}\mathfrak{X}}$, then $\text{rank}(T) = \text{rank}(L_A) = \text{ColRank}(A)$.

HINT: For (c) recall the commutative diagram in Figure 2.3 of Chapter 2; the vertical maps are isomorphisms and isomorphisms preserve dimensions of subspaces.

EXERCISE 3.32. Let A be an $n \times m$ matrix,

(a) Prove that $\text{rank}\,(A^\mathsf{T}) = \text{rank}\,(L_{A^\mathsf{T}})$ is equal to $\text{rank}\,(L_A)^\mathsf{T}$.

(b) This would follow immediately if it were true that $(L_A)^\mathsf{T} = L_{A^\mathsf{T}}$. Explain why this statement does not make sense.

HINT: Keep in mind the setting for this (and the next) exercise. If $V = \mathbb{K}^m$, $W = \mathbb{K}^n$, and A is $n \times m$, we get a map $L_A : \mathbb{K}^m \to \mathbb{K}^n$. Since the transpose matrix A^T is $m\times n$, it determines a linear map $L_{A^\mathsf{T}} : \mathbb{K}^n \to \mathbb{K}^m$ in the opposite direction:

$$V \xrightarrow{L_A} W, \quad V \xleftarrow{L_{A^\mathsf{T}}} W \quad (V = \mathbb{K}^m \text{ and } W = \mathbb{K}^n),$$

while, by definition, the transpose $(L_A)^\mathsf{T}$ maps $V^* \xleftarrow{(L_A)^\mathsf{T}} W^*$.

EXERCISE 3.33. If A is an $n \times m$ matrix, use the previous exercises to prove the following:

THEOREM. For any $n\times m$ matrix A, $\text{RowRank}(A) = \text{ColRank}(A)$.

A direct proof of this identity is given in Chapter 4 on determinants; see Proposition 4.45 and Corollary 4.47.

Additional Exercises

Section 3.1. Definitions and Examples

1. TRUE/FALSE QUESTIONS: ("True" if the statement is *always* true.)
   (a) If $\mathfrak{X} = \{e_1, \dots, e_n\}$ is the standard basis in $\mathbb{K}^n$, the dual basis $\mathfrak{X}^* = \{e_j^*\}$ consists of the maps $\phi_i : \mathbb{K}^n \to \mathbb{K}$ such that $\phi(x_1, \dots, x_n) = x_i$ for $1 \le i \le n$.
   (b) If V is a finite-dimensional vector space over a field $\mathbb{K}$ and $v_1 \neq v_2$ in V, there is some dual vector $\ell \in V^*$ that can distinguish between v_1 and v_2 in the sense that $\langle \ell, v_1\rangle \neq \langle \ell, v_2\rangle$.
   (c) The sum $\phi(A) = A_{1,n} + A_{2,n-1} + \cdots + A_{n,1}$ of the *counterdiagonal* entries in an $n \times n$ matrix (terms A_{ij} such that $i + j = n + 1$) is an element in the dual space of $V = \mathrm{M}(n, \mathbb{K})$.
   (d) Same as (c), but with $\phi(A) = \sum_{i,j=1}^n A_{ij}$, the sum of all entries in A.
   (e) Same as (c) but with $\phi(A) = \prod_{i,j=1}^n A_{ij}$, the product of all entries in A.
   (f) If $T : V \to W$ is linear and $v \in V, \ell \in W^*$, the map $\phi(\ell) = \langle \ell, T(v)\rangle$ is in V^*.
2. Which of the following maps $f : V \to \mathbb{R}$ are linear functionals on the given vector space?
   (a) The *trace* $\mathrm{Tr}(A) = \sum_{i=1}^n a_{ii}$ (sum of the diagonal entries) for matrices A in $V = \mathrm{M}(n, \mathbb{K})$.
   (b) The *determinant* $\det(A) = a_{11}a_{22} - a_{12}a_{21}$ of a matrix A in $V = \mathrm{M}(2, \mathbb{C})$.
   (c) The map $f\colon \mathbb{R}[x] \to \mathbb{R}$ given by $f(p) = 2\,Dp(0) + D^2p(1)$, where $p(x) \in \mathbb{R}[x]$, and D is the differentiation operator acting on polynomials in $\mathbb{R}[x]$.
   (d) For fixed indices $1 \le i, j \le n$, $f(A) = (A^2)_{ij}$ = the i, j entry of A^2, for matrices $A \in \mathrm{M}(n, \mathbb{K})$.
3. Explain why every $(n-1)$-dimensional subspace (hyperplane through the origin) in $\mathbb{K}^n$ is the null space $\ker(\ell)$ of some nonzero linear functional $\ell \in V^*$. Is the dual vector ℓ unique?

Section 3.2. Dual Bases in V^*

1. TRUE/FALSE QUESTIONS: ("True" if the statement is *always* true.)
   (a) Every finite-dimensional vector space V is isomorphic to its dual space V^*.

(b) If W_1, W_2 are subspaces in a finite-dimensional vector space, then $(W_1 \cap W_2)^\circ = W_1^\circ + W_2^\circ$.
(c) If T is an isomorphism from vector space V to its dual V^* and $\mathfrak{X}$ is a basis for V, then $T(\mathfrak{X})$ is the dual basis $\mathfrak{X}^*$ in V^*.
(d) If $V = \mathcal{C}^\infty(\mathbb{R})$ is the space of functions that have continuous derivatives $D^n f$ of all orders ($n \in \mathbb{N}$), and we fix a base point $p \in \mathbb{R}$, then for any $n \in \mathbb{N}$ the map $\phi(f) = D^n f(p)$ is a linear functional on V.
(e) If $V = \mathcal{C}[a, b]$ is the space of continuous functions on a closed interval $[a, b] \subseteq \mathbb{R}$, the map $I(f) = \int_a^b f(x)\,dx$ is a linear functional in V^*.

2. Check that the set $\mathfrak{X}$ of vectors

$$\mathbf{h}_1 = (1, 0, 1), \quad \mathbf{h}_2 = (1, 2, 1), \quad \mathbf{h}_3 = (0, 0, 1),$$

is a basis in $\mathbb{R}^3$. Describe elements of the dual basis $\mathfrak{X}^* = \{\mathbf{h}_i^*\}$ as functions $\mathbf{h}_i(x_1, x_2, x_3)$ on $\mathbb{R}^3$. (Recall Example 3.12.)

3. Define functionals $\mathbf{f}_i$ on $V = \mathbb{R}^3$, letting

$$\mathbf{f}_1(x, y, z) = x - 2y, \quad \mathbf{f}_2(x, y, z) = x + y + z, \quad \mathbf{f}_3(x, y, z) = y - 3z.$$

(a) Prove that $\mathfrak{Y} = \{\mathbf{f}_1, \mathbf{f}_2, \mathbf{f}_3\}$ is a basis for the dual space V^*.
(b) Find a basis $\mathfrak{X} = \{\mathbf{u}_1, \mathbf{u}_2, \mathbf{u}_3\}$ in V with $\mathfrak{Y} = \mathfrak{X}^*$, the dual basis to $\mathfrak{X}$ in V^*.

3. Let $W \subseteq \mathbb{R}^4$ be the subspace spanned by $(1, 2, -3, 4)$ and $(1, 4, -1, 8)$. If $\mathfrak{X} = \{\mathbf{e}_i\}$ is the standard basis in $V = \mathbb{R}^4$ and $\mathfrak{X}^* = \{\mathbf{e}_j^*\}$ is the dual basis in V^*, find basis vectors $\mathbf{f}_1, \mathbf{f}_2$ for the annihilator W°, described as linear combinations of the dual basis vectors $\mathbf{e}_j^*$.

4. The "matrix units" E_{ij} in matrix space $V = \mathrm{M}(n, \mathbb{K})$ were defined in Chapter 2:

$$E_{ij} \text{ is the matrix such that } (E_{ij})_{k,\ell} = \delta_{ik} \cdot \delta_{jk} \text{ for } 1 \leq k, \ell \leq n,$$

where δ_{rs} is the Kronecker delta symbol (recall Additional Exercise 4 in Chapter 2). The E_{ij} are obviously a basis in V. Describe the action of the dual basis vector E_{ij}^* on a typical matrix $A \in V$.

Section 3.3. The Transpose Operation

1. TRUE/FALSE QUESTIONS: ("True" if the statement is *always* true.)
(a) *Reflexivity* of a finite-dimensional vector space means the double dual $V^{**} = (V^*)^*$ is equal to the original space V.
(b) If A is an $n \times m$ matrix, we have

$$(L_A)^\mathsf{T} = (L_{A^\mathsf{T}})$$

for the multiplication operator $L_A : \mathbb{K}^m \to \mathbb{K}^n$.
(c) If $T : V \to W$ is linear, then $(T^\mathsf{T})^\mathsf{T}$ maps $W^{**} \to V^{**}$.
(d) If $T : V \to W$ is linear, then $T^\mathsf{T} = \mathbf{0}$ (zero operator on W^*) $\Leftrightarrow$ $T = \mathbf{0}$ on V.

(e) The transpose T^{T} of a linear operator $T : V \to W$ is well-defined even if V and W are infinite dimensional.

2. If $T : V \to W$ is an *isomorphism* (bijective linear map) between infinite-dimensional vector spaces, prove that $T^{\mathrm{T}} : W^* \to V^*$ is

(a) one-to one, (b) surjective,

so T^{T} is an isomorphism between the dual spaces W^* and V^*.

3. Prove that $(W_1 + W_2)^\circ = (W_1)^\circ \cap (W_2)^\circ$ for subspaces W_1, W_2 in a finite-dimensional vector space V.

4. If V is a finite-dimensional vector space and W a subspace, provide the details needed to prove the dimension formula for annihilators:

DIMENSION FORMULA: $\dim(W) + \dim(W^\circ) = \dim(V)$,

where $W^\circ = \{\ell \in V^* : \langle \ell, W \rangle = 0\}$.

5. For any nonempty subset S in a finite-dimensional vector space V, show that

$$\mathbb{K}\text{-span}(S) = (S^\circ)^\circ$$

when we identify $V^{**} \cong V$ in the usual way.

NOTE: Here we define $S^\circ = \{\ell \in V^* : \langle \ell, S \rangle = 0\}$, which is easily seen to be a subspace in V^*.

6. If $T : V \to W$ is a linear map between finite-dimensional vector spaces over a field $\mathbb{K}$, prove that $K(T^{\mathrm{T}}) = R(T)^\circ$.

7. If $T : V \to W$ is a linear map between finite-dimensional vector spaces over a field $\mathbb{K}$, prove that

$$|R(T^{\mathrm{T}})| = |R(T)| \quad (\text{so } \operatorname{rank}(T^{\mathrm{T}}) = \operatorname{rank}(T)).$$

Thus the ranges of T and T^{T} always have the same dimension even though they live in different spaces W and V^*.

HINT: Use the dimension formula and Additional Exercise 6.

8. Prove that we *do not* have $R(T^{\mathrm{T}}) = K(T)^\circ$ unless $|V| = |W|$ (in particular, if $V = W$).

NOTE: Compare this with the identity of Additional Exercise 6.

9. If $T : V \to W$ is a linear map between finite-dimensional vector spaces over a field $\mathbb{K}$, prove the following:

(a) T is onto if and only if T^{T} is one-to-one.
(b) T^{T} is onto if and only if T is one-to-one.

10. If $T : V \to V$ is a linear operator on a finite-dimensional vector space, we often say that "$(T^{\mathrm{T}})^{\mathrm{T}} = T$" even though this involves some abuse of terminology. In your own words, explain how the "identity" $T^{\mathrm{TT}} = T$ should be interpreted.

11. If $\mathfrak{X} = \{\mathbf{e}_i\}$ is the standard basis in $\mathbb{R}^{m+n} = \mathbb{R}^m \oplus \mathbb{R}^n$, and P is the projection onto $E = \mathbb{R}^m \times (0)$ along $F = (0) \times \mathbb{R}^n$, compute the matrix $[P]_{\mathfrak{X}\mathfrak{X}}$.

12. Let $\mathfrak{X} = \{\mathbf{e}_i\}$ be the standard basis in $V = \mathbb{R}^3$ and $P : V \to V$ the projection of V onto H along L associated with the direct sum $V = H \oplus L$, where

$$L = \mathbb{R} \cdot (\mathbf{e}_1 + \mathbf{e}_2 + 2\mathbf{e}_3) \quad \text{and} \quad H = \mathbb{R}\text{-span}\{\mathbf{e}_1, \mathbf{e}_2\}.$$

If $\mathfrak{Y} = \{\mathbf{e}_1, \mathbf{e}_2, (\mathbf{e}_1 + \mathbf{e}_2 + 2\mathbf{e}_3)\}$, compute the matrices $[P]_{\mathfrak{Y}\mathfrak{Y}}$ and $[P]_{\mathfrak{X}\mathfrak{X}}$.

HINT: The computations in Example 3.12 are relevant here.

13. Same as Additional Exercise 12, but for the projection Q of V onto L along H.

14. If V is a finite-dimensional vector space and W a subspace, explain why

$$h(W) = \{v \in V : \langle \ell, v \rangle = 0 \text{ for all } \ell \in W^\circ\}$$

is equal to W.

NOTE: The "hull" $h(W) \supseteq W$ resembles $(W^\circ)^\circ$ but the "double annihilator" lives in V^{**}, while $h(W)$ lives in the original space V. But if V is finite dimensional and we make the natural identification $V^{**} \cong V$ discussed in this section, the identity $h(W) = W$ becomes $(W^\circ)^\circ = W$. A similar pattern will be encountered in Chapter 6 when we discuss the orthogonal complement $W^\perp$ of a subspace in a finite-dimensional inner product space V and show that $(W^\perp)^\perp = W$.

CHAPTER 4

Determinants

4.1. The Permutation Group S_n

DEFINITION 4.1. The *permutation group* S_n consists of all bijections $\sigma : [1, n] \to [1, n]$ where $[1, n] = \{1, ..., n\}$, with composition of operators

$$\sigma_1 \circ \sigma_2(k) = \sigma_1(\sigma_2(k)) \quad \text{for } 1 \le k \le n$$

as the group operation. The identity element e is the identity map such that $e(k) = k$, for all $k \in [1, n]$.

Recall that a *group* is any set G equipped with a binary operation $(*)$ satisfying the following axioms:

1. ASSOCIATIVITY: $x * (y * z) = (x * y) * z$.
2. IDENTITY ELEMENT: There is an $e \in G$ such that $e * x = x = x * e$ for all $x \in G$.
3. INVERSES: Every $x \in G$ has a "two-sided inverse," an element x^{-1} in G such that $x^{-1} * x = x * x^{-1} = e$.

We do not assume that the system $(G, *)$ is commutative, with $x * y = y * x$; a group with this extra property is a *commutative group*, also referred to as an *abelian group*.

Here are some examples of familiar groups.

1. The integers $(\mathbb{Z}, +)$ become a commutative group when equipped with $(+)$ as the group operation. Multiplication $(\cdot)$ *does not* make $\mathbb{Z}$ a group. (Why?)
2. Any vector space equipped with its $(+)$ operation is a commutatve group, for instance $(\mathbb{K}^n, +)$. There is no scaling operation in a group.
3. The set $(\mathbb{C}^\times, \cdot) = \mathbb{C} \sim \{0\}$ of nonzero complex numbers equipped with complex multiplication $(\cdot)$ is a commutative group. So is the subset $\mathbb{S}^1 = \{z \in \mathbb{C} : |z| = 1\}$ (unit circle in the complex plane) because $|z|, |w| = 1 \Rightarrow |zw| = |z| \cdot |w| = 1$ and $|1/z| = 1/|z| = 1$. The $(+)$ operation is not defined on $\mathbb{C}^\times$.
4. GENERAL LINEAR GROUP. The set $\mathrm{GL}(n, \mathbb{K}) = \{A \in \mathrm{M}(n, \mathbb{K}) : \det(A) \neq 0\}$ of *invertible* $n \times n$ matrices is a group when equipped with matrix multiply as the group operation. It is noncommutative when $n \ge 2$. Validity of the group axioms for $(\mathrm{GL}, \cdot)$ follows because

$$\det(AB) = \det(A) \cdot \det(B), \quad \det(I) = 1, \quad \det(A^{-1}) = \frac{1}{\det(A)},$$

and a matrix A has a two-sided inverse $\Leftrightarrow$ $\det(A) \neq 0$.

5. SPECIAL LINEAR GROUP. These properties of the determinant imply that the subset $\mathrm{SL}(n, \mathbb{K}) = \{A \in \mathrm{M}(n, \mathbb{K}) \,:\, \det(A) = 1\}$ equipped with matrix multiply is also a (noncommutative) group;
6. The set of permutations $(\mathrm{Per}(X), \circ)$ = all bijections on a set X of n distinct objects is also a group when equipped with composition ($\circ$) as its product operation. No matter what the nature of the objects being permuted, we can restrict attention to permutations of the set of integers $[1, n]$ by labeling the original objects, and then we have the group S_n.

Cycles and the Cycle Decomposition Theorem. The implest permutations in S_n are the *k-cycles.*

DEFINITION 4.2. An ordered list $(i_1, \dots, i_k)$ of k distinct indices in $[1, n] = \{1, \dots, n\}$ determines a *k-cycle* in S_n, a permutation that acts in the following way on the set $X = [1, n]$:

$$\text{(4.1)} \quad \sigma \text{ maps } \begin{cases} i_1 \to \cdots \to i_k \to i_1 & \text{(a one-step "cyclic shift" of list entries),} \\ j \to j & \text{for all } j \text{ not in the list } \{i_1, \dots, i_k\}. \end{cases}$$

A 1-cycle (i_1) is just the identity map id_X, so we seldom indicate 1-cycles explicitly, though it is permissible and sometimes quite useful to do so. The *support* of a k-cycle is the set of entries $\mathrm{supp}(\sigma) = \{i_1, \dots, i_k\}$, in no particular order. The support of a one-cycle (i_1) is the one-point set $\{i_1\}$.

The order of the entries in the symbol $\sigma = (i_1, \dots, i_k)$ matters, but cycle notation is ambiguous because k different symbols

$$(i_1, \dots, i_k) = (i_2, \dots, i_k, i_1) = (i_3, \dots, i_k, i_1, i_2) = \cdots = (i_k, i_1, \dots, i_{k-1}),$$

obtained by "cyclic shifts" of the list entries in σ, all describe the same operation in S_n. Thus a k-cycle might best be descibed by a "cyclic list" of the sort shown in Figure 4.1, rather than a linearly ordered list, but such diagrams are typographically awkward for the printed page. If we change the cyclic order of the indices we get a new operator: $(1, 2, 3) = (2, 3, 1) = (3, 1, 2) \neq (1, 3, 2)$ because $(1, 2, 3)$ sends $1 \to 2$ while $(1, 3, 2)$ sends $1 \to 3$. Thus σ is a "1-shift" of the cyclically ordered entries in Figure 4.1. (Similarly, a "2-shift" would move points 2 steps clockwise in the cyclic order, sending $i_1 \to i_3 \to \cdots$, etc.) Obviously if σ is a k-cycle then $\sigma^k = e$ (the identity map); furthermore, the powers $e, \sigma, \sigma^2, \dots, \sigma^{k-1}$ are distinct. (They have different actions on the set $X = [1, n]$ because $e(i_1) = i_1$, $\sigma(i_1) = i_2$, $\sigma^2(i_1) = i_3$, ..., $\sigma^{k-1}(i_1) = i_k$.)

One (cumbersome) way to describe general elements $\sigma \in S_n$ employs a data array to show where each $k \in [1, n]$ ends up:

$$\sigma = \begin{pmatrix} 1 & 2 & 3 & \cdots & n \\ j_1 & j_2 & j_3 & \cdots & j_n \end{pmatrix},$$

so that $\sigma(i) = j_i$ for all i. More efficient notation is provided by the fact that every permutation σ can be uniquely written as a product of cycles with *disjoint* supports, which means that the factors commute.

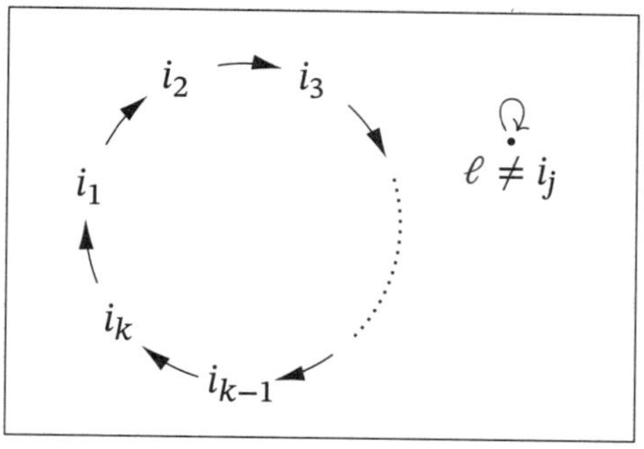

$X = \{1, 2, \dots, n\}$

FIGURE 4.1. Action of the k-cycle $\sigma = (i_1, \dots, i_k)$ on $X = \{1, 2, \dots, n\}$. Points ℓ not in the "support set" $\text{supp}(\sigma) = \{i_1, \dots, i_k\}$ remain fixed; those in the support set are shifted one step clockwise in this cyclically ordered list.

EXERCISE 4.3. If cycles $\sigma = (i_1, \dots, i_r)$, $\tau = (j_1, \dots, j_s)$ act on *disjoint* sets of indices, explain why these operators commute. This is no longer true if the sets of indices overlap. Check this by computing the effect of the following products $\sigma\tau(k) = \sigma(\tau(k))$ of permutations in S_5:

(a) $(1, 2, 4)(2, 3)$
(b) $(2, 4)(1, 2, 3)$

Is either product a cycle?

Thus the order of factors in a product of cycles is irrelevant if the cycles are disjoint.

The product of two cycles $\sigma\tau = \sigma \circ \tau$ is a composite of operators, so the action of $\sigma\tau = \sigma \circ \tau$ on an element $k \in [1, n]$ is evaluated by feeding k into the product *from the right* as below. Taking $\sigma = (1, 2)$ and $\tau = (1, 2, 3)$ in S_5, we have

$$\sigma\tau : k \xrightarrow{(1,2,3)} (1, 2, 3) \cdot k \xrightarrow{(1,2)} (1, 2) \cdot ((1, 2, 3) \cdot k) = ((1, 2)(1, 2, 3) \cdot k)$$

To determine the net effect we track what happens to each k:

Action	Net Effect
$1 \xrightarrow{(1,2,3)} 2 \xrightarrow{(1,2)} 1$	$1 \to 1$
$2 \longrightarrow 3 \longrightarrow 3$	$2 \to 3$
$3 \longrightarrow 1 \longrightarrow 2$	$3 \to 2$
$4 \longrightarrow 4 \longrightarrow 4$	$4 \to 4$
$5 \longrightarrow 5 \longrightarrow 5$	$5 \to 5$

Thus the product $(1, 2)(1, 2, 3)$ is equal to $(2, 3) = (1)(2, 3)(4)(5)$, when we include redundant 1-cycles. On the other hand, $(1, 2, 3)(1, 2) = (1, 3)$, which shows that cycles need not commute if their supports overlap.

As another example we have

$$(1, 2, 3, 4)^2 = (1, 3)(2, 4), \quad (1, 2, 3, 4)^3 = (1, 4, 3, 2), \quad (1, 2, 3, 4)^4 = e,$$

which shows that a power σ^k of a cycle need not itself be a cycle, although it is a product of disjoint cycles.

THEOREM 4.4 (Cycle Decomposition of Permutations). *Every $\sigma \in S_n$ is a product $\tau_1 \cdot \cdots \cdot \tau_s$ of disjoint cycles such that* $\mathrm{supp}(\tau_1) \cup \cdots \cup \mathrm{supp}(\tau_s) = [1, n]$, *if we include the* 1*-cycles needed to account for all indices $k \in [1, n]$. This decomposition is unique (up to order of the commuting factors).*

PROOF. Look at the iterated action of σ on the first $n_1 = 1$ in $[1, n]$,

$$1 = \sigma^0(1) \to \sigma^1(1) = \sigma(1) \to \sigma^2(1) \to \cdots \to \sigma^r(1) \to \cdots .$$

After a certain number of steps a repeat must occur, say $\sigma^i(1) = \sigma^r(1)$ with $1 \le i < r$. If this is the *first* repeat, we must actually have $\sigma^r(1) = 1$. In fact, if $\sigma^i(1) = \sigma^r(1)$ for some exponent $0 < i < r$, the law of exponents $\sigma^k \sigma^\ell = \sigma^{k+\ell}$ implies

$$1 = \sigma^{-i}(\sigma^r(1)) = \sigma^{r-i}(1),$$

so $r - i$ would be a positive exponent $< r$ such that $\sigma^{r-i}(1) = 1$, contrary to minimality of r. Then the pattern repeats with $\sigma^{r+1}(1) = \sigma(\sigma^r(1)) = \sigma(1)$, etc. This determines the first cycle in the decomposition of σ, namely

$$\tau_1 = (1, \sigma(1), \sigma^2(1), \dots, \sigma^{r-1}(1)) \quad (\text{with } \sigma^r(1) = 1)$$

Next let $n_2 > n_1 = 1$ be the smallest index in $[1, n]$ that does not appear in the cycle τ_1, and repeat the process to get a new cycle $\tau_2 = (n_2, \sigma(n_2), \sigma^2(n_2), \dots, \sigma^{r_2-1}(n_2))$ with $\sigma^{r_2}(n_2) = n_2$. Indices appearing in τ_2 are disjoint from those in τ_1 because if $\sigma^p(1) = \sigma^q(n_2)$ we would have $n_2 = \sigma^{p-q}(1)$. Thus n_2 would appear in the list of elements appearing in τ_1, violating the definition of n_2.

Continuing this process eventually yields cycles $\tau_1, \dots, \tau_s$ with disjoint supports $\mathrm{supp}(\tau_i)$ whose union is all of $[1, n]$. Since each τ_i leaves fixed all points not in its support, it is easy to verify that these cycles commute, with $\sigma_i\sigma_j(k) = \sigma_j\sigma_i(k)$ for all k, because they act independently on disjoint sets of indices $I_i = \mathrm{supp}(\tau_i)$ in $[1, n]$, each invariant under the action of σ by construction.

The action of σ on each support set $\mathrm{supp}(\tau_i) = \{n_i, \sigma(n_i), \dots, \sigma^{r_i-1}(n_i)\}$ is a cyclic shift of entries one space to the right, with $\sigma^{r_i}(n_i) = n_i$. But that is precisely the action of the cycle τ_i on its support. We conclude that $\sigma = \tau_1 \cdot \cdots \cdot \tau_s$, proving the existence of disjoint cycle decompositions.

Uniqueness follows because the support sets have a geometric interpretation: they are the "orbits" in $X = [1, n]$ under the iterated action of σ, and once a base point n_i is specified in each orbit the action of σ determines a cyclic ordering of points in the orbit such that $\tau_i = (n_i, \sigma(n_i), \dots, \sigma^{r_i-1}(n_i))$, with $r_i =$ the size of the support set. □

EXERCISE 4.5. Write

$$\sigma = \begin{pmatrix} 1 & 2 & 3 & 4 & 5 & 6 \\ 2 & 4 & 6 & 5 & 1 & 3 \end{pmatrix}$$

as a product of disjoint commuting cycles.

HINT: Start by tracking $1 \to 2 \to 4 \to \cdots$ until a cycle is completed; then feed σ the first integer not included in the previous cycle.

EXERCISE 4.6. Evaluate the net action of the following products of cycles:

(a) $(1,2)(1,3)$ in S_3 (b) $(1,2,3,4,5)(1,2)$ in S_5
(c) $(1,2)(1,3)$ in S_6 (d) $(1,2)^2$ in S_5
(e) $(1,2)(1,2,3,4,5)$ in S_5 (f) $(1,2,3)^2$ in S_5.

Write each as a product of disjoint cycles.

EXERCISE 4.7. Determine the inverses σ^{-1} of the following elements in S_5:

(a) $(1,2)$ (b) Any 2-cycle (i_1, i_2) with $i_1 \neq i_2$
(c) $(1,2,3)$ (d) Any k-cycle $(i_1, \cdots, i_k)$

EXERCISE 4.8. Evaluate the following products in S_n as products of disjoint cycles:

(a) $(1,5)(1,4)(1,3)(1,2)$
(b) $(1,2)(1,3)(1,4)(1,5)$
(c) $(1,k)(1,2,\ldots,k-1)$

EXERCISE 4.9. The *order* $o(\sigma)$ of a permutation $\sigma \in S_n$ is the smallest integer $m \geq 1$ such that $\sigma^m = \sigma \cdot \cdots \cdot \sigma = e$ (in particular, $o(e) = 1$).

(a) If $\sigma \in S_n$ and k is the smallest exponent such that the list $e, \sigma, \sigma^2, \ldots, \sigma^k$ contains a repeated entry (so $\sigma^k = \sigma^\ell$ for some $0 \leq \ell < k$), prove $\sigma^k = e$; i.e., σ^i returns to the starting point of the cyclic list at the first repeat, so $o(\sigma) = k$.
(b) Explain why every $\sigma \in S_n$ has finite order $o(\sigma) \leq n!$.
(c) Explain why $o((1,2)(3,4)) = 2$ while $o((1,2)(3,4,5)) = 6$.

EXERCISE 4.10. The rth power $\sigma^r = \sigma \cdots \sigma$ of a k-cycle $\sigma = (i_1, \ldots, i_k)$ is obviously an "r-shift" that moves every entry clockwise r steps in Figure 4.1. Powers of a cycle need not themselves be cycles (although sometimes they are).

(a) If σ is a 6-cycle, its square $\sigma^2 = \sigma \circ \sigma$ is a cyclic 2-shift of the entries $(i_1, \ldots, i_6)$. Is σ^2 a cycle? What is the order of this element in S_n?
(b) What happens if σ is a 7-cycle?

HINT: By relabeling, it suffices to consider the "standard 6-cycle" $(1,2,3,4,5,6)$ in answering (a).

EXERCISE 4.11. Some powers of the standard 6-cycle $\sigma = (1,2,\ldots,6)$ have nontrivial disjoint cycle decompositions.

(a) Determine the disjoint cycle decompositions of $\sigma, \sigma^2, \sigma^3, \ldots, \sigma^5$. (The decompositions of $\sigma^5 = \sigma^{-1}$ and $\sigma^4 = (\sigma^2)^{-1}$ can be written down easily once you know the decompositions of σ and σ^2; σ^3 is its own inverse.)
(b) Find the order $o(\sigma^k)$ for each of the powers $e, \sigma, \sigma^2, \ldots, \sigma^5$.

Do you notice any connection between the orders $o(\sigma^k)$ and the lengths of the cycles appearing in the cycle decomposition of σ^k?

EXERCISE 4.12. Prove the following:

> THEOREM. *A power* σ^j $(1 \leq j \leq k-1)$ *of a k-cycle is again a k-cycle* $\Leftrightarrow$ *j is relatively prime to k, so* $\gcd(j,k) = 1$ *and* j, k *have no prime divisors in common.*

HINT: $\gcd(a,b)$ = the smallest positive element $ra+sb$ in the lattice $\mathbb{Z}\cdot a+\mathbb{Z}\cdot b$.

EXERCISE 4.13. Let $\sigma = (i_1, \dots, i_k)$ be a k-cycle in S_n.

(a) Explain why the inverse of σ is the k-cycle $\sigma^{-1} = (i_1, i_k, i_{k-1}, \dots, i_2)$.
(b) Explain why we must have $\gcd(k-1, k) = 1$ for any $k \geq 2$.

EXERCISE 4.14. The order of any k-cycle σ is $o(\sigma) = k$. Prove:

> THEOREM. (ORDER OF A PRODUCT OF DISJOINT CYCLES). *If* σ *and* τ *are* disjoint *cycles (perhaps of different lengths), then* $o(\sigma\cdot\tau) = \text{lcm}(o(\sigma), o(\tau))$ (lcm = *least common multiple).*

NOTE: Similarly, a product $\sigma = \prod_{i=1}^{m} \tau_i$ of several pairwise disjoint (commuting) cycles has order lcm $(o(\tau_1), \dots, o(\tau_m))$.
HINT: Examine the geometric action on supports of σ and τ. It is not enough that σ and τ commute. Recall that $\text{lcm}(a,b) = ab/\gcd(a,b)$ for $a, b \in \mathbb{N}$.

EXERCISE 4.15. The "long cycle" $\sigma = (1, 2, \dots, n)$ in S_n has order n.

(a) Determine the cycle decomposition of σ^2. (The answer will depend on whether n is even or odd.)
(b) The other long cycles have the form $\tau = (k_1, k_2, \dots, k_n)$ obtained by rearranging the entries $1, 2, \dots, n$ in σ. Explain how the cycle decomposition of τ^2 can be found from decomposition for σ^2 in (a).
(c) Is every element in S_n with order $= n$ a long cycle? Explain.

NOTE: Use the theorem in Exercise 4.14.

Parity of a Permutation. In a different direction we now show that the 2-cycles (i, j) generate the entire group S_n in the sense that every $\sigma \in S_n$ can be written as a product $\sigma = \tau_1 \cdot \dots \cdot \tau_r$ of 2-cycles. However, these factors are not necessarily disjoint and need not commute, and such decompositions are far from unique since we have, for example,

$$e = (1,2)^2 = (1,2)^4 = (1,2)(2,3)(3,1)(2,3).$$

Nevertheless, an important aspect of such factorizations is unique, namely its *parity*

$$\text{sgn}(\sigma) = (-1)^r,$$

where r = #(2-cycles in the factorization $\sigma = \tau_1 \cdots \tau_r$). That means the elements $\sigma \in S_n$ fall into two disjoint classes: *even permutations* that can be written as a product of an even number of 2-cycles, and *odd permutations*. It is not obvious that all 2-cycle decompositions of a given permutation have the same parity. We prove that next, and then show how to compute $\text{sgn}(\sigma)$ effectively.

First observe that a decomposition into 2-cycles always exists. By Theorem 4.4 it suffices to show that any k-cycle can be so decomposed. For 1-cycles this is obvious since $(k) = e = (1, 2) \cdot (1, 2)$; in particular, $\text{sgn}(e) = 1$ and the identity is an even permutation. When $k > 1$ it is easy to check that

$$(1, 2, \cdots, k) = (1, k) \cdot \cdots \cdot (1, 3) \cdot (1, 2)$$

(with $k - 1$ factors).

EXERCISE 4.16. Verify the preceding factorization of the standard k-cycle $(1, 2, \dots, k)$. By relabeling deduce that $(i_1, \dots, i_k) = (i_1, i_k)(i_1, i_{k-1}) \cdots (i_1, i_2)$ for any k-cycle.

NOTE: This is an example of "proof by relabeling."

Once we verify that the parity is well defined, this tells us how to recognize the parity of any k-cycle.

$$\text{sgn}(i_1, i_2, \dots, i_k) = (-1)^{k-1} \quad \text{for all } k > 0. \tag{4.2}$$

We claim that these nonunique factorizations all assign the same *parity* to a permutation.

THEOREM 4.17. *All decompositions $\sigma = \tau_1 \cdots \tau_r$ of a permutation as a product of 2-cycles have the same* $\text{sgn}(\sigma) = (-1)^r$.

PROOF. The group S_n acts on the space of polynomials $\mathbb{K}[\mathbf{x}] = \mathbb{K}[x_1, \dots, x_n]$ by permuting the variables

$$(\sigma \cdot f)(x_1, \dots, x_n) = f(x_{\sigma(1)}, \dots, x_{\sigma(n)})$$

For instance $(1, 2, 3) \cdot f(x_1, x_2, x_3, x_4, x_5) = f(x_2, x_3, x_1, x_4, x_5)$. This is a "covariant group action" in the sense that

$$(\sigma\tau) \cdot f = \sigma \cdot (\tau \cdot f) \quad \text{and} \quad e \cdot f = f$$

for all f and all $\sigma, \tau \in S_n$. The notation makes this a bit tricky to prove; one way to convince yourself is to write

$$\begin{aligned}
\sigma \cdot (\tau \cdot f)(x_1, \dots, x_n) &= \tau \cdot f(x_{\sigma(1)}, \dots, x_{\sigma(n)}) \\
&= \tau \cdot f(w_1, \dots, w_n)\big|_{w_1 = x_{\sigma(1)}, \dots, w_n = x_{\sigma(n)}} \\
&= f(w_{\tau(1)}, \dots, w_{\tau(n)})\big|_{w_k = x_{\sigma(k)}} \\
&= f(x_{\sigma(\tau(1))}, \dots, x_{\sigma(\tau(n))}) \\
&= f(x_{(\sigma\tau)(1)}, \dots, x_{(\sigma\tau)(n)}) = (\sigma\tau) \cdot f(x_1, \dots, x_n).
\end{aligned}$$

Now consider the particular polynomial in n unknowns $\phi \in \mathbb{K}[x_1, \dots, x_n]$,

$$\phi(x_1, \dots, x_n) = \prod_{i<j} (x_i - x_j).$$

We now show that $\sigma \cdot \phi = (-1)\phi$ for any 2-cycle $\sigma = (i, j)$; then by "covariance" it will follow that $\sigma \cdot \phi = (-1)^r \phi$ if σ is a product $\tau_1 \cdots \tau_r$ of r two-cycles. Since the definition of $\sigma \cdot \phi$ makes no reference to 2-cycle decompositions of σ, we will conclude that $(-1)^r$ must be the same for all such decompositions of σ, completing the proof of Theorem 4.17.

To show that $\tau \cdot \phi = (-1)\phi$ for a 2-cycle (i, j), we may assume $i < j$. Note that the terms $x_k - x_\ell$ $(k < \ell)$ not involving i or j are unaffected when we switch $x_i \leftrightarrow x_j$. The remaining terms are of three types.

Case 1. Terms involving both i and j. The only such term is $x_i - x_j$, which becomes

$$\sigma \cdot (x_i - x_j) = x_j - x_i = (-1)(x_i - x_j),$$

suffering a change of sign.

Case 2. Terms involving only i. The possibilities upon switching $x_i \leftrightarrow x_k$ (for $k \neq j$) are

Terms	$x_k - x_i$ $1 \leq k < i$	$x_i - x_k$ $i < k < j$	$x_i - x_k$ $j < k \leq n$
#(Terms)	$i - 1$	$j - i - 1$	$n - j$
Effect of $x_i \leftrightarrow x_j$ on sign of term	No change (since $k < i < j$)	$x_i - x_k \to x_j - x_k$ $= (-1)(x_k - x_j)$	No change (since $i < j < k$)

Case 3. Terms involving only j. These are (for $k \neq i$).

Terms	$x_k - x_j$ $1 \leq k < i$	$x_k - x_j$ $i < k < j$	$x_j - x_k$ $j < k \leq n$
#(Terms)	$j - 1$	$j - i - 1$	$n - j$
Effect of $x_i \leftrightarrow x_j$ on sign of term	No change (since $i < j < k$)	$x_k - x_j \to x_k - x_i$ $= (-1)(x_i - x_k)$	No change (since $i < j < k$)

The effect of switching $x_i \leftrightarrow x_j$ is to permute the terms in $\prod_{k<l}(x_k - x_l)$, changing the sign of some, so the product gets multiplied by $+1$ or -1. Counting the number of sign changes in each case, we see that in all cases we have

$$(-1)^{\#(\text{changes})} = (-1)^{1+\text{even}} = -1$$

for any 2-cycle, as claimed. ☐

COROLLARY 4.18. *The parity map* $\mathrm{sgn} : S_n \to \{\pm 1\}$, *defined by* $\mathrm{sgn}(\sigma) = (-1)^r$ *if* σ *can be written as a product of* r *two-cycles, has the following algebraic properties*:

1. $\mathrm{sgn}(e) = +1$
2. $\mathrm{sgn}(\sigma\tau) = \mathrm{sgn}(\sigma) \cdot \mathrm{sgn}(\tau)$

3. $\text{sgn}(\sigma^{-1}) = (\text{sgn}(\sigma))^{-1} = \text{sgn}(\sigma)$ (*since* $\text{sgn} = \pm 1$).

PROOF. Obviously $\text{sgn}(e) = 1$ since we may write $e = (1, 2)^2$. If $\sigma = c_1 \cdots c_r$ and $\tau = c'_1 \cdots c'_s$ where c_i, c'_j are 2-cycles, then $\sigma\tau = c_1 \cdots c_r c'_1 \cdots c'_s$ is a product of $r + s$ cycles, proving (2.). The third property follows because

$$1 = \text{sgn}(e) = \text{sgn}(\sigma\sigma^{-1}) = \text{sgn}(\sigma) \cdot \text{sgn}(\sigma^{-1})$$

since the only values of sgn are ± 1. □

(Optional) An Alternative Proof of the Parity Theorem 4.17. We have given a self-contained proof of the parity theorem, but there is another proof based on linear algebra. It is not quite self-contained (it presumes a rudimentary knowledge of determinants), but is shorter and provides a more concrete interpretation of the signature $\text{sgn}(\sigma)$. We include this discussion for the benefit of interested readers.

ALTERNATIVE PROOF OF THEOREM 4.17. Each $\sigma \in S_n$ can be thought of as a permutation of vectors in the standard basis $\mathfrak{X} = \{\mathbf{e}_1, \dots, \mathbf{e}_n\}$ in $\mathbb{K}^n$, with $\sigma : \mathbf{e}_k \mapsto \mathbf{e}_{\sigma(k)}$. That action induces a linear operator $\tilde{\sigma} : \mathbb{K}^n \to \mathbb{K}^n$ given by

$$\tilde{\sigma}\Big(\sum_{k=1}^{n} a_k \mathbf{e}_k\Big) = \sum_{k=1}^{n} a_k \mathbf{e}_{\sigma(k)} \quad \text{where } a_k \in \mathbb{K},\ 1 \leq k \leq n.$$

Thus $\tilde{\sigma}_1 \circ \tilde{\sigma}_2 = \widetilde{(\sigma_1\sigma_2)}$. The matrix $[\tilde{\sigma}]_{\mathfrak{X}\mathfrak{X}}$ of a 2-cycle $\tilde{\sigma}$ with respect to the standard basis $\mathfrak{X} = \{\mathbf{e}_1, \dots, \mathbf{e}_n\}$ is shown in Figure 4.2. Such matrices are examples of *permutation matrices*, which are characterized by the following properties:

(i) Each entry is 0 or 1.
(ii) Each column contains exactly one "1".
(iii) Each row contains exactly one "1".

To illustrate, we show the permutation matrix corresponding to the 2-cycle $\sigma = (i, j)$ in Figure 4.2.

The correspondence $\Phi : \sigma \to \Phi(\sigma) = \tilde{\sigma}$ is a homomorphism mapping S_n into the group $(\text{GL}(n), \circ)$ of invertible linear operators on $\mathbb{K}^n$, so that $\Phi(e) = \text{id}_{[1,n]}$ and $\Phi(\sigma\tau) = \Phi(\sigma) \circ \Phi(\tau) = \tilde{\sigma} \circ \tilde{\tau}$. Since determinants are multiplicative, it follows that

(i) $\det \Phi(e) = 1$.
(ii) $\det \Phi(i, j) = -1$ for any 2-cycle $(i \neq j)$.
(iii) $\det(\Phi(\sigma)) = \prod_{i=1}^{r} \det \Phi(\sigma_i) = (-1)^r = \text{sgn}(\sigma)$ if σ is a r-fold product $\sigma = \sigma_1 \cdots \sigma_r$ of 2-cycles.

But the value of $\det \Phi(\sigma)$ is determined without reference to any factorization of σ into 2-cycles, so we get the same number $\text{sgn}(\sigma) = (-1)^r$ no matter how σ is factored. Thus, the number of 2-cycles in any factorization of σ is either always even or always odd.

Finally, since $\det(AB) = \det(A) \cdot \det(B)$ for any pair of linear operators, we see that $\text{sgn}(\sigma\tau) = \text{sgn}(\sigma) \cdot \text{sgn}(\tau)$. □

$$\begin{bmatrix}
1 & & & 0 & & & & 0 & & & 0 \\
 & \ddots & & \vdots & & & & \vdots & & & \\
 & & 1 & & & & & & & & \\
0 & \cdots & & \boxed{0} & & \cdots & & \boxed{1} & & \cdots & 0 \\
 & & & & 1 & & & & & & \\
 & & & \vdots & & \ddots & & \vdots & & & \\
 & & & & & & 1 & & & & \\
0 & \cdots & & \boxed{1} & & \cdots & & \boxed{0} & & \cdots & 0 \\
 & & & & & & & & 1 & & \\
 & & & \vdots & & & & \vdots & & \ddots & \\
0 & & & 0 & & & & 0 & & & 1
\end{bmatrix}$$

FIGURE 4.2. The permutation matrix associated with the 2-cycle $\sigma = (i, j)$ differs from the identity matrix only in rows $i, 2$ and columns i, j.

For obvious reasons, we say a permutation is *even* if it can be written as a product of an even number of 2-cycles, so that $\mathrm{sgn}(\sigma) = +1$, and otherwise it is *odd*. Note that the identity e is even since $e = (12)(12)$ and all 2-cycles are odd.

EXERCISE 4.19. Verify that the map Φ from $S_n \to \mathrm{GL}(n)$ defined above has the properties:

(a) $\Phi(e) = I$ and $\Phi(\sigma\tau) = \Phi(\sigma) \circ \Phi(\tau)$.
(b) Φ is a one-to-one mapping—i.e., that the kernel $\ker(\Phi)$ of the group homomorphism Φ is trivial.
(c) Explain why the matrix $[\widetilde{\sigma}]_{\mathfrak{X}\mathfrak{X}}$ with respect to the standard basis has a single "1" in each column, each lying in a different row, with all other entries equal to "0".
(d) Then explain why any matrix A having the properties in (c) must be $[\widetilde{\sigma}]_{\mathfrak{X}\mathfrak{X}}$ for a unique permutation $\sigma \in S_n$.

EXERCISE 4.20. Verify that the permutation matrix $[\widetilde{\sigma}]_{\mathfrak{X}\mathfrak{X}}$ associated with a two-cycle $\sigma = (i, j)$, $i \neq j$, has the form shown in Figure 4.2.

4.2. Determinants

The previous digression about the permutation group S_n is needed to formulate the basic definition of $\det(A)$ for an $n \times n$ matrix $A \in \mathrm{M}(n, \mathbb{K})$ or the determinant $\det(T)$ of a linear operator $T : V \to V$ on a finite-dimensional vector space.

Any discussion that defines $\det(A)$ in terms of "expansion by minors" is confusing the basic definition of det with a commonly used *algorithm* for computing its value. Here is the real definition.

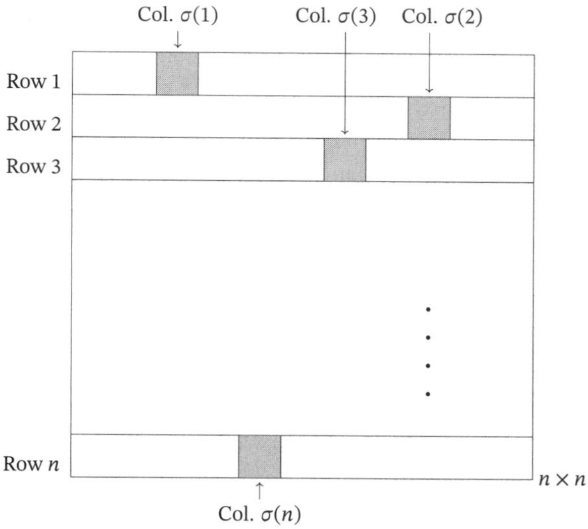

FIGURE 4.3. A permutation $\sigma \in S_n$ determines a "template" for selecting matrix entries by marking the address $(i, \sigma(i))$ – the one in Row_i, $\text{Column}_{\sigma(i)}$. Each row contains exactly one marked spot, and likewise for each column.

DEFINITION 4.21. If $A \in \text{M}(n, \mathbb{K})$, its *determinant* is defined to be

$$\det(A) = \sum_{\sigma \in S_n} \text{sgn}(\sigma) \cdot a_{1,\sigma(1)} \cdot \cdots \cdot a_{n,\sigma(n)} = \sum_{\sigma \in S_n} \text{sgn}(\sigma) \cdot \prod_{i=1}^{n} a_{i,\sigma(i)}. \tag{4.3}$$

Each product in this sum is obtained by using $\sigma \in S_n$ to select one entry from each row, taking each entry from a different column. Thus each σ determines a "template" for selecting matrix entries that are to be multiplied together (the product then being weighted by the signature $\text{sgn}(\sigma)$ of the permutation). The idea is illustrated in Figure 4.3.

Proving the Basic Properties of det(A). Many properties can be read directly from this definition; however, the all-important multiplicative property $\det(AB) = \det(A) \cdot \det(B)$ is tricky no matter what definition we start from. We begin with several easy properties:

THEOREM 4.22. *If $A \in \text{M}(n, \mathbb{K})$ and $c \in \mathbb{K}$, we have the following*:

1. $\det(I_{n \times n}) = 1$.
2. $\det(cA) = c^n \cdot \det(A)$.
3. $\det(A^{\text{T}}) = \det(A)$.
4. *When $\mathbb{K} = \mathbb{C}$ we have* $\det(\overline{A}) = \overline{(\det(A))}$ *where $\overline{z}$ = the complex conjugate of $z = x + iy$.*

5. *If A is upper triangular, so*

$$A = \begin{pmatrix} a_{11} & \cdots & * \\ & \ddots & \vdots \\ 0 & & a_{nn} \end{pmatrix},$$

then $\det(A) = \prod_{k=1}^{n} a_{kk}$ *is the product of the diagonal entries.*

PROOF. Assertions (1.), (2.), and (4.) are trivial. In (5.) the typical product $\pm a_{1,\sigma(1)} \cdots a_{n,\sigma(n)}$ in the definition of $\det(A)$ will equal 0 if any factor is zero. But unless $\sigma(k) = k$ for all k, there will be some row such that $\sigma(k) > k$ and some other row such that $\sigma(\ell) < \ell$. The resulting template includes a matrix entry below the diagonal, making the product zero for this template. Therefore the only permutation contributing a term to the sum (4.3) is $\sigma = e$, and that term is equal to $a_{11} \cdots a_{nn}$ as in (5.)

For (3.) we note that

$$\det(A^{\mathsf{T}}) = \sum_{\sigma \in S_n} \operatorname{sgn}(\sigma) \cdot (b_{1,\sigma(1)} \cdots b_{n,\sigma(n)})$$

if $B = A^{\mathsf{T}} = [b_{ij}]$. By definition of A^{T}, $b_{ij} = a_{ji}$, so the typical term becomes

$$b_{1,\sigma(1)} \cdots b_{n,\sigma(n)} = a_{\sigma(1),1} \ldots a_{\sigma(n),n}$$

and

$$\det(A^{\mathsf{T}}) = \sum_{\sigma \in S_n} \operatorname{sgn}(\sigma) b_{1,\sigma(1)} \cdots b_{n,\sigma(n)} = \sum_{\sigma \in S_n} \operatorname{sgn}(\sigma) a_{\sigma(1),1} \cdots a_{\sigma(n),n}.$$

However, we may write $a_{\sigma(j),j} = a_{\sigma(j),\sigma^{-1}(\sigma(j))}$ for each j, and then

$$\prod_{i=1}^{n} a_{\sigma(i),i} = \prod_{i=1}^{n} a_{\sigma(i),\sigma^{-1}(\sigma(i))},$$

so if we replace the dummy index i in the product with $j = \sigma(i)$, the product becomes $\prod_{j=1}^{n} a_{j,\sigma^{-1}(j)}$, and we get

$$\det(A^{\mathsf{T}}) = \sum_{\sigma \in S_n} \operatorname{sgn}(\sigma) \cdot \prod_{j=1}^{n} a_{j,\sigma^{-1}(j)}.$$

Next, let $\tau = \sigma^{-1}$, which runs through all of S_n as σ runs through S_n because S_n is a group. (This is our first encounter with the "group" property of S_n.) Furthermore, $\operatorname{sgn}(\tau) = \operatorname{sgn}(\sigma^{-1}) = \operatorname{sgn}(\sigma)$, so that

$$\det(A^{\mathsf{T}}) = \sum_{\tau \in S_n} \operatorname{sgn}(\tau) \cdot \prod_{j=1}^{n} a_{j,\tau(j)} = \det(A). \qquad \square$$

The following observation is a fundamental property of matrix determinants.

LEMMA 4.23. *If B is obtained from A by interchanging two rows, then*

$$\det(B) = (-1) \cdot \det(A).$$

Likewise, by part (3.) *of Theorem* 4.22, *this is also true when two columns are interchanged.*

PROOF. We do the case of which columns k and ℓ are interchanged. If $A = [a_{ij}]$ then $B = [b_{ij}]$ with $b_{ij} = a_{i\tau(j)}$, where τ is the two-cycle $\tau = (k, \ell)$ that switches the column indices when we interchange $\text{Col}_\ell(A) \leftrightarrow \text{Col}_k(A)$. Then $\text{Col}_j(B) = \text{Col}_{\tau(j)}(A)$ for $1 \leq j \leq n$, and for every $\sigma \in S_m$, we have

$$b_{1,\sigma(1)} \cdots b_{n,\sigma(n)} = a_{1,\tau\sigma(1)} \cdots a_{n,\tau\sigma(n)}.$$

But S_n is a *group*, so $\tau S_n = S_n$ and the elements $\tau\sigma$ run through all of S_n as σ runs through S_n; furthermore, because τ is a 2-cycle we have $\text{sgn}(\tau) = -1$ and $\text{sgn}(\tau\sigma) = \text{sgn}(\tau)\,\text{sgn}(\sigma) = (-1) \cdot \text{sgn}(\sigma)$. Thus

$$\begin{aligned}
\det(B) &= \sum_{\sigma \in S_n} \text{sgn}(\sigma) \cdot \prod_{i=1}^{n} b_{i,\sigma(i)} = \sum_{\sigma \in S_n} \text{sgn}(\sigma) \cdot \prod_{i=1}^{n} a_{i,\tau\sigma(i)} \\
&= \sum_{\sigma \in S_n} \text{sgn}(\tau)\ \text{sgn}(\tau\sigma) \cdot \prod_{i=1}^{n} a_{i,\tau\sigma(i)} \\
&= \text{sgn}(\tau) \cdot \sum_{\mu \in S_n} \text{sgn}(\mu) \cdot \prod_{i=1}^{n} a_{i,\mu(i)} = (-1) \cdot \det(A).
\end{aligned}$$

□

EXERCISE 4.24. Use the previous results to show that $\det(A) = 0$ if either

(a) A has two identical rows (or columns), or
(b) A has a row (or column) consisting entirely of zeros.

Recall the definition in Chapter 1 of the "elementary row operations" on a matrix A:

- TYPE I: $R_i \to \lambda \cdot R_i$: multiply Row_i by $\lambda \neq 0$ ($\lambda \in \mathbb{K}$)
- TYPE II: $R_i \leftrightarrow R_j$: interchange Row_i and Row_j
- TYPE III: $R_i \to R_i + \lambda R_j$: add to Row_i any scalar multiple of a different row R_j (leaving Row_j unaltered).

The effect of the first two operations on the determinant of a square matrix is easy to evaluate. We have just seen that Type II operations only cause a sign change.

EXERCISE 4.25. Prove that if B has $R_i(B) = \lambda \cdot R_i(A)$ for some index $1 \leq i \leq n$, with all other rows unchanged, then $\det(B) = \lambda \cdot \det(A)$.

To deal with Type III operations, first observe that the map $\det : \text{M}(n, \mathbb{K}) \to \mathbb{K}$ is a *multilinear function* of the rows or columns of A.

LEMMA 4.26. *If the* i^{th} *row of a matrix* A *is a linear combination* $R_i = aR_i' + bR_i''$ *of two other rows, then*

$$\det(A) = \det\begin{pmatrix} R_1 \\ \vdots \\ aR_i' + bR_i'' \\ \vdots \\ R_n \end{pmatrix} = a \cdot \det\begin{pmatrix} R_1 \\ \vdots \\ R_i' \\ \vdots \\ R_n \end{pmatrix} + b \cdot \det\begin{pmatrix} R_1 \\ \vdots \\ R_i'' \\ \vdots \\ R_n \end{pmatrix}$$
$$= a \cdot \det(A') + b \cdot \det(A'')$$

where A' *and* A'' *are the matrices obtained by replacing* $\mathrm{Row}_i(A)$ *with* R_i' *or* R_i''*, respectively. In other words,* $\det(A)$ *is a multilinear function of the rows in* A*: If we vary only* R_i*, holding the other rows fixed, the determinant is a linear function of* R_i*.*

PROOF. If $R_i' = (x_1, \dots, x_n)$ and $R_i'' = (y_1, \dots, y_n)$, then $A_{ij} = ax_j + by_j$ and

$$\begin{aligned}
\det(A) &= \sum_{\sigma\in S_n} \mathrm{sgn}(\sigma) \cdot \big(a_{1,\sigma(1)} \cdot \dots \cdot (ax_{\sigma(i)} + by_{\sigma(i)}) \cdot \dots \cdot a_{n,\sigma(n)}\big) \\
&= a \cdot \sum_{\sigma\in S_n} \mathrm{sgn}(\sigma) \cdot (a_{1,\sigma(1)} \cdot \dots \cdot x_{\sigma(i)} \cdot \dots \cdot a_{n,\sigma(n)}) \\
&\quad + b \cdot \sum_{\sigma\in S_n} \mathrm{sgn}(\sigma) \cdot (a_{1,\sigma(1)} \cdot \dots \cdot y_{\sigma(i)} \cdot \dots \cdot a_{n,\sigma(n)}) \\
&= a \cdot \det(A') + b \cdot \det(A'')
\end{aligned}$$

as claimed. □

COROLLARY 4.27. *If* B *is obtained from* A *by a* Type III *row operation* $R_i \to R_i + cR_j$ $(j \neq i)$*, then* $\mathrm{Row}_i(B) = R_i + cR_j$ *and*

$$\det(B) = \det\begin{pmatrix} R_1 \\ \vdots \\ R_i \\ \vdots \\ R_j \\ \vdots \\ R_n \end{pmatrix} + c \cdot \det\begin{pmatrix} R_1 \\ \vdots \\ R_j \\ \vdots \\ R_j \\ \vdots \\ R_n \end{pmatrix} = \det(A) + 0 = \det(A)$$

because the second matrix has a repeated row.

Row Operations, Determinants, and Matrix Inverses. Assuming A is a square matrix, we show that every row operation on A can be implemented by multiplying A on the left by a suitable $n \times n$ "elementary matrix" E; the corresponding column operation is achieved by multiplying A on the right by the transpose E^{T}. We leave verification as one of the Additional Exercises to Section 4.2 at the end of this chapter. There we also indicate how elementary row and column operations on a nonsquare $n \times m$ matrix can be realized by multiplying A on the left or right by "elementary matrices" that differ only slightly from those defined below for square matrices.

For the three basic types of elementary row operations (and of column operations), we have:

- TYPE I. $(\text{Row}_i) \to \lambda \cdot (\text{Row}_i)$ is equivalent to sending A to $E_{\text{I}}A$ where

$$E_{\text{I}} = \begin{pmatrix} 1 & & & & 0 \\ & \ddots & & & \\ & & \lambda & & \\ & & & \ddots & \\ 0 & & & & 1 \end{pmatrix}$$

  Obviously $\det(E_{\text{I}}) = \lambda$ and by Exercise 4.25

$$\det(E_{\text{I}}A) = \lambda \cdot \det(A) = \det(E_{\text{I}}) \cdot \det(A).$$

- TYPE II. $(\text{Row}_i) \leftrightarrow (\text{Row}_j)$: Now the result is achieved using the matrix

$$E_{\text{II}} = \begin{pmatrix} 1 & & & \text{Col}_i & & \text{Col}_j & & 0 \\ & \ddots & & & & & & \\ & & & \boxed{0} & \cdots & \boxed{1} & & \\ & & & \vdots & \ddots & \vdots & & \\ & & & \boxed{1} & \cdots & \boxed{0} & & \\ & & & & & & \ddots & \\ 0 & & & & & & & 1 \end{pmatrix}$$

  Since E_{II} is $I_{n\times n}$ with two rows interchanged, $\det(E_{\text{II}}) = -1$ by Lemma 4.23, and

$$\det(E_{\text{II}}A) = (-1) \cdot \det(A) = \det(E_{\text{II}}) \cdot \det(A).$$

- TYPE III. $(\text{Row}_i) \to (\text{Row}_i) + \lambda(\text{Row}_j)$, with $j \neq i$. Assuming $i < j$, the appropriate matrix is

$$E_{\text{III}} = \begin{pmatrix} 1 & & & \text{Col}_i & & \text{Col}_j & & 0 \\ & \ddots & & & & & & \\ & & & \boxed{1} & \cdots & \boxed{\lambda} & & \\ & & & \vdots & \ddots & \vdots & & \\ & & & \boxed{0} & \cdots & \boxed{1} & & \\ & & & & & & \ddots & \\ 0 & & & & & & & 1 \end{pmatrix}$$

  and we have $\det(E_{\text{III}}) = 1$. But then by Lemma 4.26 we also have

$$\det(E_{\text{III}}A) = \det(A) = \det(E_{\text{III}}) \cdot \det(A).$$

It is obvious that the elementary row operations are invertible, with the inverse of any Type III operation being the elementary operation

$$(\text{Row}_i) \to (\text{Row}_i) - \lambda(\text{Row}_j).$$

The inverse maps are then represented by invertible matrices that are easily calculated; for instance, E_{III}^{-1} is obtained by replacing $\lambda \to -\lambda$ in the off-diagonal entry of E_{III}. Note that the inverse of an elementary matrix is always of the same type as the original matrix.

As an immediate consequence of this discussion we have the following:

COROLLARY 4.28. *If E is any $(n \times n)$ elementary matrix, then*

$$\det(EA) = \det(E) \cdot \det(A)$$

for any $n \times n$ matrix A.

It is also evident that for a suitably chosen product $P = E_1 \cdots E_s$, the product $E = PA$ will be the echelon form (or row-reduced echelon form) of A.

This lemma allows us to compute determinants using row operations, exploiting the fact that $\det(A)$ can be calculated by inspection when A is upper triangular. First observe that a sequence of row operations sends $A \mapsto E_m \cdots E_1 \cdot A$ (the echelon form of A), but then by induction on n

$$\begin{aligned}\det(E_m \cdot \cdots \cdot E_1 A) &= \det(E_m) \cdot \det(E_{m-1} \cdots E_1 \cdot A)\\ &= \cdots = \left(\prod_{i=1}^{m} \det(E_i)\right) \cdot \det(A).\end{aligned}$$

Thus

$$\det(A) = \left(\prod_{i=1}^{m} \det(E_i)^{-1}\right) \cdot \det(E_1 \cdots E_m A),$$

and calculating $\det(A)$ reduces to calculating the determinant of the upper triangular row reduced form, which can be read by inspection. (You must also keep track of the row operations used, and the scalar values of their determinants.)

Computing Matrix Inverses. Suitably chosen row operations will put an $n \times n$ matrix into echelon form. If we only allow elementary operations of Type II or Type III, we can achieve nearly the same result except that the pivot entries may be nonzero scalars λ_i rather than 1's, as shown in Figure 4.4. Next recall that matrix space $\mathrm{M}(n, \mathbb{K})$ and the space of linear operators $\mathrm{Hom}(\mathbb{K}^n, \mathbb{K}^n)$ are isomorphic as *associative algebras* under the correspondence

$$A \mapsto L_A \quad (L_A(\mathbf{x}) = A \cdot \mathbf{x} = (n \times n) \cdot (n \times 1) \text{ matrix product}),$$

as we showed in the discussion surrounding Exercise 4.13. That means the following statements are equivalent:

(4.4)

1. *A matrix inverse A^{-1} exists in* $\mathrm{M}(n, \mathbb{K})$.
2. $L_A : \mathbb{K}^n \to \mathbb{K}^n$ *is an invertible linear operator.*
3. $\ker(L_A) = (0)$.
4. *The matrix equation $A\mathbf{x} = \mathbf{0}$ has only the trivial solution $\mathbf{x} = \mathbf{0}_{n \times 1}$.*

We say that a matrix is *nonsingular* if any of these conditions holds; otherwise it is *singular*.

EXERCISE 4.29. If A, B are square matrices prove the following:

(a) The product AB is *singular* if at least one of the factors is singular.

(b) The product AB is *nonsingular* if both factors are nonsingular.

Useful facts about matrix inverses can be deduced by using row operations. One consequence is an elegant invertibility criterion: a matrix A is invertible if and only if $\det(A) \neq 0$.

PROPOSITION 4.30. *The following statements regarding an $n \times n$ matrix are equivalent*:

1. $\det(A) \neq 0$.
2. *A has a multiplicative inverse A^{-1} in* $\mathrm{M}(n, \mathbb{K})$.
3. *The multiplication operator $L_A : \mathbb{K}^n \to \mathbb{K}^n$ is an invertible* (*bijective*) *linear operator on coordinate space.*

PROOF. We already know (2.) $\Leftrightarrow$ (3.). Row operations of Type II and III reduce A to one of the two "modified echelon forms" A' (see Figure 4.4(a)–(b)), in which the step corners contain nonzero scalars $\lambda_1, \ldots, \lambda_r$ that need not equal 1, and $r = \mathrm{rank}(A)$. Obviously if some column fails to meet a step-corner as in Figure 4.4(a), then the product of diagonal entries $\det(A)$ is zero; at the same time, the matrix equations $A'\mathbf{x} = \mathbf{0}$ and $A\mathbf{x} = \mathbf{0}$ will involve free variables and will have nontrivial solutions. Thus the left multiplication operator $L_A : \mathbb{K}^n \to \mathbb{K}^n$ fails to be invertible (because $\ker(L_A) \neq (0)$) and a matrix inverse A^{-1} does not exist. The situation in Figure 4.4(b) is better: since Type II and Type III operations can only change $\det(A)$ by a $\pm$ sign, $\det(A) = \pm\det(A') = \pm\prod_{i=1}^{n} \lambda_i$ is nonzero. Concurrently, $A\mathbf{x} = \mathbf{0}$ has only the trivial solution, L_A is an invertible linear operator on $\mathbb{K}^n$, and a matrix inverse A^{-1} exists. □

To summarize, we have proved the following result (and a little more).

THEOREM 4.31. *If $A \in \mathrm{M}(n, \mathbb{K})$ then A^{-1} exists if and only if* Type II *and* Type III *row operations yield a modified echelon form that is upper triangular, with all diagonal entries nonzero*:

$$E_m \cdot \cdots \cdot E_1 A = \begin{pmatrix} \boxed{\lambda_1} & * & & * \\ 0 & \boxed{\lambda_2} & & \\ & & \ddots & \\ 0 & & 0 & \boxed{\lambda_n} \end{pmatrix} \qquad (\lambda_i \neq 0)$$

Here the E_i are the matrices corresponding to the elementary row operations that are needed to put A in echelon form.

Then the determinant is

$$\det(A) = \prod_{k=1}^{m} \det(E_k)^{-1} \cdot \prod_{i=1}^{n} \lambda_i = (-1)^r \cdot \prod_{i=1}^{n} \lambda_i\,,$$

where r is the number of Type II operations involved in the row reduction. Note that the factor $(-1)^r = \prod_{k=0}^{n} \det(E_k)^{-1}$ attributed to the row operations can only

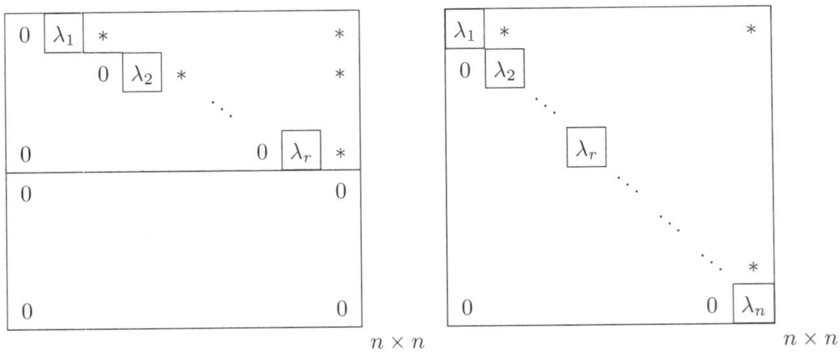

FIGURE 4.4. Row operations of Type II and III reduce an $n \times n$ matrix A to one of the two "modified echelon forms" A' shown in (a) and (b); in both the step corners contain nonzero scalars $\lambda_1, \dots, \lambda_r$ that need not $= 1$, and $r = \text{rank}(A)$ with $r = n$ in (b). If there are columns that do not meet a step corner as in (a), then some diagonal entries in A' are zero and $\det(A) = \pm\det(A') = 0$. In the situation of (b) $\det(A) = \pm\det(A') = \pm(\lambda_1 \cdots \lambda_n)$ because Type II and III elementary operations have determinant $= \pm 1$. In this case $\det(A)$ is nonzero and its value can be determined by inspection, except for a $(\pm)$ sign.

be ± 1 *since no* Type I *operations are involved. On the other hand, if the modified echelon form contains columns that do not meet a step corner, then* $\det(A) = 0$ *and* A^{-1} *does not exist.*

Computational Issues. The basic definition (4.3) of the determinant is computationally very costly. Below we will give an algorithm (*expansion by minors*), which is often useful in studying the algebraic properties of determinants, but in numerical calculations it is still pretty costly compared to the row reduction method developed above. To illustrate:

n = Matrix Size	Expansion by Minors		Row Reduction	
	Adds	Multiplies	Adds	Multiplies
2	1	2	1	3
4	23	40	14	23
5	119	205	30	45
10	3.6×10^6	6.2×10^6	285	339

The technique used above also yields a fairly efficient algorithm for computing A^{-1} (which at the same time determines whether A is in fact invertible). Allowing all three types of row operations, an invertible matrix can be driven into its *reduced echelon form*, which is just the identity matrix $I_{n\times n}$. In this case

$$E_m \cdots E_1 \cdot A = I_{n\times n} \quad \text{and} \quad A^{-1} = E_1^{-1} \cdots E_m^{-1} \cdot I_{n\times n} \tag{4.5}$$

Each inverse E_k^{-1} is easily computed; it is just another elementary matrix of the same type as E_k. This can be codified as an explicit algorithm:

> COMPUTING $\boldsymbol{A^{-1}}$: THE GAUSS-SEIDEL ALGORITHM. *Starting with the augmented $n \times 2n$ matrix $[A : I_{n\times n}]$, perform row operations to put A into reduced echelon form (upper triangular with zeros above all step corners). If* rank$(A) < n$ *and A is not invertible, this will be evident—not all columns include a step-corner—so the algorithm reports that* $\det(A) = 0$ *and A is not invertible. Otherwise, every column is a pivot column and the reduced echelon form of A is just the identity matrix. Applying the same operations to the entire augmented matrix transforms $[A : I_{n\times n}] \to [I_{n\times n} : B]$ in which $B = A^{-1}$. (Why?)*

Proving the Multiplicative Property $\det(AB) = \det(A)\det(B)$. An additional consequence of the preceding discussion is the very important multiplicative property of determinants.

THEOREM 4.32 (Multiplicative Property). *If $A, B \in \mathrm{M}(n, \mathbb{K})$, then*

$$\det(AB) = \det(A) \cdot \det(B).$$

PROOF. If A is singular, then AB is singular (Exercise 4.29) so $\det(A) = 0$ and $\det(AB) = 0$ by Proposition 4.30. Thus

$$\det(AB) = 0 = 0 \cdot \det(B) = \det(A) \cdot \det(B),$$

and similarly if B is the singular factor.

Otherwise A and B are nonsingular and so is AB; thus we can find elementary matrices such that $E_m \cdots E_1 A = I_{n\times n}$, which implies $A = E_1^{-1} \cdots E_m^{-1}$. By repeated application of Corollary 4.28, we see that

$$\det(A) = \prod_{i=1}^{m} \det(E_i^{-1}),$$

hence

$$\begin{aligned} \det(AB) &= \det(E_1^{-1}) \cdot \det\big(E_2^{-1} \cdots E_m^{-1} B\big) \\ &= \cdots = \prod_i \det(E_i^{-1}) \cdot \det(B) = \det(A) \cdot \det(B). \end{aligned}$$

□

EXERCISE 4.33. If $A \in \mathrm{M}(n, \mathbb{K})$ is invertible, then $\det(A^{-1}) = \det(A)^{-1}$. If $A, B \in \mathrm{M}(n, \mathbb{K})$ and S is an invertible matrix such that $B = SAS^{-1}$, then $\det(B) = \det(A)$.

Thus the function $\phi(A) = \det(A)$ is a "*similarity invariant*" on matrix space $\mathrm{M}(n, \mathbb{K})$, having the same value for all matrices in a similarity class. We will encounter other similarity invariants of matrices in the following discussion.

EXERCISE 4.34. Prove that rank(A) of an $n \times n$ matrix is a similarity invariant.

EXERCISE 4.35. An $n \times n$ matrix A is said to be *orthogonal* if $A^{\mathrm{T}}A = I_{n\times n}$. Prove the following:

(a) $A^{\mathrm{T}}A = I \Rightarrow AA^{\mathrm{T}} = I$, so A is orthogonal $\Leftrightarrow A^{\mathrm{T}} = A^{-1}$ (two-sided inverse).
(b) $\det(A) = \pm 1$ for any orthogonal matrix, over any field.

HINT: Recall the comments posted in (4.4). For (a) it suffices to show $A^{\mathrm{T}}A = I \Rightarrow$ the operator $L_A : \mathbb{K}^n \to \mathbb{K}^n$ is one-to-one.

EXERCISE 4.36. Use Type II and III row operations to find the determinant of the following matrix:

$$A = \begin{pmatrix} 1 & 2 & 1 & 2 \\ 2 & 1 & 2 & 1 \\ 1 & 3 & 3 & 1 \\ 1 & 3 & 3 & 4 \end{pmatrix}$$

EXERCISE 4.37. Use Type II and III row operations to show that $\det(A) = -16i$ for the following matrix in $\mathrm{M}(4, \mathbb{C})$, where $i = \sqrt{-1}$:

$$\begin{pmatrix} 1 & 1 & 1 & 1 \\ 1 & i & -1 & -i \\ 1 & -1 & 1 & -1 \\ 1 & -i & -1 & i \end{pmatrix}$$

EXERCISE 4.38. Apply the Gauss-Seidel algorithm to find A^{-1} for the matrices

$$\text{(a)}\quad A = \begin{pmatrix} 1 & 3 & 1 \\ 2 & 8 & 4 \\ 0 & 4 & 7 \end{pmatrix} \qquad \text{(b)}\quad A = \begin{pmatrix} 1 & 3 & 2 \\ 2 & 4 & 1 \\ 0 & 4 & 2 \end{pmatrix}$$

EXERCISE 4.39. Consider the set of matrices H_n of the form

$$A = \begin{pmatrix} 1 & x_1 & \cdots & x_n & z \\ & 1 & & & y_n \\ & & \ddots & & \vdots \\ & & & 1 & y_1 \\ 0 & & & & 1 \end{pmatrix} \qquad \text{(blank spots} = 0\text{)}$$

with x_i, y_j, z in $\mathbb{K}$. When $\mathbb{K} = \mathbb{R}$ this is the n-dimensional *Heisenberg group* of quantum mechanics.

(a) Prove that H_n is closed under matrix product.
(b) Prove that the inverse A^{-1} of any matrix in H_n is also in H_n (compute it explicitly in terms of the parameters x_i, y_j, z).

Since the identity matrix is also in H_n, that means H_n is a *matrix group* contained in $\mathrm{GL}(n+2, \mathbb{K})$.

EXERCISE 4.40. For $n \geq 2$ let

$$A = \begin{pmatrix} 0 & 1 & & & & & 0 \\ 1 & 0 & 1 & & & & \\ 0 & 1 & 0 & 1 & & & \\ & & \ddots & \ddots & \ddots & & \\ & & & 1 & 0 & 1 \\ 0 & & & & 1 & 0 \end{pmatrix}.$$

Use row operations to

(a) Calculate $\det(A)$.

(b) Calculate the inverse A^{-1} if it exists.

NOTE: The outcome will depend on whether n is even or odd.

EXERCISE 4.41. Given a diagonal matrix $D = \text{diag}(\lambda_1, \dots, \lambda_n)$ with distinct entries, find an invertible matrix S such that conjugation $D \mapsto SDS^{-1}$ interchanges the ith and j^{th} diagonal entries ($i \neq j$):

$$S \begin{pmatrix} \lambda_1 & & & & & 0 \\ & \ddots & & & & \\ & & \boxed{\lambda_i} & & & \\ & & & \ddots & & \\ & & & & \boxed{\lambda_j} & \\ & & & & & \ddots \\ 0 & & & & & \lambda_n \end{pmatrix} S^{-1} = \begin{pmatrix} \lambda_1 & & & & & 0 \\ & \ddots & & & & \\ & & \boxed{\lambda_j} & & & \\ & & & \ddots & & \\ & & & & \boxed{\lambda_i} & \\ & & & & & \ddots \\ 0 & & & & & \lambda_n \end{pmatrix}$$

HINT: Think row *and* column operations on D. Note that if E_{II} is a Type II elementary matrix then $E^{-1} = E = E^{\mathrm{T}}$, and right multiplication by E^{T} effects the corresponding column operation. (Recall Exercise 1.43.)

Defining Determinants of Linear Operators. A determinant $\det(T)$ can be unambiguously assigned to any linear operator $T : V \to V$ on a finite-dimensional space. Given a basis $\mathfrak{X} = \{e_i\}$ in V, we get a matrix $[T]_{\mathfrak{X}\mathfrak{X}}$ and could entertain the idea of assigning

$$\det(T) = \det([T]_{\mathfrak{X}\mathfrak{X}}), \tag{4.6}$$

but for this to make sense the outcome must be independent of the choice of basis. This actually works. If $\mathfrak{Y}$ is any other basis, we know there is an invertible matrix $S = [\text{id}_V]_{\mathfrak{Y}\mathfrak{X}}$ such that $[T]_{\mathfrak{Y}\mathfrak{Y}} = S\,[T]_{\mathfrak{X}\mathfrak{X}}S^{-1}$, and then by Theorem 4.32

$$\begin{aligned} \det\big([T]_{\mathfrak{Y}\mathfrak{Y}}\big) &= \det(S) \cdot \det\big([T]_{\mathfrak{X}\mathfrak{X}}\big) \cdot \det(S^{-1}) \\ &= \det(SS^{-1}) \cdot \det\big([T]_{\mathfrak{X}\mathfrak{X}}\big) = \det(I_{n\times n}) \cdot \det\big([T]_{\mathfrak{X}\mathfrak{X}}\big) \\ &= \det\big([T]_{\mathfrak{X}\mathfrak{X}}\big) \end{aligned}$$

as required. Thus the determinant (4.6) of a linear operator is well defined.

The *trace* $\text{Tr}(T)$ is another well-defined attribute of an operator $T : V \to V$ when $\dim(V) < \infty$. Recall Exercise 2.46 of Chapter 2: For $n \times n$ matrices the trace $\text{Tr}(A) = \sum_{i=1}^{n} A_{ii}$ is a linear operator $\text{Tr} : \text{M}(n, \mathbb{K}) \to \mathbb{K}$ such that $\text{Tr}(I_{n\times n}) = n$ and $\text{Tr}(AB) = \text{Tr}(BA)$. If $\mathfrak{X}$, $\mathfrak{Y}$ are bases for V, we get

$$\text{Tr}([T]_{\mathfrak{Y}\mathfrak{Y}}) = \text{Tr}(S\,[T]_{\mathfrak{X}\mathfrak{X}}S^{-1}) = \text{Tr}(S^{-1}S \cdot [T]_{\mathfrak{X}\mathfrak{X}}) = \text{Tr}([T]_{\mathfrak{X}\mathfrak{X}})$$

Thus for any basis $\mathfrak{X}$ in V,

$$\text{Tr}(T) = \text{Tr}([T]_{\mathfrak{X}\mathfrak{X}}) \tag{4.7}$$

determines a well-defined trace on operators. However, if $T : V \to W$ with $V \neq W$, there is no natural way to assign a "determinant" or "trace" to T, even if $\dim(V) = \dim(W)$. The problem is philosophical. In the absence of some way to identify vectors in V with those in W, there is no natural way to say that a basis $\mathfrak{X}$ in V is the "same as" another basis $\mathfrak{Y}$ in W.

The operator trace has the same algebraic properties as the matrix trace.

EXERCISE 4.42. If $A, B : V \to V$ are linear operators on a finite-dimensional space V, prove the following:

(a) $\text{Tr} : \text{Hom}_{\mathbb{K}}(V, V) \to \mathbb{K}$ is a $\mathbb{K}$-linear map between vector spaces:

$$\text{Tr}(A + B) = \text{Tr}(A) + \text{Tr}(B) \quad \text{and} \quad \text{Tr}(\lambda \cdot A) = \lambda \cdot \text{Tr}(A).$$

(b) $\text{Tr}(\text{id}_V) = n \cdot \dim(V)$.

(c) $\text{Tr}(AB) = \text{Tr}(BA)$ (composition product of operators).

(d) If S is an invertible operator and $B = SAS^{-1}$, then $\text{Tr}(B) = \text{Tr}(A)$.

The last statement shows that Tr is a similarity invariant for linear operators; as is the determinant det.

EXERCISE 4.43. If $T : V \to V$ is a linear operator on a finite-dimensional space, prove that

$$\text{(a) } \text{Tr}(T) = \text{Tr}(T^{\mathsf{T}}) \quad \text{and} \quad \text{(b) } \det(T) = \det(T^{\mathsf{T}})$$

NOTE: A conceptual issue arises in (b): T maps $V \to V$ while the transpose $T^{\mathsf{T}} : V^* \to V^*$ acts on an entirely different vector space. But if you take a basis $\mathfrak{X}$ in V and the dual basis $\mathfrak{X}^*$ in V^*, the definitions (4.6) and (4.7) still have something useful to say.

EXERCISE 4.44. Let $P : V \to V$ be a projection (associated with some direct sum decomposition $V = E \oplus F$) that projects vectors onto E along F. Prove that $\text{Tr}(P) = \dim_{\mathbb{K}}(E)$.

HINT: Pick a suitable basis compatible with the decomposition $E \oplus F$.

More on Rank, RowRank, and ColumnRank. In Chapter 1 we defined $\text{rank}(A)$ for any $n \times m$ matrix to be the dimension of the range space $R(L_A) = L_A(\mathbb{K}^m)$ of the linear operator $L_A : \mathbb{K}^m \to \mathbb{K}^n$ and explained why this is equal to the column rank of A (the dimension of $\text{ColSpace}(A)$ spanned by the columns in A, regarded as vectors in $\mathbb{K}^n$). Our purpose here is to show that (*row rank*) = (*column rank*) for any matrix. This will follow as a corollary to a very efficient way to find a basis for $R(L_A)$ by inspecting the *row-reduced* echelon form of A.

PROPOSITION 4.45. *If an $n \times m$ matrix A is row reduced to echelon form E, then $r = \operatorname{rank}(A) = \dim(R(L_A)) = \operatorname{ColRank}(A)$ is the number of pivot columns $C_{j_1}(E), \dots, C_{j_r}(E)$ in E. Moreover, the corresponding columns $C_{j_1}(A), \dots, C_{j_r}(A)$ in the original matrix A are a basis for the range space $R(L_A)$.*

PROOF. We have seen that left multiplication of A by a suitable product $P = E_1 \cdots E_s$ of elementary matrices yields the row-reduced echelon form $E_* = PA$ of A, in which all entries of the pivot columns $C_{j_1}(E_*), \dots, C_{j_r}(E_*)$ are zero except for 1's in the step corners. We claim that the corresponding columns $C_{j_1}(A), \dots, C_{j_r}(A)$ in A are linearly independent. If not, there would be coefficients c_i such that

$$c_1 C_{j_1}(A) + \cdots + c_r C_{j_r}(A) = 0_{n\times 1} \quad (n \times 1 \text{ zero matrix}), \tag{4.8}$$

and multiplying (4.8) on the left by P yields

$$c_1 P C_{j_i}(A) + \cdots + c_r P C_{j_r}(A) = 0_{n\times 1}.$$

Because E_* is in echelon form all entries below the rth row in E_* are zero (recall Figure 4.4), so by (4.8) and Exercise 4.46 below we get

$$PC_{j_i}(A) = C_{j_i}(PA) = C_{j_i}(E_*) = \operatorname{col}(0, \dots, 1, \dots, 0, 0, \dots, 0),$$

with a "1" in the ith spot and zeros in the last $n - r$ spots of this matrix.

EXERCISE 4.46. If B is $n \times n$ and C is an $n \times m$ matrix, prove that $C_j(BC) = B \cdot C_j(C)$ for $1 \le j \le m$.

Thus all coefficients $c_j = 0$ and the columns $C_{j_i}(A)$ are independent.

Because the last $n - r$ rows of E_* consist of zeros, every column in E_* has the form

$$C_j(E_*) = \operatorname{col}(d_1, \dots, d_r, 0, \dots, 0) = \sum_{i=1}^{r} d_i C_{j_i}(E_*)$$

for $1 \le j \le m$, and entries in pivot columns $C_{j_i}(E_*)$ are zero except for a "1" in the ith place. By Exercise 4.46 the corresponding columns of A have the form

$$\begin{aligned} C_j(A) = C_j(P^{-1}E_*) = P^{-1} \cdot C_j(E_*) &= P^{-1}\Big(\sum_{i=1}^{r} d_i C_{j_i}(E_*)\Big) \\ &= \sum_{i=1}^{r} d_i P^{-1} C_{j_i}(E_*) \\ &= \sum_{i=1}^{r} d_i C_{j_i}(A). \end{aligned}$$

Thus the columns $C_{j_i}(A)$ span $\operatorname{ColSpace}(A) = R(L_A)$ and are a basis. □

As an immediate consequence we have the following:

COROLLARY 4.47. $\operatorname{RowRank}(A) = \operatorname{ColRank}(A)$ *for any $n \times m$ matrix.*

NOTE: There is a coordinate-independent proof that (*column rank*) = (*row rank*) that does not involve matrix wrangling but does involve a multistep chain of reasoning. Such a proof has already been been outlined in Exercises 3.28–3.32 of Chapter 3.

Expansion by Minors and Cramer's Rule. The following result gives a recursive procedure for computing an $n \times n$ determinant in terms of $(n-1) \times (n-1)$ determinants. Although it is handy for small matrices and important in determining algebraic properties of determinants, it is prohibitively expensive in computing time for large n. This expansion is keyed to a particular row (or column) of A and requires evaluation of $(n-1) \times (n-1)$ subdeterminants (the "minors" of the title), one for each row entry.

THEOREM 4.48 (Expansion by Minors). *If we fix a row* Row_i *in* A*, we can "expand by minors" along this row, writing*

$$\det(A) = \sum_{j=1}^{n} (-1)^{i+j} a_{ij} \cdot \det(M_{ij}) \tag{4.9}$$

where M_{ij} = *the* $(n-1) \times (n-1)$ *submatrix obtained by deleting* Row_i *and* Col_j *from* A*. Similarly, for any column* $1 \le j \le n$ *we have an expansion along* Col_j*:*

$$\det(A) = \sum_{i=1}^{n} (-1)^{i+j} a_{ij} \cdot \det(M_{ij}).$$

The $(n-1) \times (n-1)$ matrices M_{ij} are called the "minors" of A.

PROOF. Since $\det(A) = \det(A^{\mathrm{T}})$, it is enough to prove the result for expansion along a row. Each term in the sum

$$\det(A) = \sum_{\sigma \in S_n} \mathrm{sgn}(\sigma) \cdot (a_{1\sigma(1)} \cdots a_{n\sigma(n)})$$

contains just one factor from $\mathrm{Row}_i(A) = (a_{i1} \cdots a_{in})$, so by gathering together terms involving the same factor a_{ik}, $k = 1, 2, \ldots, n$, we may write

$$\det(A) = a_{i1}A^*_{i1} + \cdots + a_{in}A^*_{in}$$

in which A^*_{ij} involves no entry from $\mathrm{Row}_i(A)$.

Our task is to show $A^*_{ij} = (-1)^{i+j} \det(M_{ij})$. One approach is to use the case when $i = j = n$ as a guide. In that special situation we have

$$a_{nn}A^*_{nn} = \sum_{\sigma \in S'_n} \mathrm{sgn}(\sigma) \cdot (a_{1\sigma(1)} \cdots a_{n\sigma(n)})$$

where $S'_n \subseteq S_n$ is the subgroup of permutations (isomorphic to S_{n-1}) such that $\sigma(n) = n$.

EXERCISE 4.49. If $\widetilde{\sigma} \in S_{n-1}$ is identified with the permutation σ in S_n such that $\sigma(n) = n$ and $\sigma(k) = \widetilde{\sigma}(k)$ for $1 \le k \le n-1$, show that $\mathrm{sgn}(\widetilde{\sigma}) = \mathrm{sgn}(\sigma)$.

In view of this, the sum $\sum_{\sigma\in S_n'}(\cdots)$ becomes $\sum_{\widetilde{\sigma}\in S_{n-1}}(\cdots)$, and

$$A^*_{nn} = (-1)^{n+n}\det(M_{nn}) = \det(M_{nn}).$$

Now consider any i and j. Interchange $\mathrm{Row}_i(A)$ with the row below it. This "flip" of adjacent rows does not affect the value of $\det(M_{ij})$ except for a change of sign because the relative positions of the other rows and columns are not affected. Each flip switches the sign attached to a_{ij} in (4.9), and if we flip successive rows until the original i^{th} row of $\widetilde{A}$ is at the bottom, there will be $n-i$ such changes in (4.9). Next, flip adjacent columns moving the jth column to become the nth column, incurring $n-j$ sign changes. Thus

$$A^*_{ij} = (-1)^{n-i+n-j}\det(M_{ij}) = (-1)^{i+j}\det(M_{ij})$$

for all i and j, proving the theorem. □

We post without proof the following formula for A^{-1}.[1]

THEOREM 4.50 (Cramer's Rule). *If matrix $A \in \mathrm{M}(n,\mathbb{K})$ is invertible we have*

$$A^{-1} = \frac{1}{\det(A)}\cdot(\mathrm{Cof}(A))^{\mathrm{T}}$$

where the $n\times n$ cofactor matrix $\mathrm{Cof}(A)$ has i,j entry $= (-1)^{i+j}\det(M_{ij})$, and M_{ij} is the $(n-1)\times(n-1)$ submatrix obtained by deleting (Row_i) and (Col_j) from A, and relabeling the entries that remain.

EXERCISE 4.51. When the transposition $\sigma = (i,j)$ acts by permuting the standard basis vectors $e_1,\dots,e_n$ in $\mathbb{K}^n$, the corresponding linear operator on $\mathbb{K}^n$ is represented by the matrix $A = [\widetilde{\sigma}]_{\mathfrak{X}\mathfrak{X}}$ shown in Figure 4.2. Show that $\det(A) = -1$.

EXERCISE 4.52. Explain why the calculation posted in Exercise 4.51 yields the crucial formula

$$\mathrm{sgn}(\sigma) = (-1)^r$$

where r = the number of factors τ_i in any decomposition $\sigma = \tau_1\cdots\tau_r$ as a product of 2-cycles.

A combination of expansion by minors and row operations of Type II and III (which all have det $E = 1$), yields a fairly effective way to evaluate determinants. One could develop a formal algorithm for doing this, but the general idea should be clear from the following example.

EXAMPLE 4.53. Evaluate the determinant of the 5×5 matrix A,

$$A = \begin{pmatrix} 6 & 2 & 1 & 0 & 5\\ 2 & 1 & 1 & -2 & 1\\ 1 & 1 & 2 & -2 & 3\\ 3 & 0 & 2 & 3 & -1\\ -1 & -1 & -3 & 4 & 2\end{pmatrix} \xrightarrow{\text{Row Ops}} A' = \begin{pmatrix} 2 & 0 & -1 & 4 & 3\\ 2 & \boxed{1} & 1 & -2 & 1\\ -1 & 0 & 1 & 0 & 2\\ 3 & 0 & 2 & 3 & -1\\ 1 & 0 & -2 & 2 & 3\end{pmatrix}$$

[1]For a proof see *Linear Algebra*, 2nd edition, by Seymour Lipschutz, Schaum's Outline Series, McGraw-Hill, 1991, pp. 267–268.

DISCUSSION: First, look for an entry $a_{ij} = 1$; if you can't find one, scale some row (or column) to create one, and keep track of the effect on $\det(A)$. In our example $a_{22} = 1$. Next, clear all other entries in $\text{Col}_2(A)$ by adding suitable multiples of $\text{Row}_2(A)$ to get the matrix A' such that $\det(A') = \det(A)$. Now expand by minors along $\text{Col}_2(A')$. Since all entries except $a'_{22} = 1$ are zero the expansion has just one term, the determinant of the minor M_{22} obtained by deleting the row and column of A' that pass through the "pivot entry" a'_{22}. We are reduced to finding the determinant of the single 4×4 matrix M_{22}.

$$M_{22} = \begin{pmatrix} 2 & -1 & 4 & 3 \\ -1 & 1 & 0 & 2 \\ 3 & 2 & 3 & -1 \\ 1 & -2 & 2 & 3 \end{pmatrix} \xrightarrow{\text{Col Ops}} B' = \begin{pmatrix} 1 & 1 & 4 & 5 \\ 0 & \boxed{1} & 0 & 0 \\ 5 & 2 & 3 & -5 \\ -1 & -2 & 2 & 7 \end{pmatrix},$$

which already has an entry $b'_{22} = 1$ that will serve as the pivot for the next step.

This time it is easier to apply column operations to clear the other entries from $\text{Row}_2(M_{22})$ to get the matrix B' shown above. Expansion by minors along $\text{Row}_2(B')$ yields a single 3×3 determinant that is easily evaluated to get:

$$\det(A) = \det \begin{pmatrix} 1 & 4 & 5 \\ 5 & 3 & -5 \\ -1 & 2 & 7 \end{pmatrix}$$

We omit the routine last steps that yield $\det(A) = -24$.

Additional Exercises

Section 4.1. The Permutation Group S_n

1. TRUE/FALSE QUESTIONS: ("True" if the statement is *always* true.)

(a) If $\sigma \in S_n$ is a product $\sigma_1 \cdots \sigma_r$ of cycles with disjoint supports, then $o(\sigma) \geq \max\{o(\sigma_1), \ldots, o(\sigma_r)\}$.

(b) If p is a prime and σ is a p-cycle in S_n, then all powers σ^j for $1 \leq j \leq p-1$ are p-cycles, but $\sigma^p = e$.

(c) For any $\sigma \in S_n$, $\text{sgn}(\sigma^{-1}) = (-1) \cdot \text{sgn}(\sigma)$.

(d) If σ and τ are cycles of length 10 and 12, respectively, the order of their product is $o(\sigma \cdot \tau) = 60$.

(e) Any element in S_n of order $o(\sigma) = k$ is a k-cycle.

(f) An element in S_n of order n is a "long cycle" (a cycle of length $= n$).

2. We know (see Exercise 4.14) that cycles σ, τ commute if their supports are disjoint. How about the converse: If two cycles commute must their supports be disjoint? Prove or give a counterexample.

3. Find the disjoint cycle decompositions of the following permutations:

(a) $(1, 3, 5)(2, 4, 6)$ in S_8

(b) $(1, 5, 3)(2, 4, 5)(3, 8)$ in S_8

(c) $(1, 2)(2, 3)(3, 4)(4, 5)$ in S_5

(d) $\sigma = \begin{pmatrix} 1 & 2 & 3 & 4 & 5 & 6 & 7 \\ 2 & 3 & 7 & 6 & 4 & 5 & 1 \end{pmatrix}$ in S_{10}

(e) $\sigma = \begin{pmatrix} 1 & 2 & 3 & 4 & 5 & 6 & 7 & 8 \\ 3 & 4 & 5 & 6 & 7 & 8 & 1 & 2 \end{pmatrix}$ in S_8 (a cyclic 2-shift)

4. Determine $\text{sgn}(\sigma)$ for the permutation $\sigma = (1, 5, 3, 7)(4, 6, 3)$ by exhibiting a decomposition as a product of 2-cycles.

5. Given a fixed k-cycle σ in S_n, identify *all* nontrivial cycles $\tau = (m_1, \dots, m_r)$ that commute with σ.

HINTS: By relabeling we may assume σ is the standard k-cycle $(1, 2, \dots, k)$. We already know that cycles with disjoint supports commute, so we may as well proceed by assuming

$$\text{supp}(\tau) \cap \text{supp}(\sigma) = \{m_1, \dots, m_r\} \cap \{1, 2, \dots, k\} \neq \emptyset.$$

6. A *flip* is a particular kind of 2-cycle $\sigma = (i, i + 1)$ that interchanges adjacent integers in the cyclic ordering of $\{1, 2, \dots, n\}$; note that $\sigma = (n, 1)$ is also a flip. Prove that every permutation $\tau \in S_n$ can be written as a product of flips and that its signature is $\text{sgn}(\tau) = (-1)^{\#(\text{flips})}$.

NOTE: In particular, the group of permutations on $[1, n]$ is generated by the flips it contains.

7. If G is a finite group and $x, y \in G$ are elements that *commute*, so that $xy = yx$,

(a) Prove: $o(xy) \leq \text{lcm}\,(o(x), o(y))$ (lcm = least common multiple).

(b) Is the relation in (a) actually an equality? Prove it or find a group G and elements $x, y \in G$ for which equality fails. (Compare this question with Exercise 4.14, which holds only for *cycles* in $G = S_n$.)

NOTES: Every element of a finite group has finite order $o(x) \leq |G|$. In (b) we may restrict attention to *abelian* G because all the action occurs within the (abelian) subgroup $G_{x,y}$ generated by x and y.

HINT: In (b) consider elements x, y in a suitably chosen *cyclic* group $(\mathbb{Z}_n, +)$.

Section 4.2. Determinants.

1. TRUE/FALSE QUESTIONS: ("True" if the statement is *always* true.)

(a) If $A \in \text{M}(n, \mathbb{K})$, $\det(A) \neq 0 \Leftrightarrow$ its columns are linearly independent.

(b) If A is an $n \times m$ matrix with $n \neq m$, then its determinant is zero.

(c) If $\lambda \neq 0$ in $\mathbb{K}$ and $A \in \text{M}(n, \mathbb{K})$, then $\det(\lambda \cdot A) = \lambda \cdot \det(A)$.

(d) A determinant $\det(T)$ can be defined for any linear operator $T : V \to W$ between finite-dimensional vector spaces.

(e) Let $T : V \to V$ be a linear operator on a finite-dimensional space. If $\mathfrak{X} = \{e_1, \dots, e_n\}$ is an ordered basis in V and basis $\mathfrak{Y} = \{f_1, \dots, f_n\}$ is obtained by listing the vectors in $\mathfrak{X}$ in a different order then

$$\det[T]_{\mathfrak{X}\mathfrak{X}} = \det[T]_{\mathfrak{Y}\mathfrak{Y}}.$$

2. Explain in your own words why the following statements about a matrix A in $\mathrm{M}(n, \mathbb{K})$ are equivalent.

(a) A has a two-sided inverse A^{-1}.

(b) $\det A \neq 0$.

(c) The range of the left multiplication operator $L_A : \mathbb{K}^n \to \mathbb{K}^n$ has dimension n.

(d) The column rank of A, the dimension of $\mathbb{K}$-span in $\mathbb{K}^n$ of the columns $C_1(A), \dots, C_n(A)$, is equal to n.

NOTE: Recall the discussion of *rank, row rank,* and *column rank* in Section 1.4. Since rank(A) is equal to "column rank" by definition, all these observations can be summarized by saying rank(A) $= n$.

VOLUMES OF SOLIDS IN $\mathbb{R}^\mathbf{n}$. A *parallelopiped* in $\mathbb{R}^n$, the n-dimensional analogue of a parallelogram, is the set in $\mathbb{R}^n$

$$P = \{t_1\mathbf{a}^{(1)} + \cdots + t_n\mathbf{a}^{(n)} : 0 \leq t_1, \dots, t_n \leq 1 \text{ in } \mathbb{R}\},$$

determined by n "edge vectors" $\mathbf{a}^{(i)} = (a_1^{(i)}, \dots, a_n^{(i)}) \in \mathbb{R}^n$ for $1 \leq i \leq n$. It is well-known that the n-dimensional volume of P is given by a determinant

$$\mathrm{Vol}(P) = \det(A) \quad \text{where } A = \begin{pmatrix} a_1^{(1)} & \cdots & a_n^{(1)} \\ \vdots & & \vdots \\ a_1^{(n)} & \cdots & a_n^{(n)} \end{pmatrix},$$

which will be zero unless the vectors $\mathbf{a}^{(i)}$ are linearly independent. It is an "oriented volume," which carries a $\pm$ sign that changes if any two edge vectors in the ordered list $\mathbf{a}^{(1)}, \dots, \mathbf{a}^{(n)}$ are switched. The (unsigned) "geometric volume" is the absolute value $|\det A|$.

3. Compute the volume of the parallelopipeds P in $\mathbb{R}^3$ and $\mathbb{R}^4$ determined by each of the following sets of edge vectors emanating from the origin.

(a) Volume of $P \subseteq \mathbb{R}^3$ determined by

$$\mathbf{v}_1 = (1, 2, 3), \quad \mathbf{v}_2 = (3, 4, -1), \quad \mathbf{v}_3 = (2, 1, -5).$$

(b) Volume of $P \subseteq \mathbb{R}^4$ determined by

$$\mathbf{v}_1 = (1, -1, 5, -1), \quad \mathbf{v}_2 = (2, 1, -2, 1),$$
$$\mathbf{v}_3 = (3, 0, 1, -2), \quad \mathbf{v}_4 = (1, -1, , 4, -1)$$

NOTE: Volumes Vol(P) are translation-invariant, so if the edge vectors of P emanate from some point $\mathbf{p} \neq \mathbf{0}$, we can subtract $\mathbf{p}$ from all vertices to get a solid P' of the same volume that has the origin as one of its vertices, as in (a) and (b) above.

4. A matrix $A \in \mathrm{M}(n, \mathbb{C})$ is *skew-symmetric* if $A^{\mathrm{T}} = -A$. If n is odd, prove that such a matrix cannot be invertible. What if n is even?

5. A matrix $A \in \mathrm{M}(n, \mathbb{R})$ is *orthogonal* if $A^{\mathrm{T}}A = I$.

(a) Prove that $\det(A)$ can only have the values ± 1.

(b) Explain why we must also have $AA^{\mathrm{T}} = I$, even though A and its transpose need not commute.

Thus A^{T} is a two-sided inverse for any orthogonal matrix A.

6. A matrix $A \in \mathrm{M}(n, \mathbb{C})$ is *complex-orthogonal* if $AA^{\mathrm{T}} = I$. Prove that all such matrices have the properties (a) and (b) listed in Exercise 5 above.

ADJOINT A^* OF A MATRIX. The *matrix-adjoint* A^* and *complex conjugate* $\overline{A}$ of a matrix with complex entries are defined letting

$$(\overline{A})_{ij} = \overline{A_{ij}} \text{ (complex conjugate of } A_{ij}) \quad \text{and} \quad (A^*)_{ij} = \overline{A_{ji}},$$

so that $A^* = \overline{A^{\mathrm{T}}}$. Of course, if all entries in A are real, then $A^* = A^{\mathrm{T}}$ and $\overline{A} = A$. These operations on complex matrices will be important in Chapter 6 on inner product spaces.

7. A matrix $A \in \mathrm{M}(n, \mathbb{C})$ is *unitary* if $A^*A = I$.

(a) Prove that $|\det(A)| = 1$ for any unitary matrix.

(b) Explain why we must also have $A^*A = I$, even though A and its adjoint need not commute.

Thus A is unitary $\Leftrightarrow A^* = A^{-1}$.

8. Let $\mathcal{T}(n)$ be the upper triangular matrices in $\mathrm{M}(n, \mathbb{C})$ with $\det(A) \neq 0$. Explain why A^{-1} exists and is upper triangular.

NOTE: Thus the set $\mathcal{T}(n)$ of such matrices is a *matrix group*, which means:

(i) $I_{n \times n} \in \mathcal{T}(n)$

(ii) $A, B \in \mathcal{T}(n) \Rightarrow AB \in \mathcal{T}(n)$

(iii) $A \in \mathcal{T}(n) \Rightarrow A^{-1} \in \mathcal{T}(n)$

HINT: Property (i) is obvious and (ii) was discussed in Chapter 1. The point here is to show that the inverse A^{-1} of an upper triangular matrix is also upper triangular. Cramer's Rule might be helpful.

ROW AND COLUMN OPERATIONS: NONSQUARE MATRICES. Row and column operations can be performed on matrices of any shape. Moreover, these can be be effected by multiplying A on the left by elementary $n \times n$ matrices and on the right by elementary $m \times m$ matrices (the same matrices $E_{\mathrm{I}}, E_{\mathrm{II}}, E_{\mathrm{III}}$ as in our earlier discussion of row operations and determinants). The following exercises illustrate this for the Type II operations $R_i \leftrightarrow R_2$ and $C_i \leftrightarrow C_j$.

9. If $R(A)$ is the row operation $R_i \leftrightarrow R_j$ on a square matrix A, and $C(A)$ is the corresponding column operation, prove that

$$C(A) = (R(A^{\mathrm{T}}))^{\mathrm{T}}.$$

10. If A is a square matrix, prove that

(a) The Type II row operation $R_i \leftrightarrow R_j$ is given by the product $E_{\mathrm{II}}A$ where E_{II} is the associated Type II elementary matrix.

(b) A Type II column operation $C_i \leftrightarrow C_j$ is given by the product $AE_{\mathrm{II}}^{\mathrm{T}}$.

Similar proofs apply to the other elementary row operations on A.

11. Elementary row and column operations obviously make sense for matrices of any shape. If A is an $n \times m$ matrix prove the following:

(a) Action on A of a Type II row operation $R_i \leftrightarrow R_j$ is given by

$$A \to E_{\mathrm{II}}^{n\times n} \cdot A$$

where $E_{\mathrm{II}}^{n\times n}$ is the corresponding $n\times n$ elementary matrix of Type II.

(b) Action on A of a Type II column operation $C_i \leftrightarrow C_j$ is given by

$$A \to A \cdot (E_{\mathrm{II}}^{m\times m})^{\mathrm{T}},$$

multiplying on the right by the transpose of an $m\times m$ elementary matrix of Type II.

Similar arguments work for the other elementary row and column operations on rectangular matrices.

REMARK. Once (a) is established, part (b) of this exercise could be proved using the symmetry argument of Additional Exercise 9.

CHAPTER 5

The Diagonalization Problem

5.1. Eigenvalues, Characteristic Polynomial, and Spectrum

DEFINITION 5.1. If V is a vector space, $\lambda \in \mathbb{K}$ is an *eigenvalue* for a linear operator $T : V \rightarrow V$ if there is some $v \neq 0$ such that $T(v) = \lambda v$. For an arbitrary $\lambda \in \mathbb{K}$ the λ*-eigenspace* is

$$E_\lambda(T) = \{v \in V : (T - \lambda I)v = 0\}.$$

By definition, this vector subspace is nontrivial if and only if λ is an eigenvalue. The set of distinct eigenvalues is called the *spectrum* $\text{sp}_{\mathbb{K}}(T)$ of the operator. Eigenvalues, eigenspaces, and spectrum of a matrix $A \in \text{M}(n, \mathbb{K})$ are defined similarly, by taking T to be the left multiplication operator $L_A(\mathbf{x}) = A \cdot \mathbf{x}$ on $V = \mathbb{K}^n$. Thus, for instance,

$$E_\lambda(A) = \{\mathbf{x} \in \mathbb{K}^n : A \cdot \mathbf{x} = \lambda \mathbf{x}\}.$$

When $\lambda = 0$ the eigenspace $E_{\lambda=0}(T)$ is just $\ker(T) = \{v \in V : T(v) = 0\}$, and when $\lambda = 1$ we get the subspace of *fixed vectors* $E_{\lambda=1}(T) = \{v : T(v) = v\}$.

A connection with determinants emerges when V is finite dimensional. Then $\lambda \in \mathbb{K}$ is an eigenvalue if and only if

$$\ker(T - \lambda I) \neq (0) \Leftrightarrow (T - \lambda I) \text{ is singular} \quad \Leftrightarrow \det(T - \lambda I) = 0,$$

so the eigenvalues of T are the roots of the characteristic polynomial.

On the other hand, the *characteristic polynomial* $A \in \text{M}(n, \mathbb{K})$, defined to be

$$p_A(x) = \det(A - xI) \quad (x \text{ an indeterminate})$$

is a polynomial with $\deg(p_A) = n$ in the single unknown x because $\det(A - xI)$ is a polynomial combination of the entries $(A - xI)_{ij}$. Given a linear operator $T : V \rightarrow V$ on a finite-dimensional space and a basis $\mathfrak{X}$ in V, we have

$$[T - xI]_{\mathfrak{X}\mathfrak{X}} = [T]_{\mathfrak{X}\mathfrak{X}} - xI_{n\times n} \quad (n = \dim(V))$$

so the characteristic polynomial of T becomes the characteristic polynomial of the matrix $A = [T]_{\mathfrak{X}\mathfrak{X}}$,

$$p_T(x) = \det(T - xI) = \det\left([T]_{\mathfrak{X}\mathfrak{X}} - xI_{n\times n}\right) \quad (x \text{ an indeterminate}).$$

The discussions for operators and matrices are so similar that nothing is lost if we focus on characteristic polynomials of matrices for the time being.

Now observe what happens if we write the characteristic polynomial p_A as

$$p_A(x) = \det(A - xI) = c_0(A) + c_1(A)x + \cdots + c_n(A)x^n \tag{5.1}$$

in which the coefficients $c_i(A)$ are scalar-valued functions from $\mathrm{M}(n, \mathbb{K}) \to \mathbb{K}$.

LEMMA 5.2. *Each coefficient $c_k(A)$ in (5.1) is a similarity invariant on matrix space:*

$$c_k(SAS^{-1}) = c_k(A) \quad \textit{for } A \in \mathrm{M}(n,\mathbb{K}),\ S \in \mathrm{GL}(n,\mathbb{K}).$$

Furthermore, if we identify $\mathrm{M}(n,\mathbb{K})$ with n^2-dimensional coordinate space $\mathbb{K}^{n^2}$ via the correspondence $A \mapsto (a_{11}, \dots, a_{1n}; \dots; a_{n1}, \dots, a_{nn})$, each coefficient $c_i(A)$ is a polynomial function of the matrix entries: $c_i(A) = F_i(a_{11}, a_{12}, \dots, a_{nn})$ for some $F_i \in \mathbb{K}[\mathbf{x}] = \mathbb{K}[x_1, \dots, x_{n^2}]$.

PROOF. We have

$$\begin{aligned}\det\left(S(A - xI)S^{-1}\right) &= \det(SAS^{-1} - x\, SS^{-1}) = \det(SAS^{-1} - xI) \\ &= c_0(SAS^{-1}) + c_1(SAS^{-1})x + \cdots + c_n(SAS^{-1})x^n,\end{aligned}$$

while at the same time

$$\begin{aligned}\det\left(S(A - xI)S^{-1}\right) &= \det(S) \cdot \det(A - xI) \cdot \det(S^{-1}) \\ &= \det(A - xI) = c_0(A) + c_1(A)x + \cdots + c_n(A)x^n\end{aligned}$$

for all $x \in \mathbb{K}$. Since these are the same polynomial in $\mathbb{K}[x]$, the coefficients must agree, hence $c_i(SAS^{-1}) = c_i(A)$.

Since coefficients of $\det(A - xI)$ are polynomial combinations of entries $(A - xI)_{ij}$, the coefficients $c_k(A)$ are then polynomial functions of the a_{ij} when like powers of the unknown "x" are gathered together. □

It is interesting to examine how the coefficients $c_k(A)$ are obtained from entries in A. By the original definition of the determinant in Chapter 4, we have

$$\det(B) = \sum_{\sigma \in S_n} \operatorname{sgn}(\sigma) \cdot \Big(\prod_{i=1}^{n} b_{i,\sigma(i)}\Big),$$

so if we take $B = A - xI$ we get

$$\det B = \det \begin{pmatrix} (a_{11} - x) & a_{12} & \cdot & a_{1n} \\ a_{12} & (a_{22} - x) & \cdot & a_{2n} \\ \vdots & & \ddots & \vdots \\ a_{n1} & \cdot & \cdot & (a_{nn} - x) \end{pmatrix}.$$

It is clear that the only template yielding a product $b_{1,\sigma(1)} \cdots b_{n\sigma(n)}$ that involves x^n is the diagonal template corresponding to the trivial permutation $\sigma = e$; furthermore, in expanding the product $\prod_i (a_{ii} - x)$ we must take the term "$-x$" instead of "a_{ii}" from each factor to get the power x^n. Thus $c_n(A) \equiv (-1)^n$ is constant on matrix space (and is certainly a similarity invariant).

We claim that

$$\begin{aligned}\det(A - xI) &= (-1)^n x^n + (\text{terms of lower degree}) \\ &= (-1)^n x^n + (-1)^{n-1} \operatorname{Tr}(A)\, x^{n-1} + \cdots + \det(A) \cdot 1 \end{aligned} \tag{5.2}$$

To get the coefficient of x^{n-1}, observe that a product $\prod_i b_{i,\sigma(i)}$ involving x^{n-1} must come from a template having $(n-1)$ marked spots on the diagonal, but then *all* marked spots must lie on the diagonal, and we are again dealing with the diagonal template (for $\sigma = e$). To get the terms involving x^{n-1} in the expansion of the product $\prod_i (a_{ii} - x)$ we must now select the "$-x$" from $n-1$ factors $(a_{ii} - x)$ and the "a_{ii}" from just one. Thus

$$c_{n-1}(A) = (-1)^{n-1} \cdot \sum_{i=1}^{n} a_{ii} = (-1)^{n-1}\,\mathrm{Tr}(A)$$

as in (5.2). Determining the other coefficients is tricky business, except for the constant term, which is

$$a_0(A) = \det(A).$$

This follows because *every* template yields a product that contributes to this constant term. However, if a template marks a spot on the diagonal we must select the "a_{ii}" term rather than the "x" from that diagonal entry $(a_{ii} - x)$. It follows that the constant term in (5.2) is

$$\sum_{\sigma \in S_n} \mathrm{sgn}(\sigma) \cdot \prod_{i=1}^{n} a_{i,\sigma(i)} = \det(A)$$

as claimed. We leave discussion of other terms in the expansion (5.2) for more advanced courses.

Factoring Polynomials. It is well-known that if a nonconstant polynomial $f(x)$ in $\mathbb{K}[x]$ has a root $\alpha \in \mathbb{K}$, so $f(\alpha) = \sum_{i=0}^{n} c_i \alpha^i = 0$, then we can factor $f(x) = (x-\alpha)\cdot g(x)$ by long division of polynomials, with $\deg(g) = \deg(f) - 1$. In fact, applying the euclidean algorithm for division with remainder in $\mathbb{K}[x]$: if $P, Q \in \mathbb{K}[x]$ and $\deg(Q) \geq 1$ we can always write

$$P(x) = A(x)Q(x) + R(x) \quad \text{(with remainder } R \equiv 0 \text{ or } \deg(R) < \deg(Q))$$

Taking P to be any nonconstant polynomial in $\mathbb{K}[x]$ and $Q = (x - \alpha)$, we get $f(x) = A(x)\cdot(x-\alpha) + R(x)$ where $R(x)$ is either the zero polynomial, or $R(x)$ is nonzero with $\deg(R) < \deg(x - \alpha) = 1$, which means that $R(x)$ is a nonzero *constant* polynomial $R = c\mathbb{1}$. If $\alpha \in \mathbb{K}$ is a root of f, replacing x by α everywhere yields the identity

$$0 = f(\alpha) = A(\alpha)\cdot(\alpha - \alpha) + R(\alpha) = 0 + R(\alpha) = R(\alpha).$$

Since $R = c\mathbb{1}$ this forces $R(x) \equiv 0$ and $f(x) = A(x)(x-\alpha)$ with no remainder, so $(x - \alpha)$ divides $f(x)$ exactly.

If α_1 is a root of f we may split $f(x) = (x - \alpha_1)\cdot g_1(x)$. If $g_1(x)$ has a root α_2 in $\mathbb{K}$ we can continue this process, obtaining $f(x) = (x-\alpha_1)(x-\alpha_2)\cdot g_2(x)$. Pushing this as far as possible we arrive at a factorization

$$f(x) = \prod_{i=1}^{s}(x - \alpha_i)\cdot g(x)$$

in which $g(x)$ has *no* roots in $\mathbb{K}$. We say that f *splits completely over* $\mathbb{K}$ if g reduces to a constant polynomial, so that $f(x) = c\prod_{i=1}^{n}(x-\alpha_i)$. This product may include repeated factors, and if we gather together all factors of the same type, this becomes

$$f(x) = c\prod_{j=1}^{r}(x-\alpha_i)^{m_j} \quad (\alpha_i \in \mathbb{K},\ m_1 + \cdots + m_r = \deg f).$$

In this version the roots $\{\alpha_1, \dots, \alpha_r\}$ are *distinct* and the exponents $m_i \geq 1$ are their *multiplicities* as roots of $f(x)$; the constant c out front is the coefficient of the leading term $c_n x^n$ in $f(x)$.

COROLLARY 5.3. *A nonconstant polynomial $f(x) \in \mathbb{K}[x]$ can have at most $n = \deg(f)$ roots in any field of coefficients $\mathbb{K}$. More generally, the sum of the multiplicities of the roots in $\mathbb{K}$ is at most n.*

PROOF. If $f, g \neq 0$ in $\mathbb{K}[x]$ (so they have well-defined degrees), we know that

$$\deg(f(x)\cdot g(x)) = \deg(f(x)) + \deg(g(x)),$$

hence $\deg\big(\prod_{i=1}^{r}(x-\alpha_i)^{m_i}\big) = \sum_{i=1}^{r} m_i$ and

$$r = \#(\text{distinct roots}) \leq (m_1 + \cdots + m_r) + \deg(g) = \deg(f). \qquad \square$$

EXERCISE 5.4. If $f(x), h(x)$ are nonzero polynomials over any field, explain why the "degree formula"

$$\deg(f(x)h(x)) = \deg(f(x)) + \deg(h(x))$$

is valid.

EXERCISE 5.5. Verify that if $f(x) = \prod_{i=1}^{r}(x-\alpha_i)\cdot g(x)$ and $g(x)$ has no roots in $\mathbb{K}$, then the roots of f in $\mathbb{K}$ are $\{\alpha_1, \dots, \alpha_r\}$.
NOTE: Repetitions are allowed; $f(x)$ might even have the form $(x-\alpha)^r \cdot g(x)$.

DEFINITION 5.6. The distinct roots $\{\alpha_1, \dots, \alpha_r\}$ in $\mathbb{K}$ of a nonconstant polynomial and their multiplicities are uniquely determined. The set of distinct roots is called the *spectrum* of the polynomial f and is denoted by $\mathrm{sp}_{\mathbb{K}}(f)$.

Fundamental Theorem of Algebra. Over the field $K = \mathbb{C}$ of complex numbers we have:

THEOREM 5.7 (Fundamental Theorem of Algebra). *If f is a nonconstant polynomial in $\mathbb{C}[x]$, then f has a root $\alpha \in \mathbb{C}$ so that $f(\alpha) = 0$.*

COROLLARY 5.8. *Every nonconstant $f \in \mathbb{C}[x]$ splits completely over $\mathbb{C}$, with*

$$f(x) = c\cdot\prod_{i=1}^{r}(x-\alpha_i)^{m_i} \quad \textit{where } m_1 + \cdots + m_r = n.$$

PROOF. Since f has a root, we may factor $f = (x-\alpha_1)\cdot g_1(x)$. Unless $g_1(x)$ is a constant it also has a root, so that $f = (x-\alpha_1)(x-\alpha_2)g_2(x)$. Continue recursively. $\square$

Over $\mathbb{K} = \mathbb{R}$ or $\mathbb{Q}$, things get more complicated and $f(x)$ might not have any roots at all. For instance, this happens if $f(x) = x^2+1$ over $\mathbb{R}$, or if $f(x) = x^2-2$ over $\mathbb{Q}$ since $\mathbb{Q}$ does not contain any element α such that $\alpha^2 = 2$ (there is no "square root of 2" in $\mathbb{Q}$).

Nevertheless, since $\mathbb{R} \subseteq \mathbb{C}$ we may regard any $f \in \mathbb{R}[x]$ as a complex polynomial that happens to have all real coefficients. All real roots α remain roots $\alpha + i0$ in $\mathbb{C}$ (lying on the real axis), but enough new roots appear in the larger field to split f completely as

$$f(x) = c \cdot \prod(x - \alpha_i) \quad \text{with } c \text{ and } \alpha_i \in \mathbb{C}.$$

It is important to realize that the new nonreal roots enter in "conjugate pairs."

LEMMA 5.9. *If $f(x)$ is nonconstant in $\mathbb{R}[x]$ and $z = x + iy$ is a complex root when we identify $\mathbb{R} \subseteq \mathbb{C}$ and $\mathbb{R}[x] \subseteq \mathbb{C}[x]$, then the complex conjugate $\overline{z} = x - iy$ is also a root.*

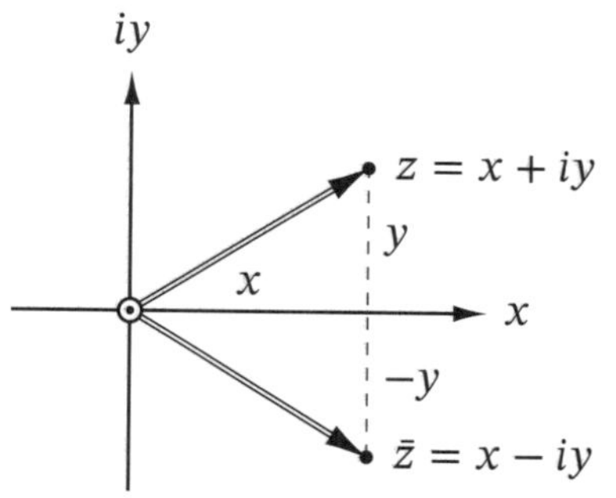

FIGURE 5.1. Nonreal roots of a polynomial with real coefficients come in conjugate pairs $z = x + iy$ and $\overline{z} = x - iy$, mirror images of each other under reflection across the x-axis.

PROOF. There is nothing to prove if z is real ($y = 0$). Otherwise, recall that conjugation has the following algebraic properties:

$$\overline{z_1 + z_2} = \overline{z}_1 + \overline{z}_2 \quad \text{and} \quad \overline{z_1 z_2} = \overline{z_1 z_2},$$

and in particular

$$\overline{z^n} = \overline{z}^n \quad \text{for all } n \in \mathbb{Z} \text{ and } z \in \mathbb{C}.$$

Hence if $0 = f(z) = \sum_{j=0} c_j z^j$ with c_j real, we have

$$\overline{(c_j z^j)} = \overline{c_j}\,\overline{(z^j)} = c_j \overline{z}^j$$

and

$$0 = \overline{0} = \overline{f(z)} = \sum_{j=0} \overline{(c_j z^j)} = \sum_{j=0} c_j (\overline{z})^j = f(\overline{z}),$$

so $\overline{z}$ is also a root in $\mathbb{C}$. □

The number of distinct nonreal roots is always *even*. The real roots of f in $\mathbb{R}[x]$ are not subject to any constraints; in fact, all the roots might be real.

The Quadratic Formula.

EXAMPLE 5.10. If $f \in \mathbb{K}[x]$ is quadratic,

$$f(x) = ax^2 + bx + c \quad \text{with } a \neq 0,$$

the quadratic formula continues to apply for all fields except those of "*characteristic* 2," in which $2 = 1 + 1$ is equal to 0 (for instance, $\mathbb{K} = \mathbb{Z}_2$). Except for this, the roots are given by

$$\text{QUADRATIC FORMULA:} \quad z_{\pm} = \frac{-b \pm \sqrt{b^2 - 4ac}}{2a}.$$

If the $\sqrt{\cdot}$ fails to exist in $\mathbb{K}$, the proper conclusion is that $f(x)$ has no roots in $\mathbb{K}$. If $\mathbb{K} = \mathbb{Q}$ or $\mathbb{R}$, this formula gives the correct roots in $\mathbb{C}$ even if there are no roots in $\mathbb{K}$.

DISCUSSION: Complete the square. By adding or subtracting a suitable constant d, we may write

$$\begin{aligned} ax^2 + bx + c = a\left(x^2 + \frac{b}{a}x + \frac{c}{a}\right) &= a\left[\left(x^2 + \frac{b}{a}x + d\right) + \left(\frac{c}{a} - d\right)\right] \\ &= a \cdot \left(x^2 + \frac{b}{a}x + d\right) + (c - ad). \end{aligned}$$

To make $x^2+(b/a)x+d$ a "perfect square" of the form $(x+k)^2 = (x^2+2kx+k^2)$, we must take $k = b/(2a)$ and $d = k^2 = (b^2/4a^2)$. Then $c - ad = c - (b^2/4a^2)$, so that

$$0 = ax^2 + bx + c = a\left(x + \frac{b}{2a}\right)^2 + \left(\frac{c}{a} - \frac{b^2}{4a^2}\right) = a \cdot \left(x + \frac{b}{2a}\right)^2 + \left(\frac{4ac - b^2}{4a}\right).$$

This happens if and only if

$$a\left(x + \frac{b}{2a}\right)^2 = \left(\frac{b^2 - 4ac}{4a}\right)$$

if and only if

$$\left(x + \frac{b}{2a}\right)^2 = \frac{b^2 - 4ac}{4a^2}.$$

Taking square roots we get

$$x = \frac{-b \pm \sqrt{b^2 - 4ac}}{2a}.$$

EXAMPLE 5.11. Here are some examples of factorization of polynomials.

1. $x^2 - 1 = (x - 1)(x + 1)$ splits over $\mathbb{R}$, with two roots $+1, -1$, each of multiplicity 1. On the other hand $x^2 + 1$ has no roots and does not split over $\mathbb{R}$, but it does split over $\mathbb{C}$, with $x^2 + 1 = (x - i)(x + i)$.
2. $x^2 + 2x + 1$ splits over $\mathbb{R}$ as $(x - 1)^2$, but there is just one root, of multiplicity 2 .

3. $x^3 - x^2 + x - 1$ has a root $x = 1$ in $\mathbb{R}$. Long division yields a quadratic,

$$x^3 - x^2 + x - 1 = (x-1)(x^2+1)$$

Over $\mathbb{R}$, there is just one root $\lambda_1 = 1$ with multiplicity $m(\lambda_1) = 1$; over $\mathbb{C}$ we get $x^2 + 1 = (x+i)(x-i)$ so there are two more roots $\lambda_2 = i$, $\lambda_3 = -i$ in the larger field $\mathbb{C}$.
4. $x^4 - 1 = (x^2-1)(x^2+1) = (x+1)(x-1)(x+i)(x-i)$.
5. $x^3 + x + 1$ has just one real root λ_1 because it is a strictly increasing function of $x \in \mathbb{R}$ whose values go to $\pm\infty$ as $x \to \pm\infty$, so its graph must cross the x-axis somewhere. But λ_1 is not so easy to write as an explicit algebraic expression involving sums, products, quotients, and cube roots. Such formulas exist, but are given by algorithms with possible branch points rather than simple expressions as in the quadratic formula. A numerical estimate yields the real root $\lambda_1 = -0.6823 + 0\,i$. There is a conjugate pair of complex roots $\lambda_2 = 0.3412 + 1.615\,i$ and $\lambda_3 = 0.3412 - 1.615\,i$, which could be found by (numerically) long dividing $f(x)$ by $(x - \lambda_1)$ and applying the quadratic formula to find the complex roots of the resulting quadratic.

5.2. Eigenvalues and the Characteristic Polynomial

We have seen that the eigenvalues of a linear operator $T : V \to V$ on a finite-dimensional vector space are the roots in $\mathbb{K}$ of the characteristic polynomial of T.

DEFINITION 5.12. If $T : V \to V$ is a linear operator on a finite-dimensional vector space, the *spectrum* $\mathrm{sp}_{\mathbb{K}}(T)$ is the set of DISTINCT roots in $\mathbb{K}$ of the *characteristic polynomial* $p_T(x) = \det(T - xI)$. We define the *geometric multiplicity* of an eigenvalue to be $\dim(E_\lambda(T))$; its *algebraic multiplicity* is the multiplicity of λ as a root of the *characteristic polynomial*, so that

$$p_T(\lambda) = \det(T - \lambda I) = 0$$

and $p_T(x) = (x - \lambda)^m \cdot g(x)$ for some polynomial $g(x)$ such that $g(\lambda) \neq 0$.

The following relation holds between geometric and algebraic multiplicities.

LEMMA 5.13. *Over any field* $\mathbb{K}$,

$$(\textit{algebraic multiplicity of } \lambda) \geq (\textit{geometric multiplicity of } \lambda).$$

Over $\mathbb{K} = \mathbb{C}$, *the sum of the algebraic multiplicities of the (distinct) in* $\mathrm{sp}_{\mathbb{C}}(T) = \{\lambda_1, ..., \lambda_r\}$ *is* $m(\lambda_1) + \cdots + m(\lambda_r) = n = \dim_{\mathbb{C}}(V)$.

PROOF. Every eigenspace E_λ is T-invariant because

$$(T - \lambda I)T(v) = T(T - \lambda I)v = 0 \quad \text{for } v \in E_\lambda.$$

This eigenspace has a basis of eigenvectors $\mathfrak{X}_\lambda = \{e_1, \dots, e_d\}$, with respect to which

$$[T]_{\mathfrak{X}_\lambda} = \begin{pmatrix} \lambda & & & 0 \\ & \lambda & & \\ & & \ddots & \\ 0 & & & \lambda \end{pmatrix}$$

(diagonal). Extending $\mathfrak{X}_\lambda$ to a basis $\mathfrak{X} = \{e_1, \dots, e_d, e_{d+1}, \dots, e_n\}$ for all of V, we get

$$[T]_{\mathfrak{X}\mathfrak{X}} = \left(\begin{array}{ccc|c} \lambda & & 0 & \\ & \ddots & & * \\ 0 & & \lambda & \\ \hline & 0 & & B \end{array}\right),$$

which implies that

$$[T - xI]_{\mathfrak{X}\mathfrak{X}} = \left(\begin{array}{ccc|c} \lambda - x & & 0 & \\ & \ddots & & * \\ 0 & & \lambda - x & \\ \hline & 0 & & B - xI \end{array}\right).$$

LEMMA 5.14. *If A has the form*

$$A = \left(\begin{array}{c|c} B & D \\ \hline 0 & C \end{array}\right)$$

where B and C are two square matrices, then $\det(A) = \det(B) \cdot \det(C)$.

PROOF. If B is $m \times m$, a sequence of Type II and III row operations on rows $R_1, \dots, R_m$ puts this block in upper triangular form; similar operations on rows $R_{m+1}, \dots, R_n$ puts block C in upper triangular form without affecting any of the earlier rows. The net result is an echelon form $A' = [B', *; 0, C']$ for which $\det A' = \det(B') \cdot \det(C')$. Each of the determinants $\det(A'), \dots, \det(C')$ differs from its counterpart by a $\pm$ sign; furthermore, the same moves that put B and C in upper triangular form also put A in upper triangular form when applied to the whole $n \times n$ matrix. We leave the reader to check that the sign changes cancel and yield $\det(A) = \det(B) \cdot \det(C)$. □

Applying Lemma 5.14 we can complete the proof of Lemma 5.13. We now see that

$$p_T(x) = \det(T - xI) = (\lambda - x)^m \cdot Q(x) \quad \text{where } Q(x) = \det(B - xI).$$

Obviously $\deg(Q(x)) = n - m$ and $p_T(x)$ has λ as a root of multiplicity at least m, so (*algebraic multiplicity of* λ) $\geq m = \dim(E_\lambda)$ as claimed. □

It might still be possible for $(x - \lambda)$ to divide $Q(x)$, making the algebraic multiplicity larger than $\dim(E_\lambda)$. A good example is $A = \left(\begin{smallmatrix} \lambda & 1 \\ 0 & \lambda \end{smallmatrix}\right)$. The operator

$L_A : \mathbb{R}^2 \to \mathbb{R}^2$ has $\dim(E_{\lambda=1}) = 1$, but $p_T(\lambda) = (\lambda - x)^2$ so the algebraic multiplicity is 2.

Finding the Eigenspaces of $T : V \to V$. The following example illustrates the complete diagonalization process.

EXAMPLE 5.15. Let $T = L_A : \mathbb{R}^3 \to \mathbb{R}^3$ with

$$A = \begin{pmatrix} 4 & 0 & 1 \\ 2 & 3 & 2 \\ 1 & 0 & 4 \end{pmatrix}.$$

If $\mathfrak{X} = \{\mathbf{e}_1, \mathbf{e}_2, \mathbf{e}_3\}$ is the standard basis in $\mathbb{R}^3$, we have $[T]_{\mathfrak{X}\mathfrak{X}} = [L_A]_{\mathfrak{X}\mathfrak{X}} = A$ as in Exercise 2.43, so

$$\begin{aligned} p_A(x) = \det(A - xI) &= \det \begin{pmatrix} 4-x & 0 & 1 \\ 2 & 3-x & 2 \\ 1 & 0 & 4-x \end{pmatrix} \\ &= [(4-x)(3-x)(4-x) + 0 + 0] - [(3-x) + 0 + 0] \\ &= -x^3 + 11x^2 - 39x + 45. \end{aligned}$$

We are looking for roots of a cubic equation. If you can guess a root α, then long divide by $x - \alpha$ to get $p_T(x) = (x - \alpha) \cdot (\textit{quadratic})$; otherwise you will have to use a numerical root-finding program. Trial and error reveals that $x = 3$ is a root and long division by $(x - 3)$ yields

$$\begin{array}{r|rrrr} & -x^2 & +8x & -15 & \\ \hline x-3) & -x^3 & 11x^2 & -39x & +45 \\ & -x^3 & +3x^2 & & \\ \hline & & 8x^2 & -39x & +45 \\ & & 8x^2 & -24x & \\ \hline & & & -15x & +45 \\ & & & -15x & +45 \\ \hline & & & & 0 \end{array}$$

Then

$$\begin{aligned} -x^3 + 11x^2 - 39x + 45 &= (x-3)(-x^2 + 8x - 15) \\ &= -(x-3)(x-5)(x-3) = -(x-3)^2(x-5)\,, \end{aligned}$$

so $\mathrm{sp}(A) = \{3, 5\}$ with algebraic multiplicities $m_{\lambda=3} = 2$, $m_{\lambda=5} = 1$. To determine the eigenspaces and geometric multiplicities, we must solve systems of equations.

EIGENVALUE $\lambda_1 = 3$: We must solve the matrix equation $(A - 3I)\mathbf{x} = \mathbf{0}$. Row operations on $[A - 3I : \mathbf{0}]$ yield

$$[A - 3I : \mathbf{0}] = \left(\begin{array}{ccc|c} 1 & 0 & 1 & 0 \\ 2 & 0 & 2 & 0 \\ 1 & 0 & 1 & 0 \end{array}\right) \to \left(\begin{array}{ccc|c} 1 & 0 & 1 & 0 \\ 0 & 0 & 0 & 0 \\ 0 & 0 & 0 & 0 \end{array}\right).$$

SOLUTIONS: x_2, x_3 are free variables and $x_1 = -x_3$, so

$$\mathbf{x} = \begin{pmatrix} -x_3 \\ x_2 \\ x_3 \end{pmatrix} \quad \text{and} \quad E_{\lambda=3} = \ker(A - 3I) = \mathbb{R}\begin{pmatrix} 0 \\ 1 \\ 0 \end{pmatrix} + \mathbb{R}\begin{pmatrix} -1 \\ 0 \\ 1 \end{pmatrix}.$$

Thus $\lambda = 3$ has geometric multiplicity $\dim(E_{\lambda=3}) = 2$.

EIGENVALUE $\lambda_2 = 5$: Solve matrix equation $(A - 5I)\mathbf{x} = \mathbf{0}$. Row operations on $[A - 5I : \mathbf{0}]$ yield

$$[A - 5I : \mathbf{0}] = \left(\begin{array}{ccc|c} -1 & 0 & 1 & 0 \\ 2 & -2 & 2 & 0 \\ 1 & 0 & -1 & 0 \end{array}\right) \to \left(\begin{array}{ccc|c} 1 & 0 & -1 & 0 \\ 0 & 1 & -2 & 0 \\ 0 & 0 & 0 & 0 \end{array}\right).$$

SOLUTIONS: x_3 is a free variable; $x_2 = 2x_3$; $x_1 = x_3$. So

$$\mathbf{x} = \begin{pmatrix} x_3 \\ 2x_3 \\ x_3 \end{pmatrix} \quad \text{and} \quad E_{\lambda=5} = \ker(A - 5I) = \mathbb{R} \cdot \begin{pmatrix} 1 \\ 2 \\ 1 \end{pmatrix}.$$

Thus $\lambda = 5$ has geometric multiplicity $\dim(E_{\lambda=5}) = 1$.

We showed earlier that the span $M = \sum_{\lambda \in \mathrm{sp}(T)} E_\lambda(T)$ of the eigenspaces of a linear operator is actually a *direct* sum $M = E_{\lambda_1} \oplus \cdots \oplus E_{\lambda_r}$. In the present situation $M = E_{\lambda=3} \oplus E_{\lambda=5}$ is all of V since the dimensions add up to $\dim(V) = 3$. Taking a basis $\mathfrak{Y} = \{\mathbf{f}_1, \ldots, \mathbf{f}_3\}$ that runs first through $E_{\lambda=3} = \mathbb{R}\mathbf{f}_1 \oplus \mathbb{R}\mathbf{f}_2$, and then through $E_{\lambda=5} = \mathbb{R}\mathbf{f}_3$, we obtain a diagonal matrix

$$[T]_{\mathfrak{Y}\mathfrak{Y}} = \left(\begin{array}{cc|c} 3 & 0 & 0 \\ 0 & 3 & 0 \\ \hline 0 & 0 & 5 \end{array}\right).$$

Once we have found the diagonalizing basis

$$\mathfrak{Y} = \{\mathbf{f}_1 = (0, 1, 0),\ \mathbf{f}_2 = (-1, 0, 1),\ \mathbf{f}_3 = (1, 2, 1)\},$$

we determine an invertible matrix Q such that $QAQ^{-1} = [T]_{\mathfrak{Y}\mathfrak{Y}} = \mathrm{diag}(3, 3, 5)$. To find Q, recall that

$$[T]_{\mathfrak{Y}\mathfrak{Y}} = [\mathrm{id}]_{\mathfrak{Y}\mathfrak{X}} \cdot [T]_{\mathfrak{X}\mathfrak{X}} \cdot [\mathrm{id}]_{\mathfrak{X}\mathfrak{Y}} = [\mathrm{id}]_{\mathfrak{Y}\mathfrak{X}} \cdot A \cdot [\mathrm{id}]_{\mathfrak{X}\mathfrak{Y}} = QAQ^{-1}.$$

Here $[\mathrm{id}]_{\mathfrak{X}\mathfrak{Y}} = Q^{-1}$ and $[\mathrm{id}]_{\mathfrak{Y}\mathfrak{X}} = [\mathrm{id}]_{\mathfrak{X}\mathfrak{Y}}^{-1}$, and by definition $[\mathrm{id}]_{\mathfrak{X}\mathfrak{Y}}$ is the transpose of the coefficient array in the system of vector identities

$$\begin{aligned} \mathbf{f}_1 &= [\mathrm{id}]\,\mathbf{f}_1 = 0 + \mathbf{e}_2 + 0, \\ \mathbf{f}_2 &= [\mathrm{id}]\,\mathbf{f}_2 = -\mathbf{e}_1 + 0 + \mathbf{e}_3, \\ \mathbf{f}_3 &= [\mathrm{id}]\,\mathbf{f}_3 = \mathbf{e}_1 + 2\mathbf{e}_2 + \mathbf{e}_3. \end{aligned}$$

Thus,

$$Q^{-1} = [\mathrm{id}]_{\mathfrak{X}\mathfrak{Y}} = \begin{pmatrix} 0 & -1 & 1 \\ 1 & 0 & 2 \\ 0 & 1 & 1 \end{pmatrix}$$

and $Q = (Q^{-1})^{-1}$ can be found efficiently via row operations:

$$\left(\begin{array}{ccc|ccc} 0 & -1 & 1 & 1 & 0 & 0 \\ 1 & 0 & 2 & 0 & 1 & 0 \\ 0 & 1 & 1 & 0 & 0 & 1 \end{array}\right) \to \left(\begin{array}{ccc|ccc} 1 & 0 & 2 & 0 & 1 & 0 \\ 0 & 1 & 1 & 0 & 0 & 1 \\ 0 & 0 & 1 & \frac{1}{2} & 0 & \frac{1}{2} \end{array}\right) \to \left(\begin{array}{ccc|ccc} 1 & 0 & 0 & -1 & 1 & -1 \\ 0 & 1 & 0 & -\frac{1}{2} & 0 & \frac{1}{2} \\ 0 & 0 & 1 & \frac{1}{2} & 0 & \frac{1}{2} \end{array}\right).$$

Thus

$$Q = \begin{pmatrix} -1 & 1 & -1 \\ -\frac{1}{2} & 0 & \frac{1}{2} \\ \frac{1}{2} & 0 & \frac{1}{2} \end{pmatrix} = \frac{1}{2}\begin{pmatrix} -2 & 2 & -2 \\ -1 & 0 & 1 \\ 1 & 0 & 1 \end{pmatrix}$$

and

$$QAQ^{-1} = \begin{pmatrix} 3 & 0 & 0 \\ 0 & 3 & 0 \\ 0 & 0 & 5 \end{pmatrix}$$

as expected. That completes the "spectral analysis" of A.

The same sort of calculations determine the eigenspaces in $\mathbb{C}^2$ when A in $\mathrm{M}(n, \mathbb{R})$ is regarded as a matrix in $\mathrm{M}(n, \mathbb{C})$.

EXAMPLE 5.16. Diagonalize the operator $L_A : \mathbb{K}^2 \to \mathbb{K}^2$ where

$$A = \begin{pmatrix} 2 & 4 \\ -1 & -2 \end{pmatrix}$$

over $\mathbb{C}$ and over $\mathbb{R}$ (insofar as this is possible).

DISCUSSION: The characteristic polynomial of A (or L_A) is

$$p_A(\lambda) = \det\begin{pmatrix} 2-\lambda & 4 \\ -1 & -2-\lambda \end{pmatrix} = -(2-\lambda)(2+\lambda) + 4 = -4 + \lambda^2 + 4 = \lambda^2.$$

The only root (real or complex) is $\lambda = 0$ so $\mathrm{sp}_{\mathbb{R}}(A) = \mathrm{sp}_{\mathbb{C}}(A) = \{0\}$. Its algebraic multiplicity is 2, but the geometric multiplicity $\dim_{\mathbb{K}}(E_{\lambda=0})$ is equal to 1. The outcome is the same over $\mathbb{C}$ and $\mathbb{R}$.

EIGENVALUE $\lambda = 0$. Here $E_{\lambda=0} = \ker(A)$. Row operations on $[A : \mathbf{0}]$ yield

$$\left(\begin{array}{cc|c} 2-\lambda & 4 & 0 \\ -1 & -2-\lambda & 0 \end{array}\right) \to \left(\begin{array}{cc|c} 2 & 4 & 0 \\ -1 & -2 & 0 \end{array}\right) \to \left(\begin{array}{cc|c} 2 & 4 & 0 \\ 0 & 0 & 0 \end{array}\right).$$

SOLUTIONS: In solving $(A-\lambda I)\mathbf{x} = A\mathbf{x} = 0$, x_2 is a free variable and $x_1 = -2x_2$, so

$$\mathbf{x} = \begin{pmatrix} -2x_2 \\ x_2 \end{pmatrix} \quad \text{and} \quad E_{\lambda=0} = \mathbb{K} \cdot \begin{pmatrix} -2 \\ 1 \end{pmatrix}.$$

Since there are no other eigenvalues, the best we can do in trying to find a simple matrix description $[T]_{\mathfrak{YY}}$ is to take a basis $\mathfrak{Y} = \{\mathbf{f}_1, \mathbf{f}_2\}$ that passes first through $E_{\lambda=0}$: let $\mathbf{f}_1 = (-2, 1)$ and then include one more vector $\mathbf{f}_2 \notin \mathbb{K}\mathbf{f}_1$ to make a basis. We have

$$[T]_{\mathfrak{XX}} = \begin{pmatrix} 2 & 4 \\ -1 & -2 \end{pmatrix}$$

with respect to the standard basis $\mathfrak{X} = \{\mathbf{e}_1, \mathbf{e}_2\}$ in $\mathbb{K}^2$ (recall Exercise 2.43). With respect to the basis $\mathfrak{Y} = \{\mathbf{f}_1, \mathbf{f}_2\}$, the matrix has block upper-triangular form,

$$[T]_{\mathfrak{Y}\mathfrak{Y}} = \begin{pmatrix} 0 & * \\ 0 & * \end{pmatrix}.$$

But this operator cannot be diagonalized by any choice of basis.

EXERCISE 5.17. In Example 5.16 we showed there is a basis $\mathfrak{Y} = \{\mathbf{f}_1, \mathbf{f}_2\}$ such that

$$A = [T]_{\mathfrak{Y}\mathfrak{Y}} = \begin{pmatrix} 0 & a \\ 0 & b \end{pmatrix}.$$

(a) Prove that b must be 0, so

$$A = \begin{pmatrix} 0 & a \\ 0 & 0 \end{pmatrix}.$$

(b) Explain how to modify the basis $\mathfrak{Y}$ to get a new basis $\mathfrak{Z}$ such that

$$[T]_{\mathfrak{Z}\mathfrak{Z}} = \begin{pmatrix} 0 & 1 \\ 0 & 0 \end{pmatrix}.$$

Example: Rotation Matrices in $\mathbb{R}^2$.

EXAMPLE 5.18. The matrix

$$A = \begin{pmatrix} \cos(\theta) & -\sin(\theta) \\ \sin(\theta) & \cos(\theta) \end{pmatrix} \quad (\theta \text{ real})$$

yields an operator $L_A : \mathbb{R}^2 \to \mathbb{R}^2$ that you will recognize as a rotation counter clockwise about the origin by θ radians. Describe its eigenspaces over $\mathbb{R}$ and over $\mathbb{C}$.

SOLUTION: Over either field we have

$$\begin{aligned} p_A(\lambda) &= \det(A - \lambda I) = \det \begin{pmatrix} \cos(\theta) - \lambda & -\sin(\theta) \\ \sin(\theta) & \cos(\theta) - \lambda \end{pmatrix} \\ &= (\cos(\theta) - \lambda)^2 + \sin^2(\theta) = \cos^2(\theta) + \sin^2(\theta) - 2\lambda\cos(\theta) + \lambda^2 \\ &= \lambda^2 - 2\lambda\cos(\theta) + 1. \end{aligned}$$

This is zero only when

$$\begin{aligned} \lambda &= \frac{2\cos(\theta) \pm \sqrt{4\cos^2(\theta) - 4}}{2} = \cos(\theta) \pm \sqrt{\cos^2(\theta) - 1} \\ &= \cos(\theta) \pm i\sqrt{1 - \cos^2(\theta)} = \cos(\theta) \pm i\sin(\theta) = e^{\pm i\theta}. \end{aligned}$$

The roots are nonreal (hence a conjugate pair as shown earlier in Figure 5.1), and they lie on the unit circle in $\mathbb{C}$ because $|e^{\pm i\theta}|^2 = \sin^2(\theta) + \cos^2(\theta) = 1$ for all θ. When $\theta = 0$ or π we have $\lambda = \pm 1 + i0$ (real), and in this case $A = I$ or $-I$. In all other cases A has no real eigenvalues at all, but it can be diagonalized as

$$[L_A]_{\mathfrak{Y}\mathfrak{Y}} = \begin{pmatrix} e^{i\theta} & 0 \\ 0 & e^{-i\theta} \end{pmatrix}$$

for a suitably chosen complex basis $\mathfrak{Y} = \{\mathbf{f}_1, \mathbf{f}_2\}$ in $\mathbb{C}^2$. To find $\mathfrak{Y}$ we need to determine the eigenspaces of $L_A : \mathbb{C}^2 \to \mathbb{C}^2$.

EIGENVALUE $\lambda_1 = e^{i\theta} = \cos(\theta) + i\sin(\theta)$ with $\theta \neq n\pi$.

$$[A - \lambda_1 I] = \begin{pmatrix} \cos(\theta) - e^{i\theta} & -\sin(\theta) \\ \sin(\theta) & \cos(\theta) - e^{i\theta} \end{pmatrix} \begin{pmatrix} -i\sin(\theta) & -\sin(\theta) \\ \sin(\theta) & -i\sin(\theta) \end{pmatrix}$$

$$= \sin(\theta) \cdot \begin{pmatrix} -i & -1 \\ 1 & -i \end{pmatrix}$$

Now, $(A - \lambda_1 I)\mathbf{x} = \mathbf{0} \Leftrightarrow B\mathbf{x} = \mathbf{0}$ where $B = \begin{pmatrix} -i & -1 \\ 1 & -i \end{pmatrix}$. Row operations yield

$$\left(\begin{array}{cc|c} -i & -1 & 0 \\ 1 & -i & 0 \end{array}\right) \to \left(\begin{array}{cc|c} 1 & -i & 0 \\ 1 & -i & 0 \end{array}\right) \to \left(\begin{array}{cc|c} 1 & -i & 0 \\ 0 & 0 & 0 \end{array}\right).$$

SOLUTIONS: Here x_2 is a free variable and $x_1 = ix_2$, so

$$\mathbf{x} = \begin{pmatrix} ix_2 \\ x_2 \end{pmatrix} \quad \text{and} \quad E_{\lambda_1 = e^{i\theta}} = \mathbb{C} \cdot \begin{pmatrix} i \\ 1 \end{pmatrix}.$$

For λ_1, *(algebraic multiplicity)* = *(geometric multiplicity)* = 1.

The discussion for the conjugate eigenvalue $\lambda_2 = e^{-i\theta} = \cos(\theta) - i\sin(\theta)$ is almost the same, with the final result that $E_{\lambda = e^{-i\theta}} = \mathbb{C} \cdot \mathrm{col}(-i, 1)$. Combining these observations we get

$$\mathbb{C}^2 = E_{\lambda = e^{i\theta}} \oplus E_{\lambda = e^{-i\theta}} = \mathbb{C} \cdot \begin{pmatrix} i \\ 1 \end{pmatrix} \oplus \mathbb{C} \cdot \begin{pmatrix} -i \\ 1 \end{pmatrix} = \mathbb{C}\mathbf{f}_1 \oplus \mathbb{C}\mathbf{f}_2.$$

Thus, with respect to the basis

$$\mathfrak{Y} = \{\mathbf{f}_1 = \mathrm{col}(i, 1), \mathbf{f}_2 = \mathrm{col}(-i, 1)\}$$

we have

$$[L_A]_{\mathfrak{Y}\mathfrak{Y}} = \begin{pmatrix} e^{i\theta} & 0 \\ 0 & e^{-i\theta} \end{pmatrix}.$$

As mentioned earlier, if $T : V \to V$ is a linear operator the span $M = \sum_{\lambda \in \mathrm{sp}_{\mathbb{K}}(A)} E_\lambda$ is a direct sum $E_{\lambda_1}(T) \oplus \cdots \oplus E_{\lambda_r}(T)$, and a suitably chosen basis partially diagonalizes T, yielding the matrix

$$[T]_{\mathfrak{Y}\mathfrak{Y}} = \left(\begin{array}{ccc|c} \boxed{\lambda_1 \cdot I_{d_1 \times d_1}} & & 0 & \\ & \ddots & & * \\ & & \boxed{\lambda_r \cdot I_{d_r \times d_r}} & \\ \hline 0 & & 0 & B \end{array}\right).$$

To proceed further and determine the structure of the lower right-hand block B we would have to develop the theory of nilpotent operators, generalized eigenspaces, and the Jordan decomposition of a linear operator over $\mathbb{C}$. We must defer that for the time being. However, the following observation can be useful.

EXERCISE 5.19. Let $L = D^2 = d^2/dx^2$ acting on the space of complex-coefficient polynomials $\mathcal{P}_n \subseteq \mathbb{C}[x]$, let $\mathfrak{X} = \{e_0, \dots, e_n\}$, with $e_0 = 1$, $e_1 = x, \dots,$ $e_n = x^n$, be a basis for this $(n+1)$-dimensional space.

(a) For arbitrary $z \in \mathbb{C}$ determine the matrix

$$[L - zI]_{\mathfrak{X}\mathfrak{X}} = -\left[(D^2 - zI)|_{\mathcal{P}_n}\right]_{\mathfrak{X}\mathfrak{X}}$$

with respect to $\mathfrak{X}$ and find all complex eigenvalues of L.

(b) Determine all nontrivial eigenspaces $E_\lambda(L)$ by finding a basis in each.

(c) When $n = 2$ and $\lambda \neq 0$, find an explicit matrix inverse for the 3×3 matrix $[L - \lambda I]_{\mathfrak{X}\mathfrak{X}}$.

HINT: The answer in (c) will of course involve the undetermined parameter $\lambda \in \mathbb{C}$, so you will have to perform symbolic row operations on the augmented matrix $[A : I]$, where $A = [L - \lambda I]_{\mathfrak{X}\mathfrak{X}}$, to find $A^{-1} = [L - \lambda I]^{-1}_{\mathfrak{X}\mathfrak{X}}$.

The Case of Distinct Eigenvalues. When a linear operator $T : V \to V$ on an n-dimensional space has n *distinct eigenvalues*, so its characteristic polynomial is a product of linear factors $p_T(x) = c \cdot \prod_{k=1}^{n}(x - \lambda_k)$, then the diagonalization problem has an easy resolution, and likewise for matrices in $\mathrm{M}(n, \mathbb{K})$ over any field.

PROPOSITION 5.20. *If* $\dim_{\mathbb{K}}(V) = n$ *and* $T : V \to V$ *has* n *distinct eigenvalues in* $\mathbb{K}$, *then* T *is diagonalizable and* V *is the direct sum* $\bigoplus_{i=1}^{n} E_{\lambda_i}(T)$ *of 1-dimensional eigenspaces.*

PROOF. Since the linear span $E = \sum_{\lambda \in \mathrm{sp}(T)} E_{\lambda_i}$ of the eigenspaces is always a direct sum (see Proposition 2.29), and each λ_i has $\dim(E_{\lambda_i}) \geq 1$, the dimension of this linear span must equal n, so $V = E = \bigoplus_{\lambda_i \in \mathrm{sp}(T)} E_{\lambda_i}$. □

In some sense (at least for complex matrices), the "n distinct eigenvalues condition" is *generic*: If entries $a_{ij} \in \mathbb{C}$ are chosen at random, then with probability 1 the matrix $A = [a_{ij}]$ would have distinct eigenvalues in $\mathbb{C}$, and the characteristic polynomial would split completely into distinct linear factors

$$p_A(x) = c \cdot \prod_{i=1}^{n}(x - \lambda_i).$$

Unfortunately, many matrices important in applications do not have n distinct eigenvalues, which is why we need the more subtle theory of "generalized eigenvalues" developed in Chapter 1 of Volume II when diagonalization fails.

EXERCISE 5.21. What happens to the spectrum $\mathrm{sp}(T)$ when you replace

(a) $T \to cT$ (b) $T \to cT + I$ (c) $T \to T + cI$

with $c \neq 0$?

5.3. Diagonalization and Limits of Operators

Here we will be concerned with limits of sequences $\{v_n : n \in \mathbb{N}\}$ in various vector spaces, particularly matrix space $\mathrm{M}(n, \mathbb{K})$. To discuss such limits we must restrict attention to spaces over $\mathbb{K} = \mathbb{R}$ or $\mathbb{C}$, and we need to equip the vector space V with additional geometric structure by imposing a *norm* $\|\ast\|$ on V that assigns a length to every vector. The distance from v to w can then be defined as $\mathrm{d}(v, w) = \|w - v\|$, and this is what we need to define the limit of a sequence of vectors in V by declaring that

$$v = \lim_{n\to\infty} v_n \quad \text{if and only if} \quad \|v_n - v\| \to 0 \text{ as } n \to \infty.$$

Norms on Finite-Dimensional Spaces.

DEFINITION 5.22. A *norm* on a vector space V over $\mathbb{K} = \mathbb{R}$ or $\mathbb{C}$ is an assignment $v \mapsto \|v\|$ with the following properties:

$$\text{POSITIVITY:} \quad \|v\| \geq 0 \text{ and } \|v\| = 0 \Leftrightarrow v = \mathbf{0}$$

$$\text{SCALING:} \quad \|cv\| = |c| \cdot \|v\| \text{ for all scalars } c \in \mathbb{K} \tag{5.3}$$

$$\text{TRIANGLE INEQUALITY:} \quad \|v \pm w\| \leq \|v\| + \|w\|$$

Then the "distance function" $d(v, w) = \|w - v\|$ satisfies the *geometric triangle inequality*

$$d(v, w) \leq d(v, z) + d(z, w) \quad \text{for } v, w, z \in V,$$

because by the scaling and triangle inequality properties in (5.3), we have

$$\|v - w\| = \|(v - z) - (z - w)\| \leq \|v - z\| + \|z - w\|.$$

The triangle inequality has a converse that is sometimes indispensable.

LEMMA 5.23 (Reverse Triangle Inequality). *For vectors v, w in a normed vector space V with norm $\|\ast\|$ we have*

$$\big|\|v\| - \|w\|\big| \leq \|v - w\|.$$

PROOF. By the Triangle Inequality

$$\|v + w\| \leq \|v\| + \|w\|.$$

Thus

$$\|v\| = \|v - w + w\| \leq \|v - w\| + \|w\|,$$

so that $\|v\| - \|w\| \leq \|v - w\|$. Reversing the roles of v, w, we also get $\|w\| - \|v\| \leq \|v - w\|$. Since the absolute value of a real number c is either $|c| = c$ or $|c| = -c$, we conclude that

$$\big|\|v\| - \|w\|\big| \leq \|v - w\|$$

□

Examples that concern us include the following

1. NORMS ON COORDINATE SPACE. *Here $V = \mathbb{K}^n$ and the norm of $\mathbf{z} = (z_1, \dots, z_n)$ is*

$$L^2\text{-NORM:} \quad \|\mathbf{z}\|_2 = \left(\textstyle\sum_{k=1}^{n} |z_k|^2\right)^{1/2} \quad \text{(the euclidean norm)}$$

These spaces support several other norms, each with its own uses in applications:

L^∞-NORM: $\|\mathbf{z}\|_\infty = \max\{|z_1|, \dots, |z_n|\}$
L^1-NORM: $\|\mathbf{z}\|_1 = |z_1| + \cdots + |z_n|$

Each measures the discrepancy between $\mathbf{z}$ and $\mathbf{w}$ in a different way. For the L^∞-norm, we take the distance $\|\mathbf{z}-\mathbf{w}\|_\infty$ to be the largest discrepancy between entries in $\mathbf{z}-\mathbf{w}$, while the L^1-distance $\|\mathbf{z}-\mathbf{w}\|_1$ is the sum of the differences $|z_k - w_k|$ between entries. The L^2-norm, $\|\mathbf{z}-\mathbf{w}\|$ is just the euclidean distance between vectors

$$\|\mathbf{z}-\mathbf{w}\|_2 = \left(\sum_{k=1}^{n} |z_k - w_k|^2\right)^{1/2},$$

which is known in engineering circles as the "RMS" or "root-mean-square" discrepancy between the entries of the vectors.

2. NORMS ON MATRIX SPACES. *Here* $V = \mathrm{M}(n \times m, \mathbb{K})$ *or* $\mathrm{M}(n, \mathbb{K})$ *and, again, various norms occur in applications. We will work mostly with the following:*

L^∞-NORM: $\|A\|_\infty = \max\{|A_{ij}| : 1 \le i \le n,\ 1 \le j \le m\}$
L^1-NORM: $\|A\|_1 = \sum\{|A_{ij}| : 1 \le i \le n,\ 1 \le j \le m\}$,

which are just variants of the L^∞- and L^1-norms for vectors. (They become the L^∞- and L^1-norms on vectors if we list the entries of A as a vector

$$\mathbf{z} = \{A_{11}, \dots, A_{1m}; \dots; A_{n1}, \dots, A_{nm}\}$$

with mn entries.)

The L^2-norm on $A \in \mathrm{M}(n, \mathbb{K})$ is referred to as the Hilbert-Schmidt norm of A. It corresponds to the L^2-norm on euclidean spaces,

HILBERT-SCHMIDT NORM ON MATRICES. $\|A\|_{\mathrm{HS}} = \|A\|_2 = \left(\sum_{i,j=1}^{n} |A_{ij}|^2\right)^{1/2}.$

These matrix norms all have the properties in (5.3).

3. NORMS ON LINEAR OPERATORS. *We have seen in Chapter 2 that once bases $\mathfrak{X}$ and $\mathfrak{Y}$ have been specified in V and W, the space of linear operators* $\mathrm{Hom}_{\mathbb{K}}(V, W)$ *can be identified with matrix space* $\mathrm{M}(n \times m, \mathbb{K})$ *via the isomorphism*

$$\Phi : \mathrm{Hom}_{\mathbb{K}}(V, W) \to \mathrm{M}(n \times m, \mathbb{K}) \quad \text{with } \Phi(T) = [T]_{\mathfrak{Y}\mathfrak{X}}.$$

Consequently norms on $\mathrm{M}(n \times m, \mathbb{K})$ *can be transferred to norms on operators, and vice versa. Thus we may speak of $L^1, \dots, L^\infty$-norms of linear operators, once bases have been specified in V and W.*

Multiplicative Properties of Norms on Matrix Space. When $m = n$ (the situation that most concerns us), the spaces $\mathrm{M}(n, \mathbb{K})$ and $\mathrm{Hom}_{\mathbb{K}}(V, V)$ are both associative algebras with identities $I = \mathrm{id}_V$ and $I_{n\times n}$, respectively, and the isomorphism Φ preserves products as well as linear combinations, with

$$\Phi(S \circ T) = \Phi(S)\cdot\Phi(T) \text{ (matrix product)} \quad \text{and} \quad \Phi(I) = I_{n\times n}.$$

All the matrix and operator norms introduced above have a multiplicative property in addition to those already listed in (5.3) and Lemma 5.23.

PROPERTY 5.24 (Multiplicative Property). $\|AB\| \le C\|A\|\cdot\|B\|$ *and* $\|A^k\| \le C^{k-1}\|A\|^k$ *for some* $C > 0$.

EXERCISE 5.25. Prove the following norm inequalities for $A, B \in \mathrm{M}(n, \mathbb{K})$.

(a) $\|AB\|_\infty \le n\|A\|_\infty\cdot\|B\|_\infty$ with $\|I\|_\infty = 1$.
(b) $\|AB\|_{\mathrm{HS}} \le \|A\|_{\mathrm{HS}}\cdot\|B\|_{\mathrm{HS}}$ with $\|I\|_{\mathrm{HS}} = \sqrt{n}$.
(c) $\|AB\|_1 \le n\|A\|_1\cdot\|B\|_1$with $\|I\|_1 = n$.

Use induction on $k = 1, 2, \ldots$ to prove the following important "geometric growth" condition on powers A^k:

(d) If

$$\|A\cdot B\| \le C\cdot\|A\|\cdot\|B\| \quad \text{for all } A, B \in \mathrm{m}(n, \mathbb{K}),$$

the powers A^k have *at most geometric growth* as $k \to \infty$:

$$\|A^k\| \le C^{k-1}\,\|A\|^k \quad \text{for } k = 0, 1, 2, \ldots.$$

NOTE: For the L^∞-norm we have $C = N$ for $N \times N$ matrices.

HINTS: In discussing the L^2-norm (aka Hilbert-Schmidt) in (b) you will want to recall the classical *Cauchy-Schwarz inequality*, which says:

$$\text{For } \mathbf{z}, \mathbf{w} \in \mathbb{C}^n \text{ we have } \left|\sum_{k=1}^n z_k w_k\right| \le \left(\sum_{k=1}^n |z_k|^2\right)^{1/2} \cdot \left(\sum_{k=1}^n |w_k|^2\right)^{1/2}$$

Definitions of these operator and matrix norms may seem rather arbitrary (except perhaps for the euclidean L^2-norms), but we will often consider operators $T : V \to V$ acting on *normed vector spaces*, and when we do there is a natural *operator norm* $\|T\|_{\mathrm{op}}$ that has nice properties and a clear geometric interpretation.

The Operator Norm $\|T\|_{\mathrm{op}}$ on Linear Operators and Matrices.

DEFINITION 5.26 (Operator Norm). Let $T : V \to V$ be a linear operator on a finite-dimensional vector space over $\mathbb{K} = \mathbb{R}$ or $\mathbb{C}$. If $\|*\|$ is a norm on V we define a corresponding *operator norm* on linear operators in $\mathrm{Hom}(V, V)$,

$$\text{(5.4)} \qquad \text{OPERATOR NORM:} \quad \|T\|_{\mathrm{op}} = \max\left\{\frac{\|Tv\|}{\|v\|} : v \ne 0 \text{ in } V\right\} = \max\{\|Tv\| : \|v\| = 1 \text{ in } V\}.$$

Thus $\|T(v)\| \le \|T\|_{\mathrm{op}}\cdot\|v\|$ and $k = \|T\|_{\mathrm{op}}$ is the smallest constant such that $\|T(v)\| \le k\,\|v\|$ for all $v \in V$.

The second equality in Definition 5.26 follows from the first via the scaling law for norms (5.3). The geometric interpretation of $\|*\|_{\mathrm{op}}$ is:

> *The operator norm $\|T\|_{\mathrm{op}}$ is the largest length $\|T(v)\|$ of any image vector $T(v)$ as v runs through the "unit sphere" $\Sigma_1 = \{x \in V : \|x\| = 1\} \subseteq V$.*

One advantage of this norm is that the multiplicative property takes the simplest possible form,

$$\|ST\|_{\mathrm{op}} \leq \|S\|_{\mathrm{op}} \cdot \|T\|_{\mathrm{op}} \quad \text{and} \quad \|I\|_{\mathrm{op}} = 1, \tag{5.5}$$

which involves no extraneous constant C.

EXERCISE 5.27. For linear operators $T : V \to V$ on a normed vector space V, verify the following:

(a) The operator norm $\|T\|_{\mathrm{op}}$ defined above has the properties (5.3) required of a vector space norm on $\mathrm{Hom}_{\mathbb{K}}(V,V)$, and that the equalities in (5.4) are valid.

(b) $\|ST\|_{\mathrm{op}} \leq \|S\|_{\mathrm{op}} \cdot \|T\|_{\mathrm{op}}$ for $S, T \in \mathrm{Hom}_{\mathbb{K}}(V,V)$.

The operator norm $\| * \|_{\mathrm{op}}$ is widely used in theoretical discussions of norm convergence because of its nice properties, but as we show in Theorem 5.30 below it is equivalent to any other norm on $\mathrm{Hom}_{\mathbb{K}}(V,V)$ when $\dim(V) < \infty$. On the other hand the numerical value of $\|T\|_{\mathrm{op}}$ is more difficult to compute than $\| * \|_1, \dots, \| * \|_\infty$. (There is no free lunch.)

EXAMPLE 5.28. Compute the operator norm $\|T\|_{\mathrm{op}}$ when $T = L_A : \mathbb{R}^n \to \mathbb{R}^n$ for some $A = [a_{ij}] \in \mathrm{M}(n, \mathbb{R})$ and $\mathbb{R}^n$ is equipped with the euclidean norm $\|\mathbf{x}\|_2 = (\sum_{i=1}^n |x_i|^2)^{1/2}$.

DISCUSSION: the operator norm is a constrained optimization problem:

$$\|T\|_{\mathrm{op}} = \max\Big\{\Big(\sum_{i=1}^n |T(\mathbf{x})_i|^2\Big)^{1/2} : \|\mathbf{x}\|_2 = 1\Big\}.$$

Since it suffices to maximize $\|T\|_{\mathrm{op}}^2$, let us square both sides to get

$$\begin{aligned}\|T\|_{\mathrm{op}}^2 &= \max\Big\{\sum_{i=1}^n |T(\mathbf{x})_i|^2 : \|\mathbf{x}\|_2 = 1\Big\} \\ &= \max\Big\{\sum_{i=1}^n \Big|\sum_{j=1}^n a_{ij}x_j\Big|^2 : \|\mathbf{x}\|_2 = 1\Big\} \\ &= \Big\{\sum_{i=1}^n |\mathrm{Row}_i(A) \bullet \mathbf{x}|^2 : \sum_{k=1}^n |x_k|^2 = 1\Big\}\end{aligned}$$

where $\mathbf{x} \bullet \mathbf{y} = \sum_{k=1}^n x_k y_k$ on $\mathbb{R}^n$ is the usual dot product on $\mathbb{R}^n$. The sum of squared dot products in the last line is a quadratic function of the variables $x_1, \dots, x_n$ and the optimization problem reduces to finding its maximum on a smooth $(n-1)$-dimensional hypersurface (the unit sphere) in $\mathbb{R}^n$, the locus where $x_1^2 + \cdots + x_n^2 = 1$. In principle this could be accomplished by standard calculus methods (Lagrange multipliers), but the calculations could get ugly.

Fortunately, if our intent is to show that some limit $\lim_{k\to\infty} T_k$ involving operators $T_k : \mathbb{R}^n \to \mathbb{R}^n$ exists, we only care about the the general properties of the operator norm and not its particular numerical values.

Equivalence of Norms on Finite-Dimensional Spaces. If $\| * \|_p$ and $\| * \|_q$ are norms on the same vector space V, we say they are *equivalent norms* if there are constants $C_p, C_q > 0$ such that

$$\|v\|_p \le C_p \|v\|_q \quad \text{and} \quad \|v\|_q \le C_q \|v\|_p \quad \text{for all } v \in V. \tag{5.6}$$

This is obviously an RST equivalence relation between norms on V.

EXERCISE 5.29. Prove that the norms $\|*\|_1, \|*\|_2, \|*\|_\infty$ on $\mathbb{K}^n$ are pairwise equivalent as in (5.6) by exhibiting explicit constants $k_i > 0$ such that

$$\|\mathbf{x}\|_1 \le k_1 \|\mathbf{x}\|_2, \quad \|\mathbf{x}\|_2 \le k_2 \|\mathbf{x}\|_\infty, \quad \|\mathbf{x}\|_\infty \le k_3 \|\mathbf{x}\|_1.$$

In the next section we will discuss limits $v = \lim_{n\to\infty} v_n$ of sequences in normed vector spaces. Although limits are defined in terms of a norm on V, we will see that equivalent norms on V determine the same notions of sequence convergence and limits: we have $v_n \to v$ as $n \to \infty$ with respect to one norm if and only if it converges (and to the same limit) with respect to the other. With this in mind we close this section by proving:

THEOREM 5.30 (Equivalence of Norms on Finite-Dimensional V). *Any two vector space norms p_1 and p_2 on a finite-dimensional vector space over $\mathbb{K} = \mathbb{R}$ or $\mathbb{C}$ are equivalent as in* (5.6)*, and determine the same notions of limit and sequence convergence.*

PROOF. Since V is finite dimensional we may identify it with $\mathbb{K}^n$ by imposing a basis in V. Transferring the norms on V over to $\mathbb{K}^n$ we may hereafter assume $V = \mathbb{K}^n$. The unit sphere $\Sigma_1 = \{\mathbf{x} \in \mathbb{K}^n : \|\mathbf{x}\|_\infty = 1\}$ is a closed, bounded set in $\mathbb{K}^n$, and by the Bolzano-Weierstrass theorem it is a compact set in $\mathbb{K}^n$. It follows that any continuous function $f : \Sigma_1 \to \mathbb{R}$ is bounded from above and below

$$m \le f(\mathbf{x}) \le M \quad \text{when } \|\mathbf{x}\|_\infty = 1,$$

and achieves both its maximum and minimum values $M \ge m$ on Σ_1,

Now let $p(\mathbf{x})$ be an arbitrary norm on $\mathbb{K}^n$. We claim that the function $f : \Sigma_1 \to \mathbb{R}$ given by $f(\mathbf{x}) = p(\mathbf{x})$ is continuous with respect to the $\| * \|_\infty$ norm on $\mathbb{K}^n$. First observe that if $\mathfrak{X} = \{\mathbf{e}_i\}$ is the standard basis in $\mathbb{K}^n$, we may write any vector as $\mathbf{x} = \sum_{i=1}^n c_i \mathbf{e}_i$, and then by the triangle inequality

$$p(\mathbf{x}) \le \sum_{i=1}^n |x_i|\, p(\mathbf{e}_i) \le M_0 \cdot \|\mathbf{x}\|_1 \le nM_0 \cdot \|\mathbf{x}\|_\infty$$

where $M_0 = \max\{p(\mathbf{e}_1), \dots, p(\mathbf{e}_n)\}$. By the reverse triangle inequality we get

$$|p(\mathbf{x}) - p(\mathbf{y})| \le p(\mathbf{x} - \mathbf{y}) \le nM_0 \cdot \|\mathbf{x} - \mathbf{y}\|_\infty,$$

which means that $f(\mathbf{x}) = p(\mathbf{x})$ is a continuous function when Σ_1 is equipped with the distance function $d(\mathbf{x}, \mathbf{y}) = \|\mathbf{x}-\mathbf{y}\|_\infty$, and it has strictly positive values because $p(\mathbf{x}) > 0$ unless $\mathbf{x} = \mathbf{0}$.

Because Σ_1 is compact $p(\mathbf{x})$ achieves well-defined maxima and minima on Σ_1, with

$$0 < m = \min\{p(\mathbf{x}) : \|\mathbf{x}\|_\infty = 1\} \leq p(\mathbf{x}) \leq M = \max\{p(\mathbf{x}) : \|\mathbf{x}\|_\infty = 1\}.$$

For arbitrary $\mathbf{x}$ this implies

$$m \cdot \|\mathbf{x}\|_\infty \leq p(\mathbf{x}) \leq p\left(\|\mathbf{x}\|_\infty \cdot \frac{\mathbf{x}}{\|\mathbf{x}\|_\infty}\right) \leq M \cdot \|\mathbf{x}\|_\infty.$$

This suffices to show that an arbitrary norm $p(\mathbf{x})$ on $\mathbb{K}^n$ ($\mathbb{K} = \mathbb{R}$ or $\mathbb{C}$) is equivalent to the L^∞-norm $\|\mathbf{x}\|_\infty$ □

As an immediate consequence of this theorem and the definition of "equivalent norms," all vector space norms on a finite-dimensional vector space V make it a complete normed space, in which convergence can be verified by checking that the sequence has the Cauchy property.

EXERCISE 5.31. Use Theorem 5.30 to prove that any vector space norm $\|*\|$ on matrix space $\mathrm{M}(n, \mathbb{C})$ has the following multiplicative properties:

> There is a constant $C > 0$, which may depend on the particular norm $\|*\|$ such that
> 1. $\|A \cdot B\| \leq C\,\|A\| \cdot \|B\|$ for all $A, B \in \mathrm{M}(n, \mathbb{C})$.
> 2. $\|A^k\| \leq C^{k-1} \cdot \|A\|^k$ for all $k \in \mathbb{N}$.

HINT: We know that $\|AB\|_2 \leq \|A\|_2\|B\|_2$ (Exercise 5.25), and that every norm on $\mathrm{M}(n, \mathbb{C})$ is equivalent to the L^2-norm.

Practical Calculations with Matrix Norms. We can define an operator norm on matrix space, letting $\|A\|_{\mathrm{op}} = \|L_A\|_{\mathrm{op}}$ for the multiplication operator $L_A(\mathbf{z}) = A \cdot \mathbf{z}$, acting on $\mathbb{C}^n$ equipped with the euclidean L^2-norm. This norm on $\mathrm{M}(n, \mathbb{C})$ has properties (5.5) that streamline calculations,

$$\|I\| = 1 \quad \text{and} \quad \|AB\| \leq \|A\| \cdot \|B\| \Rightarrow \|A^k\| \leq \|A\|^k$$

for all $k \in \mathbb{N}$. And since all norms on matrix space are equivalent we can make life easier by taking $\|*\| = \|*\|_{\mathrm{op}}$ in general discussions of norm convergence in matrix space. However, when numerical calculations are required, it is advisable to work with one of the L^p-norms, $p = 1, 2, \infty$, whose numerical values are easier to compute.

Convergence of Sequences and Series in Matrix Space. We begin by defining limits of square matrices over $\mathbb{K} = \mathbb{R}$ or $\mathbb{C}$. The following discussion actually applies to all normed vector spaces, but we will focus on convergence of sequences of matrices in the operator norm defined above. We could just as well work with any of the other norms discussed previously in this section since they all yield the same notion of limits in matrix space. If V is finite-dimensional, limits $T_n \to T$ could similarly be defined in the space of linear operators $\mathrm{Hom}_{\mathbb{K}}(V, V)$.

DEFINITION 5.32. For $\mathbb{K} = \mathbb{R}$ or $\mathbb{C}$ we define *simple convergence* of matrices in $\mathrm{M}(N, \mathbb{K})$, denoted by writing

$$\lim_{n\to\infty} A_n = A \quad \text{or} \quad A_n \to A \text{ as } n \to \infty$$

to mean that as $n \to \infty$ each entry $a_{ij}^{(n)}$ in A_n converges in $\mathbb{C}$ to the corresponding entry in the limit matrix A:

$$|a_{ij}^{(n)} - a_{ij}| \to 0 \quad \text{in } \mathbb{C} \text{ as } n \to \infty \tag{5.7}$$

for each $1 \le i, j \le N$. Since there are only finitely many entries in these matrices we obviously have simple convergence $A_n \to A$ if and only if $A_n \to A$ in the L^∞-norm, so that $\|A_n - A\|_\infty \to 0$ as $n \to \infty$.

Properties of Matrix Norms. The basic properties of matrix norms are summarized below.

EXERCISE 5.33. If $\| * \|$ is any vector space norm and A, $B \in \mathrm{M}(N, \mathbb{K})$, explain why:

(a) $\|\lambda A\| = |\lambda| \cdot \|A\|$ for all $\lambda \in \mathbb{K}$
(b) TRIANGLE INEQUALITY: $\|A + B\| \le \|A\| + \|B\|$
(c) REVERSE TRIANGLE INEQUALITY: $|\,\|A\| - \|B\|\,| \le \|A - B\|$
(d) MULTIPLICATIVE PROPERTY: $\|AB\| \le C\|A\| \cdot \|B\|$ for some $C > 0$
(e) GEOMETRIC GROWTH OF POWERS: $\|A^k\| \le C^{k-1}\|A\|^k$ for $k = 0, 1, \ldots$

This last condition will be crucial in proving that various matrix-valued power series are convergent.

As an immediate consequence of the reverse triangle inequality we have the following:

COROLLARY 5.34. *If $A_n \to A$ in $\mathrm{M}(n, \mathbb{C})$, then $\|A_n\| \to \|A\|$ in $\mathbb{R}$.*

EXERCISE 5.35. If A in $\mathrm{M}(n, \mathbb{C})$ is an *invertible* matrix and $A_n \to A$ (norm convergent), prove the following:

(a) $\det(A_n) \to \det(A) \neq 0$
(b) $A_n^{-1} \to A^{-1}$ as $n \to \infty$

HINT: For (b), use the fact that norm convergence is the same as simple convergence. Recall definition (5.7) and Cramer's Rule (Theorem 4.50) for computing the inverse A^{-1} of a nonsingular matrix A.

Various theorems regarding norm convergence in $\mathrm{M}(N, \mathbb{C})$ follow from the basic properties of normed vector spaces and Exercises 5.33 and 5.35.

EXERCISE 5.36. If $A_n \to A$ and $B_n \to B$ (norm convergent) and $\lambda_n \to \lambda$ in $\mathbb{C}$, prove the following:

(a) $A_n + B_n \to A + B$.

(b) $A_nB \to AB$ and $AB_n \to AB$.
(c) $A_nB_n \to AB$. Thus matrix multiplication is a "jointly continuous" operation on its two inputs.
(d) If Q is an invertible matrix, then $QA_nQ^{-1} \to QAQ^{-1}$. Hence every similarity transformation $A \mapsto QAQ^{-1}$ is a continuous operation on matrix space.
(e) $\lambda_n A_n \to \lambda A$.

HINT: In (c) add and subtract A_nB, then apply the triangle inequality.

5.4. Application: Computing the Exponential e^A of a Matrix

Working with the L^∞-norm we will show that the exponential series

$$e^A = \sum_{k=0}^{\infty} \frac{1}{k!} A^k \quad \text{for } A \in \mathrm{M}(N, \mathbb{C})$$

converges in the L^∞-norm, which means that its finite partial sums

$$S_n = I + A + \frac{A^2}{2!} + \cdots + \frac{A^n}{n!} \quad (n \in \mathbb{N})$$

converge to a definite limit e^A in matrix space::

$$\|S_n - e^A\|_\infty \to 0 \quad \text{as } n \to \infty.$$

This is not so easy to prove, but if $D = \mathrm{diag}(\lambda_1, \dots, \lambda_N)$ is a *diagonal* matrix

$$D = \begin{pmatrix} \lambda_1 & & & 0 \\ & \lambda_2 & & \\ & & \ddots & \\ 0 & & & \lambda_N \end{pmatrix},$$

it is quite obvious that the partial sums S_n converge in the L^∞-norm because

$$\begin{aligned} S_n &= I + D + \cdots + \frac{D^n}{n!} \\ &= \begin{pmatrix} 1 + \lambda_1 + \cdots + \frac{\lambda_1^n}{n!} & & & 0 \\ & 1 + \lambda_2 + \cdots + \frac{\lambda_2^n}{n!} & & \\ & & \ddots & \\ 0 & & & 1 + \lambda_N + \cdots + \frac{\lambda_N^n}{n!} \end{pmatrix} \\ &\to e^D = \begin{pmatrix} e^{\lambda_1} & & & 0 \\ & e^{\lambda_2} & & \\ & & \ddots & \\ 0 & & & e^{\lambda_N} \end{pmatrix} \quad \text{as } n \to \infty. \end{aligned}$$

The scalar-valued exponential series $e^z = \sum_{k=0}^{\infty} z^k/k!$ is absolutely convergent for every complex number $z \in \mathbb{C}$. Therefore S_n converges to e^D in the L^∞-norm:

$$\sum_{k=0}^{\infty} \frac{D^k}{k!} = \lim_{n\to\infty} S_n = e^D = \begin{pmatrix} e^{\lambda_1} & & & 0 \\ & e^{\lambda_2} & & \\ & & \ddots & \\ 0 & & & e^{\lambda_N} \end{pmatrix}.$$

The Cauchy Convergence Criterion in Matrix Space. For matrices that are not diagonal we must prove there actually *is* a matrix e^A to which the matrix-exponential series converges in the L^∞–norm,

$$\|S_n - e^A\|_\infty \to 0 \quad \text{as } n \to \infty.$$

This is so because $\mathrm{M}(N, \mathbb{K})$ equipped with the L^∞-norm has the following *completeness* property, similar to completeness of $\mathbb{R}^n$ and $\mathbb{C}^n$ in the euclidean norm

$$\|\mathbf{z}\|_2 = \Big(\sum_{k=1}^{n} |z_k|^2\Big)^{1/2} \qquad \text{for } \mathbf{z} = (z_1, \dots, z_n) \text{ in } \mathbb{C}^n,$$

or completeness of the number fields $\mathbb{K} = \mathbb{R}$ and $\mathbb{C}$.

THEOREM 5.37 (Cauchy Convergence Criterion). *A sequence $\{A_n\}$ of matrices in $\mathrm{M}(N, \mathbb{K})$ converges to some limit $A_0 = \lim_{n\to\infty} A_n$ in the L^∞-norm if and only if the sequence has the Cauchy property*

$$\|A_m - A_n\|_\infty \to 0 \quad \textit{eventually as} \quad m, n \to \infty. \tag{5.8}$$

To be precise, this means: Given any $r > 0$ we can find a cutoff $M > 0$ such that

$$\|A_m - A_n\|_\infty < r \quad \textit{for all } m, n \geq M.$$

Statement (5.8) is much stronger than saying *successive terms* in the sequence get close, with $\|A_{n+1} - A_n\|_\infty \to 0$ as $n \to \infty$; to verify the Cauchy criterion you must show that *all* the terms far along in $\{A_n\}$ are simultaneously close together as $n \to \infty$. Incidentally, if two norms p_2 and p_2 on matrix space are equivalent as in (5.6), it is easy to see that the Cauchy criterion is valid for $p_1 \Leftrightarrow$ it holds for p_2.

If you try to prove convergence $A_n \to A$ by examining distances $\|A_n - A\|_\infty$, you must actually have the prospective limit A in hand, and that limit might be very hard to guess. The Cauchy criterion gets around this problem. You don't need to identify the value of the limit whose existence is assured in (5.8), because the Cauchy criterion can be verified by inspecting the terms of the given sequence $\{A_n\}$. For instance, in $\mathbb{R}$ the integral test of calculus shows that the partial sums

$$S_n = 1 + \frac{1}{2^2} + \cdots + \frac{1}{n^2} \quad \text{of the } \textit{harmonic series} \sum_{n=1}^{\infty} \frac{1}{n^2}$$

have the Cauchy property for $\mathbb{R}$, and hence

$$\sum_{n=1}^{\infty} \frac{1}{n^2} = \lim_{n\to\infty} \{S_n\}$$

exists, by the completeness property (5.8). It is a lot harder to identify this limit in closed form, and show it is exactly $\pi^2/6$. We will see one way to do that in Chapter 6.

Convergence of the Exponential Series. Returning to our discussion of the matrix exponential series $\sum_{n=0}^{\infty} A^k/k!$, we now show that its partial sums $S_n = \sum_{k=0}^{n} A^k/k!$ have the Cauchy property in the $\|\cdot\|_\infty$-norm, so by completeness of $\mathrm{M}(N, \mathbb{C})$ the partial sums actually have a limit, which we name "e^A":

$$e^A = \sum_{k=1}^{\infty} \frac{A^k}{k!} = \lim_{n\to\infty} S_n.$$

PROOF. To verify the Cauchy property for $\{S_n\}$ we may assume $m > n$. By the triangle inequality and the multiplicative property of Exercise 5.25 we have

$$\|S_m - S_n\|_\infty = \left\| \sum_{k=n+1}^{m} \frac{A^k}{k!} \right\|_\infty \le \sum_{k=n+1}^{m} \frac{\|A^k\|_\infty}{k!} \le \sum_{k=n+1}^{m} \frac{N^k \|A\|_\infty^k}{k!}.$$

By the ratio test of calculus the Taylor series for $f(x) = e^x$ converges (to e^x) for all $x \in \mathbb{R}$:

$$e^x = \sum_{n=0}^{\infty} \frac{D^n f(0)}{n!} x^n = \sum_{n=0}^{\infty} \frac{x^n}{n!},$$

because $D^k\{e^x\} = e^x$ for all $k \Rightarrow D^k f(0) = 1$ for all k. Taking $x = N \cdot \|A\|_\infty$ we get

$$\sum_{k=0}^{n} \frac{N^k \|A\|_\infty^k}{k!} \to \sum_{k=0}^{\infty} \frac{N^k \|A\|_\infty^k}{k!} = e^{N\|A\|_\infty} < \infty \quad \text{as } n \to \infty.$$

Hence for $m \ge n$

$$0 \le \|S_m - S_n\|_\infty \le \sum_{k=n+1}^{m} \frac{(N \cdot \|A\|_\infty)^k}{k!} \le \sum_{k=n+1}^{\infty} \frac{(N \cdot \|A\|_\infty)^k}{k!} \to 0$$

as $m, n \to \infty$. Thus $\{S_n\}$ is a Cauchy sequence for the $\|\cdot\|_\infty$-norm and the matrix-valued series $\sum_{k=0}^{\infty} A^k/k!$ converges in L^∞-norm for every matrix A. □

Explicit Computation of e^A. It can be difficult to directly compute the sum of a convergent series such as $e^A = \sum_{n=0}^{\infty} A^n/n!$ For instance, imagine trying to evaluate e^A when

$$A = \begin{pmatrix} 1 & -1 \\ -6 & 2 \end{pmatrix}.$$

Computing higher and higher powers A^k is computationally infeasible. With this in mind it is natural to ask how many terms in the series for e^A would be needed to make Error $= \|e^A - S_n\|_\infty < 1 \times 10^{-6}$ (6-place accuracy)?

On the other hand, computing e^D was easy for a diagonal matrix $D = \mathrm{diag}(\lambda_1, \dots, \lambda_N)$.

$$S_n = I + D + \cdots + \frac{D^n}{n!} \to \begin{pmatrix} e^{\lambda_1} & & & 0 \\ 0 & e^{\lambda_2} & & \\ & & \ddots & \\ 0 & & & e^{\lambda_N} \end{pmatrix} = e^D.$$

We now show that e^{tA} can be computed in closed form for all $t \in \mathbb{C}$ for any A that is diagonalizable over $\mathbb{R}$ or $\mathbb{C}$.

EXAMPLE 5.38. Compute e^{tA} ($t \in \mathbb{R}$ or $\mathbb{C}$) for the matrix

$$A = \begin{pmatrix} 1 & -1 \\ -6 & 2 \end{pmatrix}.$$

SOLUTION: First observe that A is diagonalizable, with $QAQ^{-1} = \left(\begin{smallmatrix} 4 & 0 \\ 0 & -1 \end{smallmatrix}\right) = D$ for suitably chosen Q. The eigenvalues of A are the roots of the characteristic polynomial

$$\begin{aligned} p_A(x) &= \det\begin{pmatrix} 1-\lambda & -1 \\ -6 & 2-\lambda \end{pmatrix} = (\lambda - 2)(\lambda - 1) - 6 \\ &= \lambda^2 - 3\lambda + 2 - 6 = \lambda^2 - 3\lambda - 4 = (\lambda - 4)(\lambda + 1), \end{aligned}$$

so $\mathrm{sp}(A) = \{4, -1\}$. The eigenspaces are computed by row reduction:

EIGENVALUE $\lambda = 4$:

$$(A - \lambda I) = \begin{pmatrix} -3 & -1 \\ -6 & -2 \end{pmatrix} \xrightarrow{\text{Row Ops}} \begin{pmatrix} 1 & \frac{1}{3} \\ 0 & 0 \end{pmatrix}.$$

Solutions of $(A - \lambda I)\mathbf{x} = \mathbf{0}$ are

$$\mathbf{x} \in \mathbb{K} \cdot \begin{pmatrix} -\frac{1}{3} \\ 1 \end{pmatrix} = \mathbb{K} \cdot \begin{pmatrix} 1 \\ -3 \end{pmatrix} = \mathbb{K} \cdot \mathbf{f}_1,$$

so $E_{\lambda=4} = \mathbb{K} \cdot \mathbf{f}_1$.

EIGENVALUE $\lambda = -1$:

$$(A - \lambda I) = \begin{pmatrix} 2 & -1 \\ -6 & 3 \end{pmatrix} \xrightarrow{\text{Row Ops}} \begin{pmatrix} 1 & -\frac{1}{2} \\ 0 & 0 \end{pmatrix}.$$

Solutions of $(A - \lambda I)\mathbf{x} = \mathbf{0}$ lie in

$$\mathbb{K} \cdot \begin{pmatrix} \frac{1}{2} \\ 1 \end{pmatrix} = \mathbb{K} \cdot \begin{pmatrix} 1 \\ 2 \end{pmatrix} = \mathbb{K} \cdot \mathbf{f}_2 = E_{\lambda=-1}.$$

Thus $\mathbb{K}^2 = E_{\lambda=4} \oplus E_{\lambda=-1}$ and $\mathfrak{Y} = \{\mathbf{f}_1 = (1, -3),\ \mathbf{f}_2 = (1, 2)\}$ is a diagonalizing basis in $\mathbb{K}^2$ for the multiplication operator $L_A(\mathbf{x}) = A \cdot \mathbf{x}$.

On the other hand, we may write $L_A = (\mathrm{id}) \circ L_A \circ (\mathrm{id})$, where $\mathrm{id} : \mathbb{K}^n \to \mathbb{K}^n$ is the identity operator. Applying the discussion of change of basis in Chapter 2, we get

$$\begin{aligned} D = \begin{pmatrix} 4 & 0 \\ 0 & -1 \end{pmatrix} &= [L_A]_{\mathfrak{Y}\mathfrak{Y}} = [\mathrm{id}]_{\mathfrak{Y}\mathfrak{X}} \cdot [L_A]_{\mathfrak{X}\mathfrak{X}} \cdot [\mathrm{id}]_{\mathfrak{X}\mathfrak{Y}} \\ &= [\mathrm{id}]_{\mathfrak{Y}\mathfrak{X}} \cdot A \cdot [\mathrm{id}]_{\mathfrak{X}\mathfrak{Y}}. \end{aligned}$$

If $\mathfrak{X} = \{\mathbf{e}_1, \mathbf{e}_2\}$ is the standard basis in $\mathbb{K}^2$, we have

$$\begin{cases} \mathbf{f}_1 = \mathbf{e}_1 - 3\mathbf{e}_2, \\ \mathbf{f}_2 = \mathbf{e}_1 + 2\mathbf{e}_2, \end{cases}$$

and $[\mathrm{id}]_{\mathfrak{X}\mathfrak{Y}} = \left(\begin{smallmatrix} 1 & 1 \\ -3 & 2 \end{smallmatrix}\right)$ is the transpose of the array of coefficients in this system of equations. Thus $QAQ^{-1} = D$ where

$$Q^{-1} = [\mathrm{id}]_{\mathfrak{Y}\mathfrak{X}}^{-1} = [\mathrm{id}]_{\mathfrak{X}\mathfrak{Y}} = \begin{pmatrix} 1 & 1 \\ -3 & 2 \end{pmatrix}.$$

Since $\det(Q^{-1}) = 5$ we get

$$Q = (Q^{-1})^{-1} = \frac{1}{5} \cdot \begin{pmatrix} 2 & -1 \\ 3 & 1 \end{pmatrix}.$$

Finally, observe that

$$\begin{cases} D = QAQ^{-1} \\ A = Q^{-1}DQ \end{cases} \Rightarrow A^k = (Q^{-1}DQ) \cdot (Q^{-1}DQ) \cdot \dots \cdot (Q^{-1}DQ) = Q^{-1}D^kQ$$

for $k = 0, 1, 2, \ldots$. By part (d) of Exercise 5.36 we get

$$e^A = \sum_{k=0}^{\infty} \frac{A^k}{k!} = \sum_{k=0}^{\infty} \frac{(Q^{-1}DQ)^k}{k!} = \sum_{k=0}^{\infty} \frac{Q^{-1}D^kQ}{k!}$$
$$= Q^{-1}\Big(\sum_{k=0}^{\infty} \frac{D^k}{k!}\Big) \cdot Q = Q^{-1}e^DQ.$$

We conclude that

$$e^A = Q^{-1}\begin{pmatrix} e^4 & 0 \\ 0 & e^{-1} \end{pmatrix} Q,$$

which exhibits e^A as a product of *just three explicit matrices.*

Similarly, for $t \in \mathbb{R}$ (or even $t \in \mathbb{C}$) we have $e^{tD} = \operatorname{diag}(e^{4t}, e^{-t})$, hence

$$e^{tA} = Q^{-1}\begin{pmatrix} e^{4t} & 0 \\ 0 & e^{-t} \end{pmatrix} Q = \begin{pmatrix} 1 & 1 \\ -3 & 2 \end{pmatrix} \cdot \begin{pmatrix} e^{4t} & 0 \\ 0 & e^{-t} \end{pmatrix} \cdot \frac{1}{5}\begin{pmatrix} 2 & -1 \\ 3 & 1 \end{pmatrix}$$
$$= \frac{1}{5}\begin{pmatrix} 2e^{4t} + 3e^{-t} & -e^{4t} + e^{-t} \\ -6e^{4t} + 6e^{-t} & 3e^{4t} + 2e^{-t} \end{pmatrix}$$
$$= \frac{1}{5}e^{4t} \cdot \begin{pmatrix} 2 & -1 \\ -6 & 3 \end{pmatrix} + \frac{1}{5}e^{-t} \cdot \begin{pmatrix} 3 & 1 \\ 6 & 2 \end{pmatrix}.$$

Setting $t = 0$, we get $e^0 = I$; setting $t = 1$, we obtain e^A in closed form,

$$e^A = \frac{1}{5}e^4 \cdot \begin{pmatrix} 2 & -1 \\ -6 & 3 \end{pmatrix} + \frac{1}{5}e^{-1} \cdot \begin{pmatrix} 3 & 1 \\ 6 & 2 \end{pmatrix}.$$

5.5. Application: Linear Systems of Differential Equations

This application shows why one might want to compute the matrix-valued function $\phi(t) = e^{tA}$ that maps $\mathbb{C} \to \mathrm{M}(N, \mathbb{C})$. First we must sketch some additional properties of the exponential map on matrices (mostly without proofs).

1. EXPONENT LAW: If A and B commute then

$$e^{A+B} = e^A \cdot e^B.$$

In particular, e^A is always invertible, with $(e^A)^{-1} = e^{-A}$.

2. ONE-PARAMETER GROUP LAW: For all $s, t \in \mathbb{C}$ we have

$$e^{(s+t)A} = e^{sA} \cdot e^{tA},$$

and e^{-tA} is the inverse of e^{tA} for all t.

SKETCH OF PROOF. We give an informal proof involving rearrangement of a matrix-valued double series. But beware: rearrangement and regrouping of series are delicate matters even for scalar-valued series, and a proof that would pass muster with analysts would require considerably more detail—see any text on mathematical analysis.

The series $e^A = \sum_{k=0}^{\infty} A^k/k!$ and $e^B = \sum_{\ell=0}^{\infty} B^\ell/\ell!$ are L^∞-norm convergent. Expanding the product of the two series term by term (which in itself requires some justification for infinite series; see Exercise 5.36(c)), we get

$$e^A \cdot e^B = \Big(\sum_{k=0}^{\infty} \frac{A^k}{k!}\Big) \cdot \Big(\sum_{\ell=0}^{\infty} \frac{B^\ell}{\ell!}\Big) = \sum_{k,\ell\geq 0} \frac{1}{k!}\frac{1}{\ell!} A^k B^\ell$$
$$= \sum_{k,\ell\geq 0} \frac{1}{(k+\ell)!} \cdot \frac{(k+\ell)!}{k!\,\ell!} A^k B^\ell .$$

Now group together all terms with $k + \ell = n$ for $n = 0, 1, \ldots$ to obtain

$$e^A \cdot e^B = \sum_{n=0}^{\infty} \frac{1}{n!} \cdot \Big(\sum_{k=0}^{n} \binom{n}{k} A^k B^\ell\Big) \quad \text{where } \binom{n}{k} = \text{(binomial coefficient)}$$
$$= \sum_{n=0}^{\infty} \frac{1}{n!}(A+B)^n \quad \text{(binomial formula, valid when } AB = BA\text{)}$$
$$= e^{A+B}$$

□

3. DIFFERENTIATION LAW: The derivative of $\phi(t) = e^{tA}$ exists and is continuous, with

$$\frac{d}{dt}(e^{tA}) = A \cdot e^{tA} \quad \text{for all } t \in \mathbb{R}.$$

PROOF. From the exponent law we get

$$\frac{d}{dt}(e^{tA}) = \lim_{\Delta t \to 0} \frac{e^{(t+\Delta t)A} - e^{tA}}{\Delta t}$$
$$= \lim_{\Delta t \to 0} \left(\frac{e^{(\Delta t)A} - I}{\Delta t}\right) \cdot e^{tA} \quad (\text{since } e^{(t+\Delta t)A} = e^{tA} \cdot e^{(\Delta t)A}).$$

By using the norm properties listed in Exercise 5.33 and Exercise 5.36 it is not hard to show that

$$e^{(\Delta t)A} - I = \left(I + (\Delta t)A + \frac{(\Delta t)^2}{2!}A^2 + \frac{(\Delta t)^3}{3!}A^3 + \cdots\right) - I$$
$$= \Delta t \cdot \left(A + \frac{(\Delta t)}{2!}A^2 + \cdots\right) = \Delta t \cdot (A + o(\Delta t)).$$

The notation "$o(\Delta t)$" indicates that matrix-valued term $o(\Delta t)$ becomes small compared to Δt as $\Delta t \to 0$, so that

$$\frac{\|o(\Delta t)\|_\infty}{|\Delta t|} \to 0 \quad \text{as } \Delta t \to 0. \tag{5.9}$$

It will then follow that

$$\frac{e^{(\Delta t)A} - I}{\Delta t} = \frac{\Delta t}{\Delta t}(A + o(\Delta t)) \to A$$

in the $\|\cdot\|_\infty$-norm as $\Delta t \to 0$, and that will prove the differentiation formula.

To see why (5.2) holds, observe that

$$\|o(\Delta t)\| = \Big\| \sum_{k=2}^{\infty} \frac{(\Delta t)^k}{k!} A^k \Big\| \le \sum_{k=2}^{\infty} \frac{|\Delta t|^k}{k!} n^k \|A\|^k$$
$$= |\Delta t|^2 \cdot \Big(\sum_{k=2}^{\infty} \frac{|\Delta t|^{k-2}}{k!} n^k \|A\|^k \Big).$$

Standard estimates show that the last series sum is convergent and bounded for $|\Delta t| \le 1$, hence $\|o(\Delta t)\| \le B \cdot |\Delta t|^2$ for some constant B and all small increments $\Delta t \in \mathbb{C}$. This suffices to verify that $o(\Delta t)$ satisfies the "little oh" condition of (5.9). □

Example: Solving $dx/dt = Ax$. Any linear system of first-order constant-coefficient differential equations in n unknowns can be written in matrix form as

$$\frac{d\mathbf{y}}{dt} = A \cdot \mathbf{y}(t) \quad \text{with initial condition} \quad \mathbf{y}(0) = \mathbf{c}$$

where $\mathbf{y}(t) = (y_1(t), \dots, y_n(t))$ is a vector-valued function of t, and the $n \times n$ matrix A provides the coefficients of the system. It is well-known that once an initial value $\mathbf{c}$ is specified there is a unique, infinitely differentiable vector-valued solution $\mathbf{y}(t)$ if we regard $\mathbf{y}(t)$ as an $n \times 1$ column vector. That solution can be computed explicitly as

$$\mathbf{y}(t) = e^{tA} \cdot \mathbf{y}(0) = e^{tA} \cdot \mathbf{c} \quad \text{for } t \in \mathbb{R}. \tag{5.10}$$

This is in fact a solution because

$$\frac{d\mathbf{y}}{dt} = \frac{d}{dt}(e^{tA} \cdot \mathbf{c}) = \frac{d}{dt}(e^{tA}) \cdot \mathbf{c} = Ae^{tA} \cdot \mathbf{c} = A \cdot \mathbf{y}(t),$$

and when $t = 0$ we get $\mathbf{y}(0) = \mathbf{c}$ because $e^{0 \cdot A} = I_{n \times n}$. We must of course compute e^{tA} to arrive at $\mathbf{y}(t)$, but in the previous example we have seen how that might be done, at least when the coefficient matrix can be diagonalized.

EXAMPLE 5.39. If $A = \begin{pmatrix} 1 & -1 \\ -6 & 2 \end{pmatrix}$, determine the unique solution of the first-order vector-valued differential equation

$$\frac{d\mathbf{y}}{dt} = A \cdot \mathbf{y}(t) \quad \text{with initial value} \quad \mathbf{y}_0 = \mathbf{y}(0) = \mathbf{e}_1 = \begin{pmatrix} 1 \\ 0 \end{pmatrix}.$$

Do likewise for the initial value $\mathbf{y}_0 = y(0) = \begin{pmatrix} 0 \\ 1 \end{pmatrix}$. Then find all solutions of

$$\frac{d\mathbf{y}}{dt} = A \cdot \mathbf{y}(t) \quad \text{for an arbitrary initial value } \mathbf{y}_0 = \begin{pmatrix} c_1 \\ c_2 \end{pmatrix}.$$

SOLUTION: Earlier we found that the diagonal form of A is

$$D = QAQ^{-1} = \begin{pmatrix} 4 & 0 \\ 0 & -1 \end{pmatrix} \quad \text{where} \quad Q = \frac{1}{5}\begin{pmatrix} 2 & -1 \\ 3 & 1 \end{pmatrix}.$$

Gathering together all terms involving e^4 or e^{-1}, respectively, we get

$$e^A = e^{Q^{-1}DQ} = Q^{-1} \cdot e^D \cdot Q = \frac{1}{5} e^4 \cdot \begin{pmatrix} 2 & -1 \\ -6 & 3 \end{pmatrix} + e^{-1} \cdot \begin{pmatrix} 3 & 1 \\ 6 & 2 \end{pmatrix}.$$

Taking e^{tA} in place of e^A, we get (with little additional effort):

$$e^{tA} = \frac{1}{5}e^{4t}\cdot\begin{pmatrix} 2 & -1 \\ -6 & 3 \end{pmatrix} + \frac{1}{5}e^{-t}\cdot\begin{pmatrix} 3 & 1 \\ 6 & 2 \end{pmatrix} \quad \text{for all } t \in \mathbb{R}.$$

Taking $\mathbf{y}_0 = \mathbf{e}_1 = (1,0)$ yields one solution:

$$\mathbf{y}_1(t) = e^{tA}(\mathbf{e}_1) = \frac{1}{5}e^{4t}\cdot\begin{pmatrix} 2 \\ -6 \end{pmatrix} + \frac{1}{5}e^{-t}\cdot\begin{pmatrix} 3 \\ 6 \end{pmatrix},$$

and if $\mathbf{y}_0 = \mathbf{e}_2 = (0,1)$ we get another solution:

$$\mathbf{y}_2(t) = e^{tA}(\mathbf{e}_2) = \frac{1}{5}e^{4t}\begin{pmatrix} -1 \\ 3 \end{pmatrix} + \frac{1}{5}e^{-t}\cdot\begin{pmatrix} 1 \\ 2 \end{pmatrix}.$$

For an arbitrary initial condition $\mathbf{y}(0) = \mathbf{c} = c_1\mathbf{e}_1 + c_2\mathbf{e}_2$ with $c_1, c_2 \in \mathbb{C}$, it is obvious that the solution of $d\mathbf{y}/dt = A\cdot\mathbf{y}(t)$ with this hybrid initial condition is the same linear combination of the basic solutions $\mathbf{y}_1(t)$ and $\mathbf{y}_2(t)$, namely:

$$\begin{aligned}\mathbf{y}(t) &= c_1\mathbf{y}_1(t) + c_2\mathbf{y}_2(t) \\ &= \frac{1}{5}e^{4t}\cdot\left[c_1\begin{pmatrix} 2 \\ -6 \end{pmatrix} + c_2\begin{pmatrix} -1 \\ 3 \end{pmatrix}\right] + \frac{1}{5}e^{-t}\left[c_1\begin{pmatrix} 3 \\ 6 \end{pmatrix} + c_2\begin{pmatrix} 1 \\ 2 \end{pmatrix}\right].\end{aligned}$$

(Check for yourself that $\mathbf{y}(0) = c_1\mathbf{e}_1 + c_2\mathbf{e}_2 = \mathbf{c}$.)

The full set of differentiable maps $f : \mathbb{R} \to \mathbb{C}^2$ such that $df/dt = A\cdot f(t)$ is a two-dimensional subspace M in the infinite-dimensional space $\mathcal{C}^\infty(\mathbb{R}, \mathbb{C}^2)$ over $\mathbb{C}$ whose elements are the infinitely differentiable vector-valued maps, so

$$M = \mathbb{C}\text{-span}\{\mathbf{y}_1(t), \mathbf{y}_2(t)\} = \{c_1\mathbf{y}_1(t) + c_2\mathbf{y}_2(t) : c_1, c_2 \in \mathbb{C}\}$$

and the basic solutions $\mathbf{y}_1, \mathbf{y}_2$ are a vector basis for M. One should check that $\mathbf{y}_1$, $\mathbf{y}_2$ are linearly independent vectors in $\mathcal{C}^\infty(\mathbb{R}, \mathbb{C}^2)$. But if there were coefficients α_1, α_2 such that $\alpha_1\mathbf{y}_1(t) + \alpha_2\mathbf{y}_2(t) \equiv 0$ in $\mathbb{C}^2$, and we took any convenient base point (say $t = 0$), we would then have the following vector identity in $\mathbb{C}^2$:

$$\begin{aligned}&\begin{pmatrix} 0 \\ 0 \end{pmatrix} = \frac{\alpha_1}{5}\left[\begin{pmatrix} 2 \\ -6 \end{pmatrix} + \begin{pmatrix} 3 \\ 6 \end{pmatrix}\right] + \frac{\alpha_2}{5}\left[\begin{pmatrix} -1 \\ 3 \end{pmatrix} + \begin{pmatrix} 1 \\ 2 \end{pmatrix}\right] \\ &\Rightarrow \frac{\alpha_1}{5}\begin{pmatrix} 5 \\ 0 \end{pmatrix} + \frac{\alpha_2}{5}\begin{pmatrix} 0 \\ 5 \end{pmatrix} = \begin{pmatrix} 0 \\ 0 \end{pmatrix} \Rightarrow \alpha_1\mathbf{e}_1 + \alpha_2\mathbf{e}_2 = \mathbf{0} \Rightarrow \alpha_1 = \alpha_2 = 0\end{aligned}$$

as required.

A similar discussion holds for equations $d\mathbf{y}/dt = A\cdot\mathbf{y}(t)$ whenever A is $n\times n$ and diagonalizable. If $\{\mathbf{y}_1(t), \dots, \mathbf{y}_n(t)\} \subseteq \mathcal{C}^\infty(\mathbb{R}, \mathbb{C}^n)$ are the basic solutions, those with initial values $\mathbf{y}_k(0) = \mathbf{e}_k$ (the standard basis vectors in $\mathbb{C}^n$), then a solution with arbitrary initial value $\mathbf{y}(0) = \sum_{k=1}^n c_k\mathbf{e}_k \in \mathbb{C}^n$ is obtained by taking the same linear combination

$$\mathbf{y}(t) = c_1\mathbf{y}_1(t) + \cdots + c_n\mathbf{y}_n(t)$$

of basic solutions $\mathbf{y}_k(t)$. As above, the $\mathbf{y}_k$ are linearly independent vectors in $\mathcal{C}^\infty(\mathbb{R}, \mathbb{C}^n)$, for if $\mathbf{0} = \sum c_k\mathbf{y}_k(t)$ in $\mathcal{C}^\infty(\mathbb{R}, \mathbb{C}^n)$ for all t, then (taking $t = 0$) $\sum c_k\mathbf{e}_k = \mathbf{0}$ in $\mathbb{C}^n$; thus $c_1 = c_2 = \cdots = c_n = 0$ because $\mathbf{y}_k(0) = \mathbf{e}_k$ by definition. We conclude that the $\{\mathbf{y}_k(t)\}$ are a basis for the full set of solutions (with arbitrary initial value) of the equation $d\mathbf{y}/dt = A \cdot \mathbf{y}(t)$, and

$$\begin{aligned} M &= \left\{ f \in \mathcal{C}^\infty(\mathbb{R}, \mathbb{C}^n) : \frac{df}{dt} = A \cdot f(t) \text{ for all } t \in \mathbb{R} \right\} \\ &= \mathbb{C}\text{-span}\{\mathbf{y}_1(t), \dots, \mathbf{y}_n(t)\}. \end{aligned}$$

The set M is an n-dimensional vector subspace with $\dim_{\mathbb{C}} M = n$ in the infinite-dimensional space of differentiable vector-valued functions $f \in \mathcal{C}^\infty(\mathbb{R}, \mathbb{C}^n)$.

5.6. Application: Matrix-Valued Geometric Series

The scalar *geometric series* $1 + z + z^2 + \cdots$ is absolutely convergent for $|z| < 1$ in $\mathbb{C}$, with a sum that can be written in closed form:

$$\sum_{k=0}^{\infty} z^k = \frac{1}{1-z} \quad \text{for all } z \in \mathbb{C} \text{ such that } |z| < 1. \tag{5.11}$$

(The series diverges for $|z| > 1$.) The proof of (5.11) is based on two well-known observations:

(i) $\lim_{k\to\infty} z^k = 0$ when $|z| < 1$ in $\mathbb{C}$ (because $x^n \to 0$ in $\mathbb{R}$ when $-1 < x < 1$).

(ii) The following algebraic identity holds for all $z \in \mathbb{C}$ and $n \in \mathbb{N}$:

$$(1 - z) \cdot (1 + z + z^2 + \cdots + z^n) = 1 - z^{n+1} \quad \text{for all } z,$$

because when the product is expanded we get a "telescoping sum" in which all but the first and last terms cancel.

Then for $z \neq 1$ we can divide both sides of the identity (ii) by $(1 - z)$ to get

$$\sum_{k=0}^{n} z^k = 1 + z + \cdots + z^n = \frac{1 - z^{n+1}}{1 - z} = \left(\frac{1}{1-z}\right) - z^{n+1} \cdot \left(\frac{1}{1-z}\right).$$

Since $|z| < 1 \Rightarrow z^{n+1} \to 0$ as $n \to \infty$, the entire second term above also goes to zero and the partial sums $1 + z + z^2 + \cdots + z^n$ converge to $\frac{1}{1-z}$ as $n \to \infty$.

We will now prove that the matrix-valued version of this series is norm convergent in $\mathrm{M}(N, \mathbb{C})$, with

$$\sum_{k=0}^{\infty} A^k = I + A + A^2 + \cdots = (I - A)^{-1} \quad \text{for sufficiently small } \|A\|. \tag{5.12}$$

By Theorem 5.30 and Exercise 5.31 this is true for any vector space norm $\| * \|$ on $\mathrm{M}(N, \mathbb{C})$, although the particular domain of convergence in matrix space will depend on the norm we are using.

Convergence of Matrix-Valued Power Series. With slightly more effort, the ideas needed to prove (5.12) yield a much more general result about convergence of matrix-valued power series.

PROPOSITION 5.40. *If a complex power series $f(z) = \sum_{k=0}^{n} a_k z^k$ has radius of convergence $R > 0$, it is absolutely convergent on $D_R = \{z \in \mathbb{C} : |z| < R\}$, so that*

$$\sum_{k=0}^{\infty} |a_k z^k| = \sum_{k=0}^{\infty} |a_k| \cdot |z|^k < \infty \quad \textit{for all } z \in \mathbb{C} \textit{ such that } |z| < R.$$

For any vector space norm on $\mathrm{M}(N, \mathbb{C})$ there is a constant $C > 0$, which will depend on the particular norm, such that $\|A^k\| \le C^{k-1} \cdot \|A\|^k$ (as in Theorem 5.30). Then the matrix-valued series $f(A) = \sum_{k=0}^{\infty} a_k A^k$ is norm convergent in $\mathrm{M}(N, \mathbb{C})$ for all A such that $\|A\| < R/C$. The matrix valued series is also absolutely norm convergent in the sense that

$$\sum_{k=1}^{\infty} |a_k| \cdot \|A\|^k < \infty \quad \textit{when } \|A\| < R/C.$$

PROOF. Using Exercise 5.31 we can verify absolute norm convergence directly, from which norm convergence of the matrix-valued series will follow by completeness of $\mathrm{M}(N, \mathbb{C})$. In fact,

$$\sum_{k=1}^{\infty} \|a_k A^k\| \le \sum_{k=1}^{\infty} |a_k| \cdot C^{k-1} \|A\|^k = \frac{1}{C} \sum_{k=1}^{\infty} |a_k| \cdot (C \cdot \|A\|)^k,$$

which is finite when $C\|A\| < R$ (the radius of convergence).

The partial sums $S_n = \sum_{k=1}^{n} a_k A^k$ of the matrix-valued series are then a Cauchy sequence in $\mathrm{M}(N, \mathbb{C})$ because for $m > n$ we have

$$\begin{aligned} \|S_m - S_n\| &\le \sum_{k=n+1}^{m} |a_k| \cdot \|A^k\| \le \frac{1}{C} \sum_{k=n+1}^{m} |a_k| \cdot (C \cdot \|A\|)^k \\ &\le \frac{1}{C} \sum_{k=n+1}^{\infty} |a_k| \cdot (C \cdot \|A\|)^k. \end{aligned}$$

But if $\sum_{k=1}^{n} b_k$ is a convergent series of positive real numbers, the very definition of series convergence implies that the tails of the series converge to zero as $n \to \infty$,

$$\sum_{k=n+1}^{\infty} b_k \to 0 \quad \text{as } n \to \infty.$$

Thus $\|S_m - S_n\| \to 0$ as $m, n \to \infty$, so $\{S_n\}$ is a Cauchy sequence in matrix space, and by completeness of $\mathrm{M}(N, \mathbb{C})$ it converges to a well-defined limit in the $\| * \|$ norm. □

NOTE: Even though all norms on matrix space are equivalent by Theorem 5.30, the statement "$\|A\| < 1$" means different things for various choices of the norm.

The operator norm of Definition 5.26 is often the optimal choice, as in the next result.

Small Perturbations of Invertible Matrices.

PROPOSITION 5.41. *If* $\|*\|_{\text{op}}$ *is the operator norm on* $\text{M}(N,\mathbb{C})$ *then* $(I-A)$ *is invertible with*

$$(I-A)^{-1} = \sum_{k=0}^{\infty} A^k = I + A + A^2 + \cdots \quad (\textit{convergent when } \|A\|_{\text{op}} < 1).$$

PROOF. For any $n \in \mathbb{N}$ we obviously have

$$(I-A)\cdot(I+A+\cdots+A^n) = I - A^{n+1}.$$

Since $\|A\|_{\text{op}} < 1$ we have $\|AB\|_{\text{op}} \leq \|A\|_{\text{op}} \cdot \|B\|_{\text{op}}$ and $\|A^k\|_{\text{op}} \leq \|A\|_{\text{op}}^k \to 0$ as $k \to \infty$, so the partial sums $S_n = \sum_{k=0}^{n} A^k$ converge in $\|*\|_{\text{op}}$-norm to some matrix B as $n \to \infty$. By the limit theorems of Exercise 5.36 we get

$$(I-A)\cdot B = (I-A)\cdot \lim_{n\to\infty} S_n = \lim_{n\to\infty}\left(I - A^{n+1}\right) = I,$$

because $\|A\|_{\text{op}} < 1 \Rightarrow A^{n+1} \to 0$. □

Similarly $(I+A)^{-1} = \sum_{k=0}^{\infty}(-1)^k A^k$ for $\|A\|_{\text{op}} < 1$.

EXERCISE 5.42. Consider $\text{M}(n,\mathbb{C})$ equipped with the L^∞-norm.

(a) Find an explicit $C > 0$ such that $\|A^k\|_\infty \leq C^{k-1} \cdot \|A\|_\infty^k$ for A in $\text{M}(N,\mathbb{C})$ and $k \in \mathbb{N}$.

(b) Find a value of $r > 0$ for which the geometric series $\sum_{k=0}^{\infty} A^k$ is guaranteed to converge in the L^∞-norm when $\|A\|_\infty < r$. Explain your reasoning.

HINT: For (a) find a $C > 0$ such that $\|AB\|_\infty < C \cdot \|A\|_\infty \|B\|_\infty$.

EXERCISE 5.43. Claim: *Any matrix A sufficiently close to the identity matrix I is automatically invertible.*

(a) If $\|*\|$ is any norm on $\text{M}(n,\mathbb{C})$ prove that

There is some $r > 0$ *such that* $\|A - I\| < r \Rightarrow A^{-1}$ *exists.*

Then show that the inverse is given by the $\|*\|$-convergent geometric series

$$A^{-1} = (I-(I-A))^{-1} = \sum_{k=0}^{\infty}(I-A)^k.$$

(b) When matrix space is equipped with the operator norm, find an explicit $r > 0$ such that $\|A - I\|_{\text{op}} < r$ implies that this series converges to A^{-1} in $\|*\|_{\text{op}}$ norm.

This leads to a much more general observation about small perturbations of invertible matrices.

EXERCISE 5.44. Claim: *For any norm $\|*\|$ on $M(N,\mathbb{C})$, every matrix A sufficiently close to an invertible matrix B is also invertible.*

(a) If $\|\Delta B\|$ is sufficiently small, prove that $A = B + \Delta B$ is invertible, with

$$A^{-1} = \left(I + B^{-1}\cdot\Delta B\right)\cdot B^{-1}$$

for all sufficiently small values of $\|\Delta B\|$.

(b) For the operator norm on $M(N,\mathbb{C})$ prove that this formula for A^{-1} holds whenever

$$\|\Delta B\|_{\mathrm{op}} < \frac{1}{\|B^{-1}\|_{\mathrm{op}}}.$$

(c) Answer (b) when $M(N,\mathbb{C})$ is equipped with the L^1-norm.

NOTE: The result in (a) can be interpreted as saying "*All sufficiently small perturbations $B+\Delta B$ of a invertible matrix B are invertible.*"

Geometric Series for Nilpotent N.

EXERCISE 5.45. A matrix $N \in M(n,\mathbb{K})$ is *nilpotent* if $N^k = 0$ for all sufficiently large k, and the smallest exponent d such that $N^d = 0$ is called the *nilpotence degree* of N. The geometric series for such a matrix is a *finite sum* since $N^k = 0$ for $k \geq d$.

(a) Verify that $(I - N)$ is invertible for every nilpotent matrix, with

$$(I-N)^{-1} = I + N + N^2 + \cdots + N^{d-1}.$$

(b) If B is the 4×4 matrix

$$B = \begin{pmatrix} 1 & 1 & 0 & 0 \\ 0 & 1 & 1 & 0 \\ 0 & 0 & 1 & 1 \\ 0 & 0 & 0 & 1 \end{pmatrix}$$

prove that $A = B - I$ is nilpotent and compute B^{-1} using the preceding algebraic formula (rather than row reduction or Cramer's rule).

NOTES: The formula in (a) shows that for nilpotent N the inverse $(I-N)^{-1}$ is a polynomial in N instead of an infinite sum; the norm $\|N\|$ is irrelevant.

Additional Exercises

Section 5.1. Eigenvalues and the Characteristic Polynomial

1. TRUE/FALSE QUESTIONS: ("True" if the statement is *always* true.)

(a) There are matrices in $M(n,\mathbb{K})$ whose characteristic polynomial is $p_A(x) = x^n$.

(b) Similar matrices $A, B \in M(n,\mathbb{K})$ have the same *eigenvalues*.

(c) Similar matrices $A, B \in M(n,\mathbb{K})$ have the same *eigenspaces*, so $E_\lambda(L_A) = E_\lambda(L_B)$ in $\mathbb{K}^n$ for all $\lambda \in \mathbb{K}$.

(d) Similar matrices $A, B \in M(n,\mathbb{K})$ have the same *characteristic polynomials*.

(e) The sum of two eigenvectors for a matrix in $\mathrm{M}(n, \mathbb{K})$ is an eigenvector.
(f) A square matrix with entries in a field $\mathbb{K}$ can fail to have to have any eigenvalues at all.
(g) Every square matrix with entries in $\mathbb{C}$ has an eigenvalue in $\mathbb{C}$.
(h) A linear operator $T : V \to V$ on a *finite-dimensional* vector space is invertible if and only if $\lambda = 0$ is not an eigenvalue.
(i) A matrix $A \in \mathrm{M}(n, \mathbb{K})$ is invertible if and only if its transpose A^{T} is invertible.

2. If matrices $A, B \in \mathrm{M}(n, \mathbb{K})$ are similar, with $B = SAS^{-1}$ for some invertible matrix S, prove that the eigenspaces are related via

$$E_\lambda(B) = S\,(E_\lambda(A)) \quad \text{for all } \lambda \in \mathbb{K},$$

and hence have the same dimensions $|E_\lambda(B)| = |S\,(E_\lambda(A))\,| = |E_\lambda(A)|$.

3. If matrices A, B in $\mathrm{M}(n, \mathbb{K})$ have the same characteristic polynomial, so $p_A(x) = p_B(x)$ in $\mathbb{K}[x]$, does that mean they are similar matrices with $B = SAS^{-1}$ for some invertible matrix S? Prove or provide a counterexample.

4. If $T : V \to V$ is a linear map on a finite-dimensional space,
(a) Show that the *dimensions* of $K(T)$ and $R(T)$ are similarity invariants on the space $\mathrm{Hom}_{\mathbb{K}}(V, V)$ of linear operators.
(b) Are the *spaces* $K(T)$ and $R(T)$ constant on similarity classses in the space of linear operators $\mathrm{Hom}_{\mathbb{K}}(V, V)$?

5. If $A \in \mathrm{M}(n, \mathbb{K})$,
(a) Prove that A and its transpose A^{T} have the same eigenvalues.
(b) Find a matrix $A \in \mathrm{M}(2, \mathbb{C})$ and an eigenvalue $\lambda \in \mathbb{C}$ such that $E_\lambda(A^{\mathrm{T}}) \neq E_\lambda(A)$.

HINT: For (b), consider the matrix $A = \begin{pmatrix} 0 & 1 \\ -1 & 0 \end{pmatrix}$ in $\mathrm{M}(2, \mathbb{C})$.

6. Prove that a matrix $A \in \mathrm{M}(n, \mathbb{K})$ is diagonalizable if and only if its transpose A^{T} is diagonalizable.

7. If $A \in \mathrm{M}(n, \mathbb{K})$ is invertible prove that $\lambda \in \mathrm{sp}_{\mathbb{K}}(A) \Leftrightarrow \frac{1}{\lambda} \in \mathrm{sp}_{\mathbb{K}}(A^{-1})$.

Section 5.2. Eigenvalues and the Characteristic Polynomial.

1. TRUE/FALSE QUESTIONS: ("True" if the statement is *always* true.)
(a) All $\mathbb{C}$-linear operators $T : V \to V$ on a complex finite-dimensional vector space are diagonalizable.
(b) If a basis $\mathfrak{X} = \{e_i\}$ diagonalizes a linear operator $T : V \to V$ on a finite-dimensional space, and we define a new basis by independently scaling each basis vector to get $f_i = c_i \cdot e_i$ with $c_i \neq 0$, then the new basis $\mathfrak{Y} = \{f_j\}$ also diagonalizes T.
(c) If $\lambda \in \mathbb{K}$ is an eigenvalue for a linear operator $T : V \to V$ then λ^m is an eigenvalue for T^m for every $m = 0, 1, 2, \ldots$
(d) If $\lambda \in \mathbb{K}$ is an eigenvalue for a linear operator $T : V \to V$, what is the relationship between the eigenspaces $E_{\lambda^m}(T^m)$ and $E_\lambda(T)$?

(e) A linear operator T on an n-dimensional vector space cannot be diagonalizable if it has fewer than n distinct eigenvalues in $\mathbb{K}$.
(f) Over any field $\mathbb{K}$, a linear operator on a finite-dimensional space is diagonalizable if (*algebraic multiplicity*) = (*geometric multiplicity*) for each eigenvalue.
(g) An *invertible* linear operator T on an n-dimensional vector space is diagonalizable $\Leftrightarrow T^{-1}$ is diagonalizable.

2. If $A, B \in \mathrm{M}(n, \mathbb{K})$, prove that AB and BA have the same eigenvalues, although the eigenspaces $E_\lambda(AB)$ and $E_\lambda(BA)$ may differ.

3. Find the eigenvalues in $\mathbb{C}$ and their geometric multiplicities for the 3×3 matrix

$$A = \begin{pmatrix} 3 & -1 & 1 \\ 7 & -5 & 1 \\ 6 & -6 & 2 \end{pmatrix}$$

Using this analysis, explain why A is *not* diagonalizable.

4. For the matrix $A = \left(\begin{smallmatrix} 3 & -1 \\ 1 & 4 \end{smallmatrix}\right)$, adapt the discussion of Example 5.15 to the following:
(a) Find all eigenvalues $\lambda \in \mathbb{C}$ and determine an explicit basis for each eigenspace $E_\lambda(A) \subseteq \mathbb{C}^2$.
(b) Find an explicit "conjugator" $S \in \mathrm{M}(2, \mathbb{C})$ such that $SAS^{-1} = D$ (a diagonal matrix).

TWO NATURAL QUESTIONS. If a linear operator $T : V \to V$ is diagonalizable, so $[T]_{\mathfrak{X}\mathfrak{X}} = D = \mathrm{diag}(\lambda_1, \dots, \lambda_n)$ for some basis in V

QUESTION 1: How unique is the diagonal form D?
QUESTION 2: What is the relationship between the various diagonalizing bases?

The analysis is easiest when T has the following property:

DISTINCT EIGENVALUES. *T has $n = \dim(V)$ distinct eigenvalues in $\mathbb{K}$.*

In the next few questions we consider such an operator and two diagonalizing bases $\mathfrak{X} = \{e_1, \dots, e_n\}$, $\mathfrak{Y} = \{f_1, \dots, f_n\}$, writing $D = [T]_{\mathfrak{X}\mathfrak{X}}$ and $D' = [T]_{\mathfrak{Y}\mathfrak{Y}}$. Let S be the invertible matrix that effects the change of basis, so that $D' = SDS^{-1}$.

5. If $T : V \to V$ has $n = \dim(V)$ distinct eigenvalues, explain why the entries in $D' = \mathrm{diag}(\lambda'_1, \dots, \lambda'_n)$ must be those in D, perhaps arranged in a different order, so that

$$\lambda'_i = \lambda_{\sigma(i)} \quad \text{for some permutation } \sigma \in S_n.$$

NOTE: Evidently the diagonalized form of T is not unique but, at least for operators with distinct eigenvalues, the only ambiguity is the order in which the diagonal entries appear. The outcome is more complicated when T is diagonaliazable and has some eigenvalues with multiplicities $m(\lambda) > 1$. We won't pursue this here.

6. Let $A \in \mathrm{M}(n, \mathbb{K})$ be diagonalizable and assume that the eigenvalues of A are distinct.

(a) If D is a diagonal $n \times n$ matrix with distinct entries, prove that an arbitrary matrix $B \in \mathrm{M}(n, \mathbb{K})$ commutes with D (so $BD = DB$) if and only if B is diagonal.

(b) If S_1, S_2 are invertible matrices that diagonalize A with $S_1 A S_1^{-1} = D = S_2 A S_2^{-1}$, prove that $S_2 = D' \cdot S_1$ for some diagonal matrix D' with nonzero entries.

7. Suppose $A \in \mathrm{M}(n, \mathbb{K})$ has n distinct eigenvalues $\lambda_1, \dots, \lambda_n$ and $\mathbf{f}_j \in \mathbb{K}^n$ is an arbitrary nonzero eigenvector for each λ_j, so that $L_A(\mathbf{f}_j) = A \cdot \mathbf{f}_j = \lambda_j \mathbf{f}_j$ and the vectors $\mathfrak{X} = \{f_1, \dots, f_n\}$ are a basis for $\mathbb{K}^n$ that diagonalizes the left multiplication operator $L_A(\mathbf{x}) = A \cdot \mathbf{x}$. Now assemble the vectors $\mathbf{f}_j$ as the columns of the $n \times n$ matrix $S = [\mathbf{f}_1; \dots; \mathbf{f}_n]$.

(a) Verify that S diagonalizes A, with $SAS^{-1} = \mathrm{diag}(\lambda_1, \dots, \lambda_n)$.

(b) Prove that *every* nonsingular transition matrix S that diagonalizes A is obtained by this construction.

HINT: See also the commutative diagram in Figure 2.3 associated with Proposition 2.33, which explains the connection between linear operators $T : V \to W$ and the multiplication operators $L_A : \mathbb{K}^m \to \mathbb{K}^n$ with $A = [T]_{\mathfrak{Y}\mathfrak{X}}$.

Taking $A = [T]_{\mathfrak{X}\mathfrak{X}}$ for a basis in V, this construction yields all bases $\mathfrak{Y} = S(\mathfrak{X})$ that diagonalize $T : V \to V$, once one such basis $\mathfrak{X}$ has been identified.

8. The *left shift operator* on the infinite-dimensional space V consisting of all bounded infinite sequences $\mathbf{x} = (a_1, a_2, \dots)$ of complex numbers is defined letting

$$T(a_1, a_2, \dots) = (a_2, a_3, \dots).$$

(a) Determine the kernel $K(T)$ and range $R(T)$ and their dimensions (which may or may not be infinite).

(b) Determine all eigenvalues $\lambda \in \mathbb{C}$ and the dimensions $|E_\lambda(T)| = \dim_{\mathbb{C}} E_\lambda(T)$ of the eigenspaces.

Sections 5.3–5.6. Diagonalization, Limits of Operators, and Applications

1. TRUE/FALSE QUESTIONS: ("True" if the statement is *always* true.)

(a) The exponent law $e^{(A+B)} = e^A \cdot e^B$ holds for all $A, B \in \mathrm{M}(n, \mathbb{C})$.

(b) The exponent law $e^{(s+t)A} = e^{sA} \cdot e^{tA}$ holds for all matrices in $A \in \mathrm{M}(n, \mathbb{C})$ and $s, t \in \mathbb{R}$.

(c) For any norm on matrix space $\mathrm{M}(n, \mathbb{C})$ the identity matrix $I = I_{n \times n}$ always has norm $\|I\| = 1$.

(d) For any norm $\| * \|$ on matrix space $\mathrm{M}(N, \mathbb{C})$, $\sum_{k=0}^{\infty} A^n$ is $\| * \|$-convergent to $(I - A)^{-1}$ if $\|A\| < 1$.

2. If $\| * \|$ is any vector space norm on $\mathrm{M}(n, \mathbb{C})$ and $B = \sum_{k=1}^{\infty} A_k$ is a norm-convergent series, explain why each of the following is true:

(a) Each "tail series" $T_n = \sum_{k=n+1}^{\infty} A_k$ is norm convergent for $n = 1, 2, \dots$.

(b) $\sum_{k=1}^{\infty} A_k = \sum_{k=1}^{n} A_k + \sum_{k=n+1}^{\infty} A_k$.

(c) The limit $\lim_{n\to\infty} T_n$ exists and is equal to 0.

3. Writing $\text{Exp}(A) = e^A = \sum_{k=0}^{\infty} \frac{A^k}{k!}$, explain why the following are true:

$$\text{(i) } \text{Exp}(A^{\mathsf{T}}) = (e^A)^{\mathsf{T}} \quad \text{(ii) } \text{Exp}(\overline{A}) = \overline{e^A} \quad \text{(iii) } \text{Exp}(-A) = (e^A)^{-1}$$

where A^{T} is the transpose of $A \in \text{M}(n, \mathbb{C})$ and $\overline{A}_{ij}$ = complex conjugate of A_{ij}.

4. For each of the matrices $A \in \text{M}(2, \mathbb{C})$ listed in (i)–(iv) below, answer the following:

(a) Find all eigenvalues in $\mathbb{C}$ and a basis for the eigenspace $E_\lambda(A)$.

(b) Find an explicit invertible $S \in \text{M}(2, \mathbb{C})$ such that $SAS^{-1} = D$ is diagonal.

(c) Find *closed formulas* for scalar components $E_{11}(t), \ldots, E_{22}(t)$ of the sum

$$E(t) = e^{tA} = \sum_{n=0}^{\infty} \frac{t^n A^n}{n!} \qquad (t \in \mathbb{R}),$$

The matrices of interest are:

$$\text{(i) } A_1 = \begin{pmatrix} 0 & -1 \\ 1 & 0 \end{pmatrix} \qquad \text{(ii) } A_2 = \begin{pmatrix} 0 & -i \\ i & 0 \end{pmatrix} = i \cdot A_1$$

$$\text{(iii) } A_3 = \begin{pmatrix} 0 & 1 \\ 1 & 0 \end{pmatrix} \qquad \text{(iv) } A_4 = \begin{pmatrix} 1 & 0 \\ 0 & -1 \end{pmatrix}$$

HINTS: In each case there are only two or four distinct powers I, A, A^2, and A^3 (then $A^4 = I$ and the pattern repeats). This will yield series expansions for the entries $E_{ij}(t)$ of the matrix e^{tA} that should all be familiar from calculus.

5. Answer the questions posed in Exercise 4 (a)–(c) for the following matrices A in $\text{M}(2, \mathbb{C})$:

$$\text{(i) } A_1 = \begin{pmatrix} 0 & \frac{1}{2} \\ 3 & 0 \end{pmatrix} \qquad \text{(ii) } A_2 = \begin{pmatrix} \frac{1}{2} & 0 \\ 0 & 3 \end{pmatrix}$$

NOTE: The powers A^k no longer repeat periodically, so the idea that worked in Additional Exercise 4 above will not work here. Solving (a) may require the detailed analysis employed in Example 5.15.

In Chapter 1 in Volume II we will prove an important general formula about matrix exponentials:

$$\det\left(e^A\right) = e^{\text{Tr(A)}} \quad \text{for all } A \in \text{M}(n, \mathbb{C}),$$

whether or not A is diagonalizable. This allows us to compute $\det(e^A)$ directly, for any matrix A, without the detailed calculations of Example 5.15 because the trace $\mathrm{Tr}(A)$ can be calculated immediately once A is given. In particular, $\det(e^A) = 1$ whenever $\mathrm{Tr}(A) = 0$.

6. Using the explicit formulas for e^{tA} obtained in Additional Exercise 4 above, verify that the preceding formula is valid for each matrix A discussed there.

7. Explain why we have simple convergence $A_n \to A$ as $n \to \infty$ in $\mathrm{M}(N, \mathbb{C})$ if and only if $A_n \to A$ in the L^2-norm (Hilbert-Schmidt norm).

NOTE: The same is then true for the L^1, L^2, and L^∞-norms by Theorem 5.30.

8. If $A \in \mathrm{M}(N, \mathbb{C})$ how small does $\|A\|$ have to be to insure that $\sum_{k=0}^{\infty} A^k$ converges to $(I - A)^{-1}$ for the following norms?

(a) $\|A\|_1$ (b) $\|A\|_2$ (c) $\|A\|_\infty$

NOTE: We do not necessarily have $\|A^k\| \le \|A\|^k$ for these norms. Recall Exercise 5.31.

9. If $\mathrm{M}(2, \mathbb{C})$ is equipped with the L^2-norm and B is the invertible matrix

$$B = \begin{pmatrix} 2i & 1 \\ -1 & 3 \end{pmatrix}.$$

(a) How small does $r > 0$ have to be to insure that $A = B + \Delta B$ is invertible when $\|\Delta B\|_2 < r$? (Find an explicit value of r that works.)

(b) Does the particular perturbation $\Delta B = \left(\begin{smallmatrix} 0 & 0.20 \\ -0.15 & 0 \end{smallmatrix}\right)$ satisfy the convergence criterion in (a)?

HINT: You might have to compute B^{-1} and its norm $\|B^{-1}\|_2$ by hand.

10. Now suppose $\mathrm{M}(2, \mathbb{C})$ is equipped with the $\| * \|_\infty$ norm and B is the matrix in Additional Exercise 9. Calculations performed in Exercises 5.43 and 5.44 gave an explicit formula for $(I + \Delta B)^{-1}$ which involved the geometric series

$$\left(I + B^{-1} \cdot \Delta B\right)^{-1} = I + (B^{-1} \cdot \Delta B) + (B^{-1} \cdot \Delta B)^2 + \cdots.$$

Whenever this series converges in L^∞ norm, that formula yields an actual solution for the inverse $(B + \Delta B)^{-1}$ of the perturbed matrix $A = (B + \Delta B)$.

(a) How small must $\|\Delta B\|_\infty$ be if we want this geometric series to be $\| * \|_\infty$-norm convergent?

(b) Does this condition hold for the particular matrix B and perturbation ΔB in Additional Exercise 9?

NOTE: We do not necessarily have $\|AB\|_\infty \le \|A\|_\infty \|B\|_\infty$ or $\|A^k\|_\infty \le \|A\|_\infty^k$ in the L^∞ norm. Recall Exercises 5.27 and 5.31.

ERROR ESTIMATES FOR MATRIX SERIES. The following digression on scalar-valued Taylor series will be needed to answer the following question about matrix-valued series:

> AN ERROR ESTIMATE. *If $A \in \mathrm{M}(N, \mathbb{C})$, how many terms must be taken in the partial sum $S_n = I + A + \frac{1}{2!}A^2 + \cdots + \frac{1}{n!}A^n$ to insure that the L^∞-norm error is small enough to guarantee 6-decimal-place accuracy of the approximation?*
>
> $$\mathrm{ERROR}_n = \left\| e^A - \sum_{k=0}^{n} \frac{A^k}{k!} \right\|_\infty < 1 \times 10^{-6}.$$

Given an $(n+1)$-times differentiable function $f(t)$ on an interval $[c, x] \subseteq \mathbb{R}$, Taylor's formula with remainder (at base point c) says:

$$f(x) = S_n(x) + \frac{D^{n+1} f(t_*)}{(n+1)!} \cdot (x-c)^{n+1} \quad \text{for some } c < t_* < x,$$

where S_n is the nth Taylor polynomial based at c:

$$S_n(t) = \sum_{k=0}^{n} \frac{D^k f(c)}{k!} \cdot (t-c)^k.$$

If the Taylor series converges to $f(t)$ on the interval $[c, x]$, as it does for $f(t) = e^t$, we get

$$\begin{aligned}\mathrm{ERROR}_n = |f(x) - S_n(x)| &= \left| f(x) - \sum_{k=0}^{n} \frac{D^k f(c)}{k!}(x-c)^k \right| \\ &= \left| \frac{D^{n+1} f(t_*)}{(n+1)!} \right| \cdot |x-c|^{n+1}\end{aligned}$$

for some t_* lying between c and x. For the particular function $f(t) = e^t$, we can take $c = 0$ and replace this upper bound with

$$\mathrm{ERROR}_n \leq \frac{e^{t_*}}{(n+1)!} \cdot x^{n+1} \quad \text{for some } t_* \in [c, x].$$

On the other hand, for the L^∞-norm convergent matrix-exponential series $e^A = \sum_{k=0}^{\infty} \frac{1}{k!} A^k$ we have

$$\begin{aligned}\left\| e^A - \sum_{k=0}^{n} \frac{A^k}{k!} \right\|_\infty = \left\| \sum_{k=n+1}^{\infty} \frac{A^k}{k!} \right\|_\infty &\leq \sum_{k=n+1}^{\infty} \frac{N^k \|A\|_\infty^k}{k!} \\ &\leq \left| e^{N\|A\|_\infty} - \sum_{k=0}^{n} \frac{(N \cdot \|A\|_\infty)^k}{k!} \right| \\ &= \left[e^t - S_n(t)\big|_{t = N \cdot \|A\|_\infty} \right] \\ &\leq e^{N \cdot \|A\|_\infty} \cdot \frac{(N \cdot \|A\|_\infty)^{n+1}}{(n+1)!}.\end{aligned}$$

On the right we have applied the scalar Taylor estimate to get a norm-error estimate for matrix-exponentials.

11. Apply the preceding comments to estimate ERROR_n for:

$$3 \times 3 \ (N = 3) \text{ matrices } A \text{ such that } \|A\|_\infty \leq 1.$$

Find an explicit value for n that guarantees

$$\text{ERROR}_n < 1 \times 10^{-6} \quad \text{if } \|A\|_\infty \leq 1.$$

NOTE: In the preceding discussion we derived a general estimate for the cutoff n in terms of the parameters N, $\|A\|_\infty$, and the allowable error ϵ.

SPECTRAL RADIUS OF AN OPERATOR. If $T : V \to V$ is a linear operator on a finite-dimensional space $\mathbb{K} = \mathbb{C}$, its *spectral radius* is the size of the largest eigenvalue

$$\text{SPECTRAL RADIUS:} \quad r(T) = \max\{|\lambda| : \lambda \in \mathbb{C} \text{ an eigenvalue}\},$$

which is also the radius of the smallest closed disc D_r about $z = 0$ in the complex plane that contains the spectrum $\text{sp}(T)$.

If T happens to be diagonalizable we may take a basis $\mathfrak{X} = \{e_1, \dots, e_n\}$ of eigenvectors labeled so that

$$T(e_i) = \lambda_i e_i \quad \text{with} \quad |\lambda_1| \geq \cdots \geq |\lambda_n|$$

(repeats allowed). Then $[T]_{\mathfrak{X}\mathfrak{X}} = \text{diag}(\lambda_1, \dots, \lambda_n)$ and $r(T) = |\lambda_1|$. There may be several eigenvalues with largest absolute value. Similarly we may assign a spectral radius to any matrix $A \in \text{M}(n, \mathbb{C})$ taking $r(A) = r(L_A)$. Note carefully that the definition of spectral radius makes no mention of norms on V, norms on operators, or norms on matrices. The next exercise shows how to compute $r(T)$ directly, without having to determine the eigenvalues.

12. With the preceding comments in mind, let $T : V \to V$ be diagonalizable over $\mathbb{C}$ and choose a diagonalizing basis $\mathfrak{X} = \{e_i\}$, labeling them so $|\lambda_1| \geq \cdots \geq |\lambda_m| \geq 0$. This assigns matrices $[T]_{\mathfrak{X}\mathfrak{X}}$ to operators on V, as well as an L^∞ norm

$$\|T\|_\infty = \|[T]_{\mathfrak{X}\mathfrak{X}}\|_\infty = \max\{|T_{ij}] : 1 \leq i, j \leq n\}.$$

This norm on operators is ideally suited to calculating the operator norms $\|T\|_{\text{op}}$ determined by the the $\| * \|_\infty$ norm on V,

$$\|T\|_{\text{op}} = \max\{\, \|T(\mathbf{z})\|_\infty : \|\mathbf{z}\|_\infty = 1 \text{ in } \mathbb{C}^n\} \quad \text{(recall Example 5.28)}$$

as well as the spectral radius $r(T)$.

(a) Prove that the operator norm $\|T\|_{\text{op}}$ determined by $\| * \|_\infty$ on V is precisely the spectral radius, so that

$$r(T) = \max\{\|T(\mathbf{z})\|_\infty : \|\mathbf{z}\|_\infty = 1\}.$$

(b) Prove that the spectral radius is also given by

$$r(T) = \lim_{k\to\infty} ||T^k||_\infty^{1/k},$$

and explain why this limit exists.

(c) Use equivalence of norms on matrix space $\mathrm{M}(n, \mathbb{C})$ to show that any other norm $|| * ||$ on matrix space yields the same numerical value

$$r(T) = \lim_{k\to\infty} ||A^k||^{1/k}$$

HINT: All vector space norms on matrix space (or the space of linear operators on V) are equivalent; see Theorem 5.30 and Exercise 5.31. Operator norms $||T||_{\text{op}}$ were discussed in Example 5.28.

CHAPTER 6

Inner Product Spaces

6.1. Basic Definitions and Examples

In calculus you encountered euclidean coordinate spaces $\mathbb{R}^n$ equipped with additional structure: an *inner product* $B : \mathbb{R}^n \times \mathbb{R}^n \to \mathbb{R}$.

$$\text{EUCLIDEAN INNER PRODUCT:} \quad B(\mathbf{x}, \mathbf{y}) = \sum_{i=1}^{n} x_i y_i,$$

which is often abbreviated to $B(x, y) = (x, y)$. Associated with it we have the *euclidean norm*

$$\|\mathbf{x}\| = \left(\sum_{i=1}^{n} |x_i|^2 \right)^{1/2} = (\mathbf{x}, \mathbf{x})^{1/2},$$

which represents the length of a vector, and a *distance function*

$$d(\mathbf{x}, \mathbf{y}) = \|\mathbf{y} - \mathbf{x}\|,$$

which gives the euclidean distance from $\mathbf{x}$ to $\mathbf{y}$, as in Figure 6.1. Note that $\mathbf{y} = \mathbf{x} + (\mathbf{y} - \mathbf{x})$.

This standard inner product on $\mathbb{R}^n$ has the following geometric interpretation (see Figure 6.2):

$$(\mathbf{x}, \mathbf{y}) = \|\mathbf{x}\| \cdot \|\mathbf{x}\| \cdot \cos(\theta(\mathbf{x}, \mathbf{y})),$$

where θ is the angle between x and y, measured in the plane $M = \mathbb{R}\text{-span}\{\mathbf{x}, \mathbf{y}\}$, the 2-dimensional subspace in $\mathbb{R}^n$ spanned by $\mathbf{x}$ and $\mathbf{y}$. *Orthogonality* of two vectors is then interpreted to mean $(\mathbf{x}, \mathbf{y}) = 0$; the zero vector is orthogonal to everybody, by definition. These notions of *length, distance*, and *orthogonality*

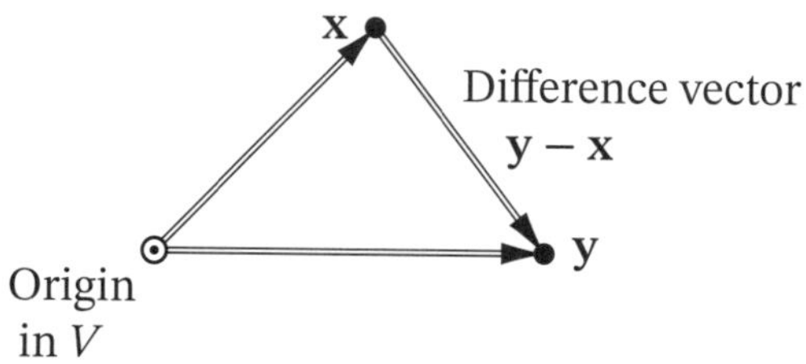

FIGURE 6.1. The distance between points $\mathbf{x}$, $\mathbf{y}$ in an inner product space is interpreted as the norm (length) $\|\mathbf{y} - \mathbf{x}\|$ of the difference vector $\Delta\mathbf{x} = \mathbf{y} - \mathbf{x}$.

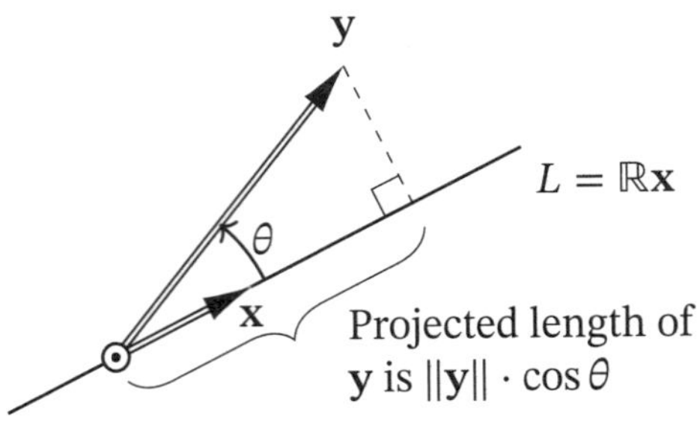

FIGURE 6.2. Geometric interpretation of the inner product $(\mathbf{x}, \mathbf{y}) = \|\mathbf{x}\|\,\|\mathbf{y}\| \cdot \cos(\theta(\mathbf{x}, \mathbf{y}))$ in $\mathbb{R}^n$. The projected length of a vector $\mathbf{y}$ onto the line $L = \mathbb{R}\mathbf{x}$ is $\|\mathbf{y}\| \cdot \cos(\theta)$. The angle $\theta(\mathbf{x}, \mathbf{y})$ is measured within the two-dimensional subspace $M = \mathbb{R}\text{-span}\{\mathbf{x}, \mathbf{y}\}$. Vectors are *orthogonal* when $\cos\theta = 0$, so $(\mathbf{x}, \mathbf{y}) = 0$. The zero vector is orthogonal to everybody.

do not exist in unadorned vector spaces. We now generalize the notion of inner product to arbitrary vector spaces, even if they are infinite-dimensional.

DEFINITION 6.1. If V is a vector space over $\mathbb{K} = \mathbb{R}$ or $\mathbb{C}$, an *inner product* is a map $B : V \times V \to \mathbb{K}$ taking ordered pairs of vectors to scalars $B(v_1, v_2) \in \mathbb{K}$ with the following properties:

1. SEPARATE ADDITIVITY IN EACH ENTRY. B is additive in each input if the other input is held fixed:
   - $B(v_1 + v_2, w) = B(v_1, w) + B(v_2, w)$
   - $B(v, w_1 + w_2) = B(v, w_1) + B(v, w_2)$

   for v, v_i, w, w_i in V.
2. HERMITIAN. For $\lambda \in \mathbb{K}$ and $v, w \in V$,
   - $B(\lambda v, w) = \lambda B(v, w)$ and $B(v, \lambda w) = \overline{\lambda} B(v, w)$.
3. POSITIVE DEFINITE. For all $v \in V$,
   - $B(v, v) \geq 0$ and $B(v, v) = 0$ if and only if $v = 0$.
4. HERMITIAN SYMMETRIC. For all $v, w \in V$,
   - $B(v, w) = \overline{B(w, v)}$ when inputs are interchanged.

   Conjugation does nothing for $x \in \mathbb{R}$ ($\overline{x} = x$ if $x \in \mathbb{R}$), so an inner product on a real vector space is simply *symmetric*, with $B(w, v) = B(v, w)$.

An inner product on a real vector space is just a *bilinear map*—one that is $\mathbb{R}$-linear in each input when the other is held fixed—because conjugation does nothing in $\mathbb{R}$.

Euclidean Norms on $\mathbb{R}^n$ and $\mathbb{C}^n$. The euclidean inner product in $\mathbb{R}^n$ is a special case of the standard euclidean inner product in complex coordinate space $V = \mathbb{C}^n$,

$$(\mathbf{z}, \mathbf{w}) = \sum_{j=1}^{n} z_j \overline{w_j}\,,$$

which is easily seen to have properties (1.)-(4.). The corresponding euclidean norm and distance functions on $\mathbb{C}^n$ are then

$$\|\mathbf{z}\| = (\mathbf{z}, \mathbf{z})^{1/2} = \Big(\sum_{j=1}^{n} |z_j|^2\Big)^{1/2} \quad \text{and} \quad d(\mathbf{z}, \mathbf{w}) = \|\mathbf{z} - \mathbf{w}\| = \Big(\sum_{j=1}^{n} |z_j - w_j|^2\Big)^{1/2}.$$

Again, properties (1.)–(4.) are easily verified.

For an arbitrary inner product B we define the corresponding norm and distance functions

$$\|v\|_B = B(v, v)^{1/2}, \quad d_B(v_1, v_2) = \|v_1 - v_2\|_B,$$

which are no longer given by such formulas.

EXAMPLE 6.2. Here are two important examples of inner product spaces.

1. On $V = \mathbb{C}^n$ (or $\mathbb{R}^n$) we can define "nonstandard" inner products by taking

$$B_\alpha(\mathbf{z}, \mathbf{w}) = \sum_{j=1}^{n} \alpha_j \cdot z_j \overline{w_j} \quad \text{with norm} \quad \|\mathbf{z}\|_\alpha = \Big(\sum_{j=1}^{n} \alpha_j \cdot |z_j|^2\Big)^{1/2},$$

   where $\alpha_j > 0$ for $1 \leq j \leq n$. This is easily seen to be an inner product. Thus the standard euclidean inner product on $\mathbb{R}^n$ or $\mathbb{C}^n$, for which $\alpha_1 = \cdots = \alpha_n = 1$, is part of a much larger family.

2. If we define

$$(f, h)_2 = \int_a^b f(t)\overline{h(t)}dt \quad \text{(Riemann integral)},$$

   the space $\mathcal{C}[a, b]$ of continuous complex-valued functions f on $[a, b]$ becomes an inner product space. The corresponding L^2-*norm* of a function is then

$$\|f\|_2 = \Big(\int_a^b |f(t)|^2\, dt\Big)^{1/2}.$$

   The inner product axioms follow from simple properties of the Riemann integral. This infinite-dimensional inner product space arises in many applications, particularly Fourier analysis.

EXERCISE 6.3. Verify that both inner products in the last example actually satisfy the inner product axioms. In particular, explain why the L^2-inner product $(f, h)_2$ has $\|f\|_2 > 0$ when f is not the zero function ($f(t) \equiv 0$ for all t).

We now take up the basic properties common to all inner product spaces.

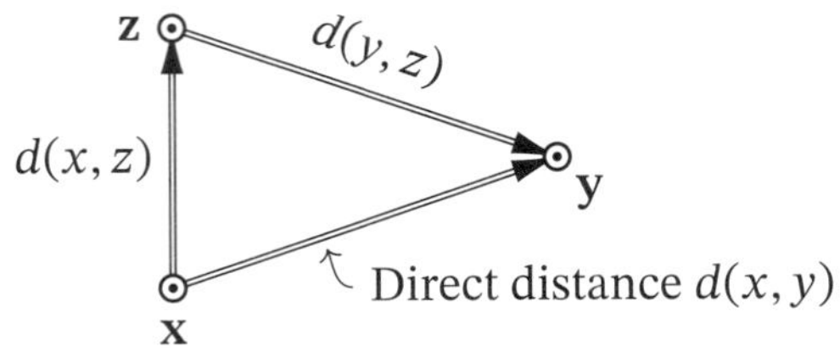

FIGURE 6.3. The meaning of the triangle inequality: direct distance from $\mathbf{x}$ to $\mathbf{y}$ is always $\leq$ the sum of distances $d(\mathbf{x},\mathbf{z}) + d(\mathbf{z},\mathbf{y})$ to any third vector $\mathbf{z} \in V$.

THEOREM 6.4. *On any inner product space V, the associated norm has the following properties:*

1. $\|x\| \geq 0$, *with* $\|x\| > 0$ *if* $x \neq 0$.
2. $\|\lambda x\| = |\lambda| \cdot \|x\|$ *(and in particular,* $\|-x\| = \|x\|$*).*
3. TRIANGLE INEQUALITY. *For* $x, y \in V$, $\|x \pm y\| \leq \|x\| + \|y\|$.

PROOF. The first two are obvious. The third is important because it implies that the distance function $d_B(x,y) = \|x - y\|$ satisfies the *geometric triangle inequalityi*

$$d_B(x,y) \leq d_B(x,z) + d_B(z,y) \quad \text{for all } x,y,z \in V$$

as indicated in Figure 6.3. This follows directly from (3.) because

$$\begin{aligned} d_B(x,y) = \|x - y\| = \|(x - z) + (z - y)\| &\leq \|x - z\| + \|z - y\| \\ &= d_B(x,z) + d_B(z,y) \end{aligned}$$

The version of (3.) involving a $(-)$ sign follows from that featuring a $(+)$ because $v - w = v + (-w)$ and $\|-w\| = \|w\|$. □

The Cauchy-Schwarz Inequality. The proof of (3.) is based on an equally important inequality:

LEMMA 6.5 (Cauchy-Schwarz Inequality). *If B is an inner product on a real or complex vector space, then*

$$|B(x,y)| \leq \|x\|_B \cdot \|y\|_B$$

for all $x, y \in V$.

PROOF. For all real t we have $\phi(t) = \|x + ty\|_B^2 \geq 0$. By the axioms governing B we can rewrite $\phi(t)$ as

$$\begin{aligned} \phi(t) &= B(x + ty, x + ty) \\ &= B(x,x) + B(ty,x) + B(x,ty) + B(ty,ty) \\ &= \|x\|_B^2 + t\,B(x,y) + t\,\overline{B(x,y)} + t^2\|y\|_B^2 \\ &= \|x\|_B^2 + 2t\,\mathrm{Re}(B(x,y)) + t^2\|y\|_B^2 \end{aligned}$$

because $B(tx, y) = tB(x, y)$ and $B(x, ty) = tB(x, y)$ for $t \in \mathbb{R}$, and $z + \overline{z} = 2\operatorname{Re}(z) = 2x$ for $z = x + iy$ in $\mathbb{C}$. Now $\phi : \mathbb{R} \to \mathbb{R}$ is a quadratic function whose minimum value occurs at t_0 where

$$\frac{d\phi}{dt}(t_0) = 2t_0\|y\|_B^2 + \operatorname{Re}(B(x, y)) = 0$$

or

$$t_0 = \frac{-\operatorname{Re}(B(x, y))}{2\|y\|_B^2}.$$

Inserting this into ϕ we find the actual minimum value of ϕ:

$$0 \leq \min\{\phi(t) : t \in \mathbb{R}\} = \frac{\|x\|_B^2 \cdot \|y\|_B^2 - 2|\operatorname{Re}(B(x, y))|^2 + |\operatorname{Re}(B(x, y))|^2}{\|y\|_B^2}.$$

Thus

$$0 \leq \|x\|_B^2 \cdot \|y\|_B^2 - |\operatorname{Re}(B(x, y))|^2,$$

which in turn implies

$$|\operatorname{Re} B(x, y)| \leq \|x\|_B \cdot \|y\|_B \quad \text{for all } x, y \in V.$$

Replacing $x \mapsto e^{i\theta}x$ does not change $\|x\|$ since $|e^{i\theta}| = |\cos(\theta) + i\sin(\theta)| = 1$ for real θ; in the inner product on the left we have $B(e^{i\theta}x, y) = e^{i\theta}B(x, y)$. We may now take $\theta \in \mathbb{R}$ so that $e^{i\theta} \cdot B(x, y) = |B(x, y)|$. For this particular choice of θ, we get

$$\begin{aligned} 0 \leq |\operatorname{Re}(B(e^{i\theta}x, y))| &= |\operatorname{Re}(e^{i\theta}B(x, y))| \\ &= \operatorname{Re}(|B(x, y)|) = |B(x, y)| \leq \|x\|_B \cdot \|y\|_B. \end{aligned}$$

That proves the Cauchy-Schwarz inequality. □

PROOF OF THE TRIANGLE INEQUALITY. The algebra is easier if we prove the (equivalent) inequality obtained when we square both sides:

$$0 \leq \|x + y\|^2 \leq (\|x\| + \|y\|)^2 = \|x\|^2 + 2\|x\| \cdot \|y\| + \|y\|^2.$$

In proving the Schwarz inequality we saw that

$$\|x + y\|^2 = (x + y, x + y) = \|x\|^2 + 2\operatorname{Re}(x, y) + \|y\|^2,$$

so our proof is finished if we can show $2\operatorname{Re}(x, y) \leq 2\|x\| \cdot \|y\|$. But

$$\operatorname{Re}(z) \leq |\operatorname{Re}(z)| \leq |z| \quad \text{for all } z \in \mathbb{C}$$

and then the Schwarz inequality yields

$$\operatorname{Re}(B(x, y)) \leq |B(x, y)| \leq \|x\|_B \cdot \|y\|_B$$

as desired. □

Hilbert-Schmidt Norm.

EXAMPLE 6.6. On $V = \mathrm{M}(n, \mathbb{K})$ we define the *Hilbert-Schmidt* inner product and norm for matrices:

$$(A,B)_{\mathrm{HS}} = \mathrm{Tr}(B^*A) \quad \text{and} \quad \|A\|_{\mathrm{HS}}^2 = \sum_{i,j=1} |a_{ij}|^2 = \mathrm{Tr}(A^*A). \tag{6.1}$$

It is easily verified that this is an inner product. First note that the trace map from $\mathrm{M}(n, \mathbb{K}) \to \mathbb{K}$,

$$\mathrm{Tr}(A) = \sum_{i=1}^{n} a_{ii},$$

is a complex linear map and $\mathrm{Tr}(\overline{A}) = \overline{\mathrm{Tr}(A)}$; then observe that

$$\|A\|_2^2 = (A,A)_{\mathrm{HS}} = \sum_{i,j=1}^{n} |a_{ij}|^2 \text{ is } > 0 \text{ unless } A \text{ is the zero matrix.}$$

Alternatively, consider what happens when we identify $\mathrm{M}(n,\mathbb{C}) \cong \mathbb{C}^{n^2}$ as complex vector spaces. The Hilbert-Schmidt norm becomes the usual euclidean norm on $\mathbb{C}^{n^2}$, and likewise for the inner products; obviously $(A,B)_{\mathrm{HS}}$ is then an inner product on matrix space.

The norm $\|A\|_{\mathrm{HS}}$ and the L^∞-norm $\|A\|_\infty$ discussed in Chapter 5 are different ways to measure the size of a matrix; the HS-norm turns out to be particularly well adapted to applications in statistics, starting with least-squares regression and moving on into analysis of variance. Each of these norms determines a notion of matrix convergence $A_n \to A$ as $n \to \infty$ in $\mathrm{M}(N,\mathbb{C})$:

$$L^2\text{-CONVERGENCE:} \quad \|A_n - A\|_{\mathrm{HS}} = \Big(\sum_{i,j} |a_{ij}^{(n)} - a_{ij}|^2\Big)^{1/2} \to 0 \text{ as } n \to \infty,$$

$$L^\infty\text{-CONVERGENCE:} \quad \|A_n - A\|_\infty = \max_{i,j}\{|a_{ij}^{(n)} - a_{ij}|\} \to 0 \text{ as } n \to \infty.$$

Despite their differences both norms *determine the same notion of matrix convergence:*

$$A_n \to A \text{ in } \|\cdot\|_2\text{-norm} \Leftrightarrow A_n \to A \text{ in } \|\cdot\|_\infty\text{-norm}.$$

The reason is explained below and was explored in more detail in Chapter 5.

EXERCISE 6.7. Show that there exist bounds $M_2, M_\infty > 0$ such that the $\|*\|_2$ and $\|*\|_\infty$ norms on matrix space mutually dominate each other,

$$\|x\|_2 \leq M_\infty \|x\|_\infty \quad \text{and} \quad \|x\|_\infty \leq M_2 \|x\|_2$$

for all $x \in \mathbb{C}^n$, so that $A_n \to A$ in the L^2-norm if and only if $A_n \to A$ in L^∞-norm.

HINT: The Cauchy-Schwarz inequality might be helpful in one direction.

Polarization Identity. The *polarization identities* below show that inner products over $\mathbb{R}$ or $\mathbb{C}$ can be reconstructed if we only know the norms of vectors in V. Over $\mathbb{C}$ we have

$$B(x,y) = \frac{1}{4}\sum_{k=0}^{3}\frac{1}{i^k}B(x+i^ky, x+i^ky) = \frac{1}{4}\sum_{k=0}^{3}\frac{1}{i^k}\,\|x+i^ky\|^2, \tag{6.2}$$

where $i=\sqrt{-1}$. Over $\mathbb{R}$ we only need two terms,

$$B(x,y) = \frac{1}{4}(B(x+y,x+y)+(-1)B(x-y,x-y)).$$

EXERCISE 6.8. Expand

$$(x+i^ky, x+i^ky) = \|x+i^ky\|^2$$

to verify the polarization identities.

Orthonormal Bases in Inner Product Spaces. Abbreviating $B(x,y) = (x,y)$, we say that a set $\mathfrak{X} = \{e_i : i \in I\}$ of vectors is *orthogonal* if $(e_i, e_j) = 0$ for $i \neq j$; it is *orthonormal* if

$$(e_i, e_j) = \delta_{ij} \text{ (Kronecker delta)} \quad \text{for all } i, j \in I.$$

In infinite-dimensional inner product spaces, an orthonormal set can be infinite and all vectors in it are nonzero; vectors in an orthogonal family could have various lengths $\|v_i\|$, or even $v_i = 0$ for some indices since $(v, 0) = 0$ for any v. The set $\mathfrak{X}$ is an *orthonormal basis* if it is orthonormal and V is spanned by $\mathfrak{X}$.

PROPOSITION 6.9. *Orthonormal sets have the following properties:*

1. *Orthonormal sets are independent.*
2. *If $\mathfrak{X} = \{e_i : i \in I\}$ is a finite orthonormal set and v is in $M = \mathbb{K}\text{-span}\{\mathfrak{X}\}$, then by* (1.) *$\mathfrak{X}$ is a basis for M and the expansion of any v in M with respect to this basis is just*

$$v = \sum_{i\in I}(v, e_i)e_i.$$

   (*Finiteness of $\mathfrak{X}$ is required for $\sum_{i\in I}(\cdot)$ to make sense; otherwise the right side is an infinite series.*)

In particular, if $\mathfrak{X} = \{e_1, ..., e_n\}$ is an orthonormal basis for a finite-dimansional inner product space V, the coefficients in the expansion

$$v = \sum_{i=1}^{n}(v, e_i)e_i \quad \textit{for every } v \in V$$

are easily computed by taking inner products.

PROOF. For (1.), if a finite sum $\sum_i c_ie_i$ equals 0 we have

$$0 = (v, e_k) = \sum_i c_i(e_i, e_k) = \sum_i c_i\delta_{ik} = c_k$$

for each k, so the e_i are independent. Part (2.) follows from (1.): we know $\{e_i\}$ is a basis, and if $v = \sum_i c_i e_i$ is its expansion, the inner product with a typical basis vector is

$$(v, e_k) = \sum_i c_i (e_i, e_k) = \sum_i c_i \delta_{ik} = c_k. \qquad \square$$

COROLLARY 6.10. *If vectors* $\{v_1, ..., v_n\}$ *are nonzero, orthogonal, and a vector basis in* V*, the renormalized vectors*

$$e_i = \frac{v_i}{\|v_i\|} \quad \text{for } 1 \leq i \leq n$$

are an orthonormal basis.

Entries in the matrix $[T]_{\mathfrak{Y}\mathfrak{X}}$ of a linear operator between inner product spaces are easily computed by taking inner products if the bases are orthonormal (but not for arbitrary bases).

EXERCISE 6.11. Let $T : V \to W$ be a linear operator between finite-dimensional inner product spaces and let $\mathfrak{X} = \{e_i\}$, $\mathfrak{Y} = \{f_i\}$ be orthonormal bases. Prove that the entries in $[T]_{\mathfrak{Y}\mathfrak{X}}$ are given by

$$T_{ij} = (T(e_j), f_i)_W = \overline{(f_i, T(e_j))_W}$$

for $1 \leq i \leq \dim(W)$, $1 \leq j \leq \dim(V)$.

Bessel's Inequality. The fundamental fact about orthonormal bases is that the coefficients in $v = \sum_{k=1}^{n} (v, e_i)\, e_i$ determine the norm $\|v\|$ via a generalization of the Pythagoras' Formula for $\mathbb{R}^n$,

$$\text{PYTHAGORAS:} \quad \text{If } \mathbf{x} = \sum_{i=1}^{n} x_i \mathbf{e}_i, \text{ then } \|\mathbf{x}\|^2 = \sum_{i=1}^{n} |x_i|^2.$$

We start by proving a fundamental inequality.

THEOREM 6.12 (Bessel's Inequality). *If* $\mathfrak{X} = \{e_1, \ldots, e_n\}$ *is any finite orthonormal set in an inner product space* V *(possibly infinite-dimensional), then*

$$\sum_{i=1}^{n} |(v, e_i)|^2 \leq \|v\|^2 \quad \text{for all } v \in V \tag{6.3}$$

Furthermore, if $v' = v - \sum_{i=1}^{n} (v, e_i)\, e_i$*, this vector is orthogonal to each* e_j *and hence is orthogonal to all the vectors in the linear span* $M = \mathbb{K}\text{-span}\{\mathfrak{X}\}$.

NOTE: *The inequality* (6.3) *becomes an equality if* $\mathfrak{X}$ *is an orthonormal basis for* V *because then* $v' = 0$.

PROOF. Since inner products are conjugate bilinear, we have

$$
\begin{aligned}
0 \le \|v'\|^2 = (v', v') &= \Big(v - \sum_{i=1}^{m} (v, e_i)\, e_i, v - \sum_{j=1}^{m} (v, e_j)\, e_j\Big) \\
&= (v, v) - \Big(\sum_i (v, e_i) e_i, v\Big) - \Big(v, \sum_j (v, e_j)\, e_j\Big) + \Big(\sum_i (v, e_i)\, e_i, \sum_j (v, e_j)\, e_j\Big) \\
&= \|v\|^2 - \sum_i (v, e_i) \cdot (e_i, v) - \sum_j \overline{(v, e_j)} \cdot (v, e_j) + \sum_{i,j} (v, e_i) \cdot \overline{(v, e_j)} \cdot (e_i, e_j) \\
&= \|v\|^2 - \sum_i |(v, e_i)|^2 - \sum_j |(v, e_j)|^2 + \sum_i |(v, e_i)|^2 \quad (\text{since } (e_k, v) = \overline{(v, e_k)}\,) \\
&= \|v\|^2 - \sum_i |(v, e_i)|^2,
\end{aligned}
$$

so that

$$\sum_{i=1} |(v, e_i)|^2 \le \|v\|^2.$$

The second statement now follows easily because

$$
\begin{aligned}
(v', e_k) = \Big(v - \sum_j (v, e_j)\, e_j, e_k\Big) &= (v, e_k) - \sum_j (v, e_j) \cdot (e_j, e_k) \\
&= (v, e_k) - (v, e_k) = 0 \quad \text{for all } k.
\end{aligned}
$$

Furthermore, if $w = \sum_{k=1}^{m} c_k e_k$ is any vector in M, we also have

$$(v', w) = \sum_k c_k (v', e_k) = 0,$$

so v' is orthogonal to M as claimed. □

COROLLARY 6.13 (Pythagoras). *If $\mathfrak{X} = \{e_1, \dots, e_n\}$ is an orthonormal basis in a finite-dimensional inner product space, then*

$$\|v\|^2 = \sum_{i=1}^{m} |(v, e_i)|^2$$

(sum of squares of the coefficients in the basis expansion $v = \sum_i (v, e_i) e_i$).

THEOREM 6.14. *Orthonormal bases exist in a finite-dimensional inner product space.*

PROOF. We argue by induction on $n = \dim(V)$, the result being trivial if $n = 1$ (any vector of length 1 is an orthonormal basis). If $\dim(V) = n+1$, let v_0 be any nonzero vector. The linear functional $\ell_0 : v \to (v, v_0)$ is nonzero, and as in Example 3.4, its kernel

$$M = \{v : (v, v_0) = 0\} = (\mathbb{K} v_0)^{\perp}$$

is a hyperplane of dimension $\dim(V) - 1 = n$. By the induction hypothesis there is an orthonormal basis $\mathfrak{X}_0 = \{e_1, , \dots, e_n\}$ in M, and every vector in M is orthogonal to v_0. If we rescale v_0 and adjoin $e_{n+1} = v_0/\|v_0\|$ to $\mathfrak{X}_0$, the enlarged set $\mathfrak{X} = \{e_1, \dots, e_n, e_{n+1}\}$ is obviously orthonormal; it is also a basis for V. (Obviously $\mathfrak{X}$ is a basis for $W = \mathbb{K}\text{-span}\{\mathfrak{X}\} \subseteq V$, and since $\dim(W) = |\mathfrak{X}| = n+1 = \dim(V)$, we must have $W = V$.) □

6.2. Orthogonal Complements and Projections

For M a subspace of a (possibly infinite-dimensional) inner product space V, its *orthogonal complement* $M^\perp$ is the set of vectors orthogonal to every vector in M,

$$M^\perp = \{v \in V : (v, m) = 0 \text{ for all } m \in M\} = \{v : (v, M) = \{0\}\}.$$

Obviously $\{0\}^\perp = V$ and $V^\perp = \{0\}$ by the axioms for the inner product.

EXERCISE 6.15. Show that $M^\perp$ is again a subspace of V and that

$$M_1 \subseteq M_2 \Rightarrow M_2^\perp \subseteq M_1^\perp.$$

PROPOSITION 6.16. *If M is a finite-dimensional subspace of a (possibly infinite-dimensional) inner product space V, then*

1. *$M \cap M^\perp = \{0\}$ and $M + M^\perp = V$, so we have a direct sum decomposition $V = M \oplus M^\perp$.*
2. *If $\dim(V) < \infty$ we also have $(M^\perp)^\perp = M$; if $|V| = \infty$ we can only say that $M \subseteq (M^\perp)^\perp$.*

PROOF. If $v \in M \cap M^\perp$, then $\|v\|^2 = (v, v) = 0$ so $v = 0$ and $M \cap M^\perp = \{0\}$. Now let $\{e_1, \dots, e_n\}$ be an orthonormal basis for M. If $v \in V$, write

$$v = \left(v - \sum_{i=0}^{m} (v, e_i)\, e_i\right) + \sum_{i=1}^{m} (v, e_i)\, e_i = v_\perp + v_\|,$$

in which $v_\perp$ is orthogonal to M and $v_\|$ is the component of v “parallel to” the subspace M (because it lies in M). Then for all $v \in V$ we have

$$(v, v_\perp) = (v_\perp + v_\|, v_\perp) = (v_\perp, v_\perp) + (v_\|, v_\perp) = \|v_\perp\|^2 + 0 = \|v_\perp\|^2.$$

Obviously $(M^\perp)^\perp \supseteq M$. If $v \in (M^\perp)^\perp$, so $(v, v_\perp) = 0$, we conclude that $\|v_\perp\|^2 = 0$ and hence $v = v_\perp + v_\| = 0 + v_\|$ is in M. That proves the reverse inclusion $M^{\perp\perp} \subseteq M$. □

The situation is illustrated in Figure 6.4.

Orthogonal Projections on Inner Product Spaces. If an inner product space is a direct sum $V = V_1 \oplus \cdots \oplus V_r$, it is an *orthogonal direct sum* if the subspaces are pairwise orthogonal,

$$(V_i, V_j) = 0 \quad \text{if } i \neq j.$$

We indicate this by writing $V = V_1 \dot{\oplus} \cdots \dot{\oplus} V_r = \dot{\bigoplus}_{i=1}^{r} V_i$. The decomposition $V = M \dot{\oplus} M^\perp$ of Proposition 6.16 was an orthogonal decomposition.

In Exercise 2.25 we defined the linear projection operators $P_i : V \to V$ associated with an ordinary direct sum decomposition $V = V_1 \oplus \cdots \oplus V_r$, and

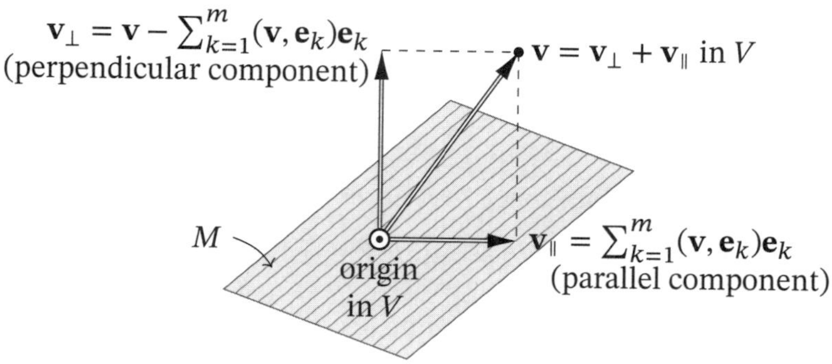

FIGURE 6.4. Given an orthonormal basis $\{e_i, \dots, e_m\}$ in a finite-dimensional subspace $M \subseteq V$, the vector $v_{\|} = \sum_{k=1}^{m}(v, e_k)\, e_k$ is in M and $v_{\perp} = v - v_{\|}$ is orthogonal to M. These are the components of $v \in V$ "parallel to M" and "perpendicular to M," with $v = v_{\perp} + v_{\|}$.

showed that these linear operators have the *idempotent property* $P^2 = P$. In fact, there is a bijective correspondence

$$(\textit{idempotent linear operators}) \longleftrightarrow (\textit{direct sum decompositions } V = R \oplus K),$$

described in Proposition 2.27, reprised below.

PROPOSITION 2.27. If a linear operator $P : V \to V$ is idempotent, so $P^2 = P$, there is a direct sum decomposition $V = R \oplus K$ such that P projects V onto R along K. In particular,

$$R = R(P) = \text{range}(P) \quad \text{and} \quad K = K(P) = \ker(P).$$

Furthermore, $Q = I - P$ is also idempotent and

$$R(Q) = K(P) \quad \text{and} \quad K(Q) = R(P).$$

When V is an inner product space we will see that the projections associated with an *orthogonal* direct sum $V = E\dot{\oplus}F$ have special properties. They are also easy to compute using the inner product. (Compare what follows with the calculations in Example 2.26, of projections associated with an ordinary direct sum decomposition $V = E \oplus F$ in a space without inner product.)

Projections associated with an orthogonal direct sum decomposition $V = V_1\dot{\oplus} \cdots \dot{\oplus}V_r$ are called *orthogonal projections.*

LEMMA 6.17. *If $V = E\dot{\oplus}F$ is an orthogonal direct sum decomposition of a finite-dimensional inner product space, then*

$$E^{\perp} = F \textit{ and } F^{\perp} = E, \quad E^{\perp\perp} = E \textit{ and } F^{\perp\perp} = F.$$

PROOF. The argument for F is the same as that for E. We proved that $E^{\perp\perp} = E$ in Proposition 6.16 and we know that $E \subseteq F^{\perp}$ by definition; based on this we will prove the reverse inequality $E \supseteq F^{\perp}$.

Since $|V| < \infty$ we have $V = F\dot{\oplus}F^{\perp}$, so that $|V| = |F|+|F^{\perp}|$; since $V = E\dot{\oplus}F$ we also have $|V| = |F|+|E|$. Therefore $|E| = |F^{\perp}|$. But $E \subseteq F^{\perp}$ in an orthogonal direct sum $E\dot{\oplus}F$, so we conclude that $E = F^{\perp}$. □

EXERCISE 6.18. Let $V = V_1\dot{\oplus}\cdots\dot{\oplus}V_r$ be an orthogonal direct sum decomposition of an inner product space (not necessarily finite dimensional).

(a) If W_i is the linear span $\sum_{j\neq i} V_j$, prove that $W_i \perp V_i$ for each i, and $V = V_i\dot{\oplus}W_i$.

(b) If $v = v_1 + \cdots + v_r$ is the unique decomposition into pairwise orthogonal vectors $v_i \in V_i$, prove that $\|v\|^2 = \sum_i \|v_i\|^2$.

The identity (b) is yet another version of Pythagoras' formula.

EXERCISE 6.19. In a finite-dimensional inner product space, prove that the *Parseval formula*

$$(v, w) = \sum_{i=1}^{n} (v, e_i)\cdot(e_i, w) = \sum_{i=1}^{n}(v, e_i)\overline{(e_i, w)}$$

holds for every orthonormal basis $\{e_1, \dots, e_n\}$.

The Gram-Schmidt Construction. We now show how any independent set of vectors $\{v_1, \dots, v_n\}$ in an inner product space can be modified to obtain an orthonormal set of vectors $\{e_1, \dots, e_n\}$ with the same linear span. This *Gram-Schmidt construction* is recursive, and at each step we have the following:

1. $e_k \in \mathbb{K}\text{-span}\{v_1, \dots, v_k\}$.
2. $M_k = \mathbb{K}\text{-span}\{e_1, \dots, e_k\}$ is equal to $\mathbb{K}\text{-span}\{v_1, .., v_k\}$.

The result is an orthonormal basis $\{e_1, \dots, e_k\}$ for $M = \mathbb{K}\text{-span}\{v_1, .., v_k\}$ (and for all of V if the $\{v_i\}$ span V). The construction proceeds inductively by constructing two sequences of vectors $\{u_i\}$ and $\{e_i\}$.

Step 1. Take

$$u_1 = v_1 \quad \text{and} \quad e_1 = \frac{v_1}{\|v_1\|}.$$

Conditions (1.) and (2.) obviously hold and $\mathbb{K}\cdot v_1 = \mathbb{K}\cdot u_1 = \mathbb{K}\cdot e_1$.

Step 2. Define

$$u_2 = v_2 - (v_2, e_1)\cdot e_1 \quad \text{and} \quad e_2 = \frac{u_2}{\|u_2\|}.$$

Obviously $u_2 \in \mathbb{K}\text{-span}\{v_1, v_2\}$ and $u_2 \neq 0$ because $v_2 \notin \mathbb{K}v_1 = \mathbb{K}e_1 = M_1$; thus e_2 is well-defined. Furthermore:

1. $u_2 \perp M_1$ because

$$(u_2, e_1) = (v_2 - (v_2, e_1)e_1, e_1) = (v_2, e_1) - (v_2, e_1)\cdot(e_1, e_1) = 0.$$

This implies $e_2 \perp M_1$, so $\{e_1, e_2\}$ is an orthonormal set of vectors.

2. $M_2 = \mathbb{K}\text{-span}\{e_1, e_2\} = \mathbb{K}u_2 + \mathbb{K}e_1 = \mathbb{K}v_2 + \mathbb{K}e_1 = \mathbb{K}v_2 + \mathbb{K}v_1$
$= \mathbb{K}\text{-span}\{v_1, v_2\}$.

If $n = 2$ we're done; otherwise we continue:

Step 3. Define

$$u_3 = v_3 - \sum_{i=1}^{2}(v_3, e_i) \cdot e_i = v_3 - \sum_{i=1}^{2} \frac{(v_3, u_i)}{\|u_i\|^2} u_i.$$

Then $u_3 \neq 0$ because the sum is in $\mathbb{K}$-span$\{v_1, v_2\}$ and the v_i are independent; thus $e_3 = \frac{u_3}{\|u_3\|}$ is well-defined. We have $u_3 \perp M_2$ because

$$\begin{aligned}(u_3, e_1) = \left(v_3 - \sum_{i=1}^{2}(v_3, e_i)e_i, e_1\right) &= (v_3, e_1) - \sum_{i=1}^{2}(v_3, e_i) \cdot (e_i, e_1) \\ &= (v_3, e_1) - (v_3, e_1) = 0,\end{aligned}$$

and similarly $(u_3, e_2) = 0$; hence $u_3 \perp M_2$ and $e_3 \perp M_2$. Finally,

$$\begin{aligned}\mathbb{K}\text{-span}\{e_1, e_2, e_3\} &= \mathbb{K}u_3 + \mathbb{K}\text{-span}\{e_1, e_2\} = \mathbb{K}v_3 + \mathbb{K}\text{-span}\{e_1, e_2\} \\ &= \mathbb{K}v_3 + \mathbb{K}\text{-span}\{v_1, v_2\} = \mathbb{K}\text{-span}\{v_1, v_2, v_3\}\end{aligned}$$

At the kth step we have produced orthonormal vectors $\{e_1, ..., e_k\}$ with $\mathbb{K}$-span$\{e_1, ..., e_k\} = \mathbb{K}$-span$\{v_1, ..., v_k\} = M_k$. Now for the induction step:

Step $k+1$. Define

$$u_{k+1} = v_{k+1} - \sum_{i=1}^{k}(v_{k+1}, e_i)\, e_i = v_{k+1} - \sum_{i=1}^{k} \frac{(v_{k+1}, u_i)}{\|u_i\|^2} u_i$$

and

$$e_{k+1} = \frac{u_{k+1}}{\|u_{k+1}\|}.$$

Again $u_{k+1} \neq 0$ because $v_{k+1} \notin M_k = \mathbb{K}$-span$\{v_1, ..., v_k\} = \mathbb{K}$-span$\{e_1, ..., e_k\}$, so e_{k+1} is well-defined. Furthermore, $u_{k+1} \perp M_k$ because

$$\begin{aligned}(u_{k+1}, e_j) = \left(v_{k+1} - \sum_{i=1}^{k}(v_{k+1}, e_i)\, e_i, e_j\right) &= (v_{k+1}, e_j) - \sum_{i=1}^{k}(v_{k+1}, e_i) \cdot (e_i, e_j) \\ &= (v_{k+1}, e_j) - (v_{k+1}, e_j) = 0,\end{aligned}$$

hence also $e_{k+1} \perp M_k$. Then

$$\begin{aligned}\mathbb{K}\text{-span}\{e_1, ..., e_{k+1}\} &= \mathbb{K}u_{k+1} + \mathbb{K}\text{-span}\{e_1, ..., e_k\} \\ &= \mathbb{K}v_{k+1} + \mathbb{K}\text{-span}\{e_1, ..., e_k\} = \mathbb{K}\text{-span}\{v_1, ..., v_{k+1}\}.\end{aligned}$$

By induction, $\{e_1, ..., e_n\}$ has the properties claimed.

Note that the outcome of Step $(k+1)$ depends only on the $\{e_1, ..., e_k\}$ and the new vector v_{k+1}; the original vectors $\{v_1, ..., v_k\}$ play no further role in the inductive process.

EXAMPLE 6.20. The standard inner product in $\mathcal{C}[-1,1]$ is the L^2 inner product

$$(f,h)_2 = \int_{-1}^{1} f(t)\overline{h(t)}dt$$

for functions $f : [-1,1] \to \mathbb{C}$. Regarding $v_1 = \mathbb{1}$, $v_2 = x$, and $v_3 = x^2$ as functions from $[-1,1] \to \mathbb{C}$, these vectors are independent. Find the orthonormal set $\{e_1, e_2, e_3\}$ produced by the Gram-Schmidt process.

SOLUTION: We have $u_1 = v_1 = \mathbb{1}$ and since $\|u_1\|^2 = \int_{-1}^{1} \mathbb{1}\, dx = 2$, we get $e_1 = \frac{1}{\sqrt{2}} \cdot \mathbb{1}$. At the next step,

$$u_2 = v_2 - (v_2, e_1)\, e_1 = v_2 - \frac{(v_2, u_1)}{\|u_1\|^2}\, u_1 = x - \frac{\int_{-1}^{1} x \cdot \mathbb{1}\, dx}{\|u_1\|^2} \cdot \mathbb{1} = x - 0 = x$$

and

$$\|u_2\|^2 = \int_{-1}^{1} x^2\, dx = 2\int_{0}^{1} x^2\, dx = 2\left[\frac{1}{3}x^3\Big|_0^1\right] = \frac{2}{3}$$

The second basis vector is

$$e_2 = \frac{u_2}{\|u_2\|} = \sqrt{\frac{3}{2}} \cdot x.$$

At the next step:

$$\begin{aligned}
u_3 &= v_3 - \left((v_3, e_1)e_1 + (v_3, e_2)e_2\right) \\
&= v_3 - \left(\frac{(v_3, u_2)}{\|u_2\|^2} \cdot u_2 + \frac{(v_3, u_1)}{\|u_1\|^2} \cdot u_1\right) \\
&= x^2 - \frac{\int_{-1}^{1} x^2 \cdot x\, dx}{\frac{2}{3}} \cdot x - \frac{\int_{-1}^{1} x^2 \cdot \mathbb{1}\, dx}{2} \cdot \mathbb{1} \\
&= x^2 - 0 - \frac{1}{3}\mathbb{1} = x^2 - \frac{1}{3}
\end{aligned}$$

Then

$$\begin{aligned}
\|u_3\|^2 &= \int_{-1}^{1} |u_3(x)|^2\, dx = \int_{-1}^{1} \left(x^2 - \frac{1}{3}\right)^2 dx \\
&= \int_{-1}^{1} \left(x^4 - \frac{2}{3}x^2 + \frac{1}{9}\right) dx \\
&= 2 \cdot \left[\frac{x^5}{5} - \frac{2}{9}x^3 + \frac{1}{9}x\Big|_0^1\right] = \frac{8}{45},
\end{aligned}$$

and the third orthonormal basis vector is

$$e_3 = \frac{u_3}{\|u_3\|} = \sqrt{\frac{45}{8}}\left(x^2 - \frac{1}{3}\right) = \sqrt{\frac{5}{8}}(3x^2 - 1).$$

If we extend the original list to include $v_4 = x^4$ we may compute e_4 knowing only e_1, e_2, e_3 (or u_1, u_2, u_3) and v_4; there is no need to repeat the previous calculations.

EXERCISE 6.21. Find u_4 and e_4 in the above situation.

The Legendre Polynomials. This process can be continued indefinitely to produce the orthonormal family $e_1(t), e_2(t), \dots, e_n(t), \dots$ of *Legendre polynomials* in the space of polynomials $\mathbb{C}[x]$ restricted to the interval $[-1, 1]$. (This is also true for $\mathbb{R}[x]$ restricted to $[-1, 1]$ since the Legendre polynomials all have real coefficients.) Clearly the $(n + 1)$-dimensional subspace M_n obtained by restricting the space of polynomials of degree $\leq n$,

$$\mathcal{P}_n = \mathbb{K}\text{-span}\{e_1, \dots, e_{n+1}\} = \mathbb{K}\text{-span}\{1, x, \dots, x^n\}$$

to the interval, so $M_n = \mathcal{P}_n|_{[-1,1]}$, has $\{e_1, \dots, e_{n+1}\}$ as an orthonormal basis with respect to the usual inner product on $\mathcal{C}[-1, 1]$,

$$(f, h)_2 = \int_{-1}^{1} f(t)\overline{h(t)}\, dt.$$

Restricting the set of Legendre polynomials $e_1(t), \dots, e_{n+1}(t), \dots$ to $[-1, 1]$ yields an orthonormal set of vectors in the infinite-dimensional inner product space $\mathcal{C}[-1, 1]$. The orthogonal projection $P_n : \mathcal{C}[-1, 1] \to M_n \subseteq \mathcal{C}[-1, 1]$ associated with the orthogonal direct sum decomposition $V = M_n \oplus (M_n)^{\perp}$ (in which $\dim(M_n)^{\perp} = \infty$) is given by the explicit formula

$$\begin{aligned} P_n f(t) &= \sum_{k=1}^{n+1} (f, e_k) e_k(t) \quad (-1 \leq t \leq 1) \\ &= \sum_{k=1}^{n+1} \left(\int_{-1}^{1} f(x)\overline{e_k(x)} dx \right) \cdot e_k(t) = \sum_{k=0}^{n} c_{k,n} t^k \quad (c_k \in \mathbb{C}) \end{aligned}$$

for any continuous function on $[-1, 1]$. The projected image $P_n f(t)$ is a polynomial of degree $\leq n$ even though $f(t)$ is continuous and need not be differentiable. Note that the coefficients of t^k change with n, unlike those in the Taylor series.

A standard result from analysis shows that the partial sums of the infinite series $\sum_{k=0}^{\infty} c_k e_k(t)$ converge in the L^2-norm to the original function $f(t)$ throughout $-1 \leq t \leq 1$,

$$\|f - P_n f\|_2 = \left(\int_{-1}^{1} |f(t) - P_n f(t)|^2\, dt \right)^{1/2} \to 0 \quad \text{as } n \to \infty$$

for all $f \in \mathcal{C}[-1, 1]$.

It must be noted that this series expansion of $f(t) \sim \sum_{k=0}^{\infty} c_k\, t^k$ is not at all the same thing as a Taylor series expansion about $t = 0$, which in any case would not make sense because $f(t)$ is only assumed continuous (the derivatives used to compute Taylor coefficients might not exist). In fact, convergence of this series in the L^2-norm is much more robust than pointwise convergence of Taylor series, which is why norm convergence is so useful in applications.

Fourier Series Expansions. The *complex trig polynomials*

$$E_n(t) = e^{2\pi int} = \cos(2\pi nt) + i\sin(2\pi nt) \quad (n \in \mathbb{Z})$$

are periodic complex-valued functions on $\mathbb{R}$; each has period $\Delta t = 1$ since

$$e^{2\pi in(t+1)} = e^{2\pi int} \cdot e^{2\pi in} = e^{2\pi int} \quad \text{for all } t \in \mathbb{R} \text{ and } n \in \mathbb{Z}.$$

If $e_n(t)$ is the restriction of $E_n(t)$ to the "period interval" $I = [0,1]$, we get an orthonormal family of vectors in $\mathcal{C}[0,1]$ with respect to the usual inner product $(f,h) = \int_0^1 f(t)\overline{h(t)}dt$ because

$$\|e_n\|^2 = \int_0^1 |e_n(t)|^2\,dt = \int_0^1 1\,dt = 1,$$

$$(e_m, e_n) = \int_0^1 e_m(t)\overline{e_n(t)}dt = \int_0^1 e^{2\pi i(m-n)t}\,dt$$

$$= \left[\frac{e^{2\pi i(m-n)t}}{2\pi i(m-n)}\Big|_0^1\right] = 0 \quad \text{if } m \neq n.$$

Thus $\{e_n : n \in \mathbb{Z}\}$ is an orthonormal family in $\mathcal{C}[0,1]$.

For $N \geq 0$ let $M_N = \mathbb{K}\text{-span}\{e_k : -N \leq k \leq N\}$. For f in this subspace we have the basis expansion

$$f = \sum_{k=-N}^{N} (f, e_k)\, e_k = \sum_{k=-N}^{N} c_k e^{2\pi ikt}$$

where c_k is the kth *Fourier coefficient*

$$c_k = (f, e_k) = \int_0^1 f(t) e^{-2\pi ikt}\,dt. \tag{6.4}$$

By Bessel's inequality:

$$\|f\|_2^2 = \int_0^1 |f(t)|^2\,dt \geq \sum_{k=-N}^{N} |c_k|^2 = \sum_{k=-N}^{N} |(f, e_k)|^2,$$

and this is true for $N = 1, 2, \ldots$. The projection P_N of $\mathcal{C}[0,1]$ onto M_N along $M_N^\perp$ is then given by

$$P_N f(t) = \sum_{k=-N}^{N} c_k e_k(t) = \sum_{k=-N}^{N} (f, e_k)\, e^{2\pi ikt}, \quad N = 0, 1, 2, \ldots$$

because $P_N(f) \in M_N$ by definition, and $(f - P_N f, e_k) = 0$ for $-N \leq k \leq N$.

The *Fourier series* of a continuous (or bounded Riemann integrable) complex-valued function $f : [0,1] \to \mathbb{C}$ is the infinite series of trigonometric polynomials

$$f \sim \sum_{k\in\mathbb{Z}} (f, e_k) \cdot e^{2\pi ikt} \tag{6.5}$$

whose coefficients $c_k = (f, e_n)$ are the Fourier coefficients defined in (6.4).

It is not immediately clear when this series converges,iiwheeries! convergence is suitably interpreted it can be proved that the series does converge, and to the initial function $f(t)$. This expansion has proved to be extremely useful in applications. Its significance is best described as follows.

If t is regarded as a time variable, and $F(t)$ is some sort of periodic "signal" or "waveform" such that $F(t+1) = F(t)$ for all t, then F is completely determined by its restriction $f = F|_{[0,1]}$ to the basic period interval $0 \le t \le 1$. The Fourier series expansion of f on this interval can in turn be regarded as representating the original waveform as a "superposition," with suitable weights, of the basic periodic waveforms $E_n(t) = e^{2\pi int}$ $(t \in \mathbb{R})$:

$$F(t) \sim \sum_{n=-\infty}^{+\infty} c_n \cdot E_n(t) \quad \text{for all } t \in \mathbb{R}.$$

For instance, this implies that any periodic sound wave $F(t)$ with period $\Delta t = 1$ can be reconstructed by superposing scalar multiples of the "pure tones" $E_n(t)$, which have frequencies $\omega_n = n$ cycles per second. This is precisely how sound synthesizers work. It is remarkable that the correct "weight" assigned to each pure tone is the Fourier coefficient $c_n = (f, e_n)$; even more remarkable is the fact that *complex-valued* weights $c_k \in \mathbb{C}$ must be allowed, even if the signal is real-valued, because the functions $E_n(t) = \cos(2\pi nt) + i\sin(2\pi nt)$ are complex-valued.

If f is piecewise differentiable the infinite series, (6.5) converges (except at points of discontinuity) to the original periodic functiom $f(t)$. Furthermore, the following results can be proved for any continuous (or Riemann integrable) function on $[0,1]$.

THEOREM 6.22. *If $f(t)$ is bounded and Riemann integrable for* $0 \le t \le 1$, *then*

1. L^2-NORM CONVERGENCE: *The partial sums of the Fourier series* (6.5) *converge to $f(t)$ in the L^2-norm*:

$$\left\| f - \sum_{k=-N}^{N} (f, e_k)\, e_k \right\|_2 \to 0 \quad \text{as } N \to \infty.$$

2. EXTENDED BESSEL: $\|f\|^2 = \int_0^1 |f(t)|^2\,dt$ *is equal to* $\sum_{k\in\mathbb{Z}} |(f, e_k)|^2$.

The norm $\|f - h\|_2 = \left(\int |f-h|^2\,dt\right)^{1/2}$ is often referred to as the "RMS = root mean square" distance between f and h.

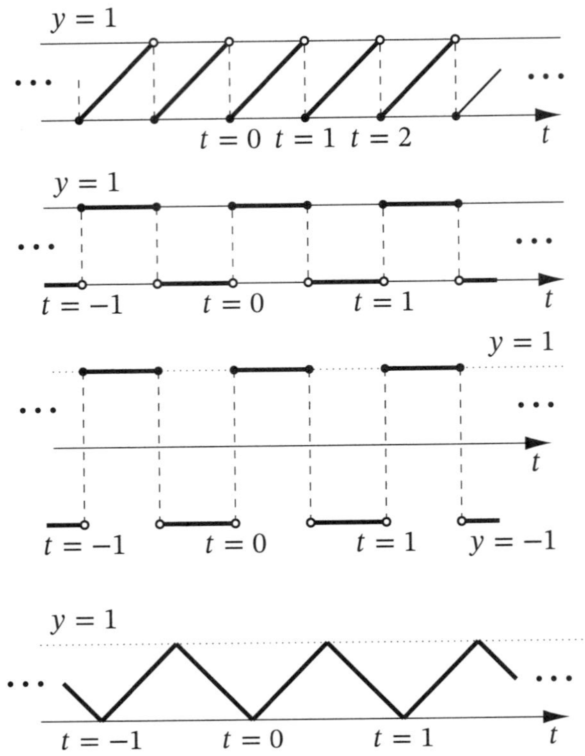

FIGURE 6.5. Various waveforms with period $\Delta t = 1$, whose Fourier transforms can be computed by calculus methods.

EXAMPLE 6.23. Let

$$f(t) = \begin{cases} t & \text{for } 0 \le t < 1, \\ 0 & \text{for } t = 1. \end{cases}$$

This is the restriction to $[0, 1]$ of the periodic "sawtooth" waveform in Figure 6.5(a). Find its Fourier series.

Solution: If $k \neq 0$ integration by parts yields

$$\begin{aligned} c_k &= \int_0^1 te^{-2\pi ikt}\,dt \\ &= \left[\frac{-1}{2\pi ik}e^{-2\pi ikt} \cdot t\bigg|_0^1\right] - \int_0^1 \frac{-1}{2\pi ik}e^{-2\pi ikt}\,dt \\ &= \frac{-1}{2\pi ik} + \frac{1}{2\pi ik}(e_k, e_0) \quad (\text{where } e_0(t) \equiv 1 \text{ for all } t) \\ &= \frac{-1}{2\pi ik} \quad \text{if } k \neq 0. \end{aligned}$$

For $k = 0$ we get a different result: by direct calculation

$$c_0 = \int_0^1 t\,dt = \frac{1}{2}.$$

By Bessel's inequality we have

$$\begin{aligned} \|f\|_2^2 &= \int_0^1 |f(t)|^2\,dt = \int_0^1 t^2\,dt = \frac{1}{3} \quad \text{(by direct calculation)} \\ &\geq \sum_{k=-N}^{N} |(f, e_k)|^2 = \sum_{k=-N}^{N} |c_k|^2, \\ &= \frac{1}{4} + \sum_{k\neq 0, -N\leq k\leq N} \frac{1}{4\pi^2 k^2} \end{aligned}$$

for any $N = 1, 2, \dots$. If we multiply both sides by $4\pi^2$, then for all N we get

$$\frac{4}{3}\pi^2 \geq \sum_{0<|k|\leq N} \frac{1}{k^2} + \pi^2 \quad \text{or} \quad \frac{1}{3}\pi^2 \geq 2 \cdot \sum_{k=1}^{N} \frac{1}{k^2},$$

$$\frac{\pi^2}{6} \geq \sum_{k=1}^{N} \frac{1}{k^2} \quad \text{for all } N = 1, 2, \dots \Rightarrow \frac{\pi^2}{6} \geq \sum_{k=1}^{\infty} \frac{1}{k^2}$$

(the infinite series converges by the Integral Test). Once we know that $\|f\|^2 = \sum_{k\in\mathbb{Z}} |c_k|^2$, we get the famed formula

$$\sum_{k=1}^{\infty} \frac{1}{k^2} = \frac{\pi^2}{6}.$$

The Fourier series associated with the sawtooth function $f(t)$ is

$$f(t) \sim \sum_{k=-\infty}^{\infty} (f, e_k)\, e_k(t) = \frac{1}{2} \cdot \mathbb{1} + \sum_{k\neq 0} \frac{-1}{2\pi i k} e^{2\pi i k t},$$

which converges pointwise for all $t \in \mathbb{R}$ except the "jump points" $t \in \mathbb{Z}$, where the series converges to the middle value $\frac{1}{2}$.

EXERCISE 6.24. Compute the Fourier transforms of the periodic functions whose graphs are shown in Figure 6.5(b)–(d).

A Geometry Problem. The following result provides further insight into the meaning of the projection $P_N(v) = \sum_{i=1}^{N} (v, e_i)\, e_i$ where $\{e_i\}$ is an orthonormal family in an inner product space V.

THEOREM 6.25. *If $\{e_1, \dots, e_n\}$ is an orthonormal family in an inner product space, and $P_M(v) = \sum_{i=1}^{n} (v, e_i)\, e_i$ the projection of v onto $M = \mathbb{K}\text{-span}\{e_1, \dots, e_n\}$ along $M^\perp$, then the image $P_M(v)$ is the point in M closest to v,*

$$\|P_M(v) - v\| = \min\{\|u - v\| : u \in M\}$$

for any $v \in V$. In particular, the minimum is achieved at the unique point $P_M(v) \in M$.

PROOF. Write $v = v_{\|} + v_{\perp}$ where $v_{\|} = P_M(v) = \sum_{i=1}^{N} (v, e_i)\, e_i$ and $v_{\perp} = v - \sum_i (v, e_i)\, e_i$. Obviously $v_{\|} \perp v_{\perp}$ and if z is any point in M we have $(v_{\|} - z) \in M$ and $(v - v_{\|}) \perp M$, so by Pythagoras

$$\|v - z\|^2 = \|(v - v_{\|}) + (v_{\|} - z)\|^2 = \|v - v_{\|}\|^2 + \|v_{\|} - z\|^2.$$

Thus

$$\|v - z\|^2 \geq \|v - v_{\|}\|^2$$

for all $z \in M$, so $\|v - z\|^2$ is minimized at $z = v_{\|} = \sum_{i=1}^{N} (v, e_i)\, e_i$. Figure 6.6 shows why the formula $\|v\|^2 = \|v_{\|}\|^2 + \|v_{\perp}\|^2$ really is equivalent to Pythagoras' formula for right triangles (see the shaded triangle). $\square$

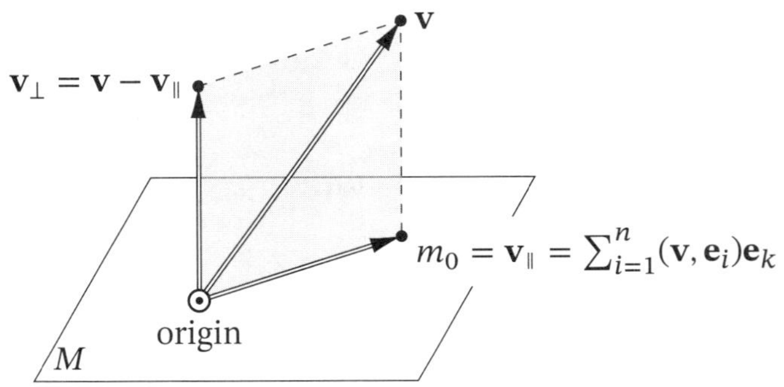

FIGURE 6.6. If M is a finite-dimensional subspace of inner product space V and $v \in V$, the unique point in M closest to v is $m_0 = v_{\|} = \sum_i (v, e_i)\, e_i$, and the minimized distance is $\|v - v_{\|}\|$. The shaded plane is spanned by the orthogonal vectors $v_{\|}$ and $v_{\perp}$, and we have $\|v\|^2 = \|v_{\|}\|^2 + \|v_{\perp}\|^2$ (Pythagoras' formula).

6.3. Adjoints and Orthonormal Decompositions

Diagonalization vs. Orthogonal Diagonalization. Let V be a finite-dimensional inner product space over $\mathbb{K} = \mathbb{R}$ or $\mathbb{C}$. A linear operator $T : V \to V$ is *diagonalizable* if there is a basis $\{e_1, \ldots, e_n\}$ of eigenvectors (so $T(e_i) = \mu_i e_i$ for some $\mu_i \in \mathbb{K}$), and we have seen that this happens if and only if $V = \bigoplus_{\lambda \in \mathrm{sp}(T)} E_\lambda(T)$ where

$$\mathrm{sp}(T) = (\text{distinct eigenvalues of } T \text{ in } \mathbb{K}) = \{\lambda \in \mathbb{K} : E_\lambda(T) \neq (0)\},$$
$$E_\lambda(T) = \{v \in V : (T - \lambda I)v = 0\} = \ker(T - \lambda I).$$

An operator T is *orthogonally diagonalizable* if there is an orthonormal basis $\{e_1, \ldots, e_n\}$ of eigenvectors, so $T(e_i) = \mu_i e_i$ for some $\mu_i \in \mathbb{K}$.

LEMMA 6.26. *A linear operator $T : V \rightarrow V$ on a finite-dimensional inner product space is orthogonally diagonalizable if and only if the eigenspaces span V and are pairwise orthogonal, so $E_\lambda(T) \perp E_\mu(T)$ for $\lambda \neq \nu$ in* $\mathrm{sp}(T)$.

PROOF ($\Leftarrow$). We have seen that the span $W = \sum_{\lambda \in \mathrm{sp}(T)} E_\lambda(T)$ is a *direct* sum whether or not $W = V$. If $W = V$ and the E_λ are orthogonal, then we have an orthogonal direct sum decomposition $V = \dot{\bigoplus}_\lambda E_\lambda(T)$. Taking an orthonormal basis in each E_λ, we get a diagonalizing orthonormal basis for all of V. □

PROOF ($\Rightarrow$). If $\mathfrak{X} = \{e_1, \dots, e_n\}$ is a diagonalizing orthonormal basis with $T(e_i) = \mu_i e_i$, each μ_i is an eigenvalue. Define

$$\mathrm{sp}' = \{\lambda \in \mathrm{sp}(T) : \lambda = \mu_i \text{ for some } i\} \subseteq \mathrm{sp}(T)$$

and for $\lambda \in \mathrm{sp}(T)$, let

$$M_\lambda = \sum \{\mathbb{K} e_i : \mu_i = \lambda\} \subseteq E_\lambda(T)$$

(which will $= (0)$ if λ does not appear among the scalars μ_i). Obviously $|M_\lambda| \leq |E_\lambda|$; furthermore, each e_i lies in some eigenspace E_λ, so

$$V = \mathbb{K}\text{-span}\{e_1, \dots, e_n\} \subseteq \sum_{\lambda \in \mathrm{sp}(T)} E_\lambda \subseteq V$$

and these subspaces coincide. Thus

$$|V| = \sum_{\lambda \in \mathrm{sp}(T)} |E_\lambda| \geq \sum_{\lambda \in \mathrm{sp}'} |E_\lambda| \geq \sum_{\lambda \in \mathrm{sp}'} |M_\lambda| \geq |V|$$

and all sums are equal. The last inequality holds because

$$\sum_{\lambda \in \mathrm{sp}'} M_\lambda \supseteq \sum_{j=1}^{n} \mathbb{K} e_j = V.$$

Now if $\mathrm{sp}(T) \neq \mathrm{sp}'$, the first inequality would be strict, and if $M_\lambda \subsetneqq E_\lambda$ the second would be strict; both situations are impossible. We conclude that $|M_\lambda| = |E_\lambda(T)|$ so $M_\lambda = E_\lambda(T)$. But the M_λ are mutually orthogonal by definition, so the eigenspaces E_λ are pairwise orthogonal as desired. □

Simple examples (discussed later) show that a linear operator on an inner product space can be diagonalizable in the ordinary sense but fail to be orthogonally diagonalizable. To explore this distinction further, we need additional background, particularly the definition of *adjoints* of linear operators.

Dual Spaces of Inner Product Spaces. There is a natural identification of any finite-dimensional inner product space V with its dual space V^*. It is implemented by a map $J : V \rightarrow V^*$ where $J(v) =$ the functional $\ell_v \in V^*$ such that

$$\langle J(v), x\rangle = \langle \ell_v, x\rangle = (x, v) \quad \text{for all } x \in V.$$

Each map ℓ_v is a linear functional because the inner product $(*, *)$ is $\mathbb{K}$-linear in its left-hand entry (but conjugate linear in the right-hand entry unless $\mathbb{K} = \mathbb{R}$). The map J is one-to-one because

$$J(v_1) = J(v_2) \Rightarrow 0 = \langle \ell_{v_1}, x\rangle - \langle \ell_{v_2}, x\rangle = (x, v_1) - (x, v_2) = (x, v_1 - v_2)$$

for all $x \in V$. Taking $x = v_1 - v_2$, we see that $0 = \|v_1 - v_2\|^2 \Rightarrow v_1 - v_2 = 0$ and $v_1 = v_2$ by positive definiteness of the inner product. To see that J is also surjective we invoke:

LEMMA 6.27. *If V is a finite-dimensional inner product space, $\{e_1, \dots, e_n\}$ an orthonormal basis, and $\ell \in V^*$, then*

$$\ell = J(v_0) \quad \textit{where } v_0 = \sum_{i=1}^{n} \overline{\langle \ell, e_i\rangle}\, e_i$$

(thus proving J surjective).

PROOF. For any $x \in V$ we have $x = \sum_i (x, e_i)\, e_i$, and by conjugate linearity of $(*, *)$

$$\begin{aligned}\langle J(v_0), x\rangle = (x, v_0) &= \Big(\sum_i x_i e_i, \sum_j \overline{\langle \ell, e_j\rangle}\, e_j\Big) = \sum_{i,j} x_i \langle \ell, e_j\rangle \cdot (e_i, e_j)\\ &= \sum_i x_i \langle \ell, e_i\rangle = \Big\langle \ell, \sum_i x_i e_i\Big\rangle = \ell(x) \quad \text{for all } x \in V.\end{aligned}$$

Therefore $J(v_0) = \ell$ as elements of V^*. □

EXERCISE 6.28. Prove that $J : V \to V^*$ is a *conjugate linear* bijection: it is additive, with $J(v + v') = J(v) + J(v')$ for all $v, v' \in V$, but $J(\lambda v) = \overline{\lambda} J(v)$ for $v \in V, \lambda \in \mathbb{C}$.

The Adjoint Operator $T^* : W \to V$. If $T : V \to W$ is a linear operator between finite-dimensional vector spaces, we showed in Chapter 2 that there is a natural transpose operator $T^{\mathrm{T}} : W^* \to V^*$. Since $V \cong V^*$ for inner product spaces, it follows that there is a natural *adjoint operator* $T^* : W \to V$ between the original vector spaces, rather than their duals.

THEOREM 6.29 (Adjoint Operator). *Let V, W be finite-dimensional inner product spaces and $T : V \to W$ a $\mathbb{K}$-linear operator. Then there is a unique $\mathbb{K}$-linear adjoint operator $T^* : W \to V$ such that*

$$(T(v), w)_W = (v, T^*(w))_V \quad \textit{for all } v \in V,\ w \in W, \tag{6.6}$$

or equivalently $(T^*(w), v)_V = (w, T(v))_W$ owing to Hermitian symmetry of the inner product.

PROOF. We define $T^*(w)$ for $w \in W$ using our observations about dual spaces. Given $w \in W$, we get a well-defined linear functional ϕ_w on V such that

$$\langle \phi_w, v\rangle = (T(v), w)_W$$

(w is fixed; the variable is v).

Obviously $\phi_w \in V^*$ because $(*,*)_W$ is linear in its left-hand entry. By the previous discussion there is a unique vector in V, which we label $T^*(w)$, such that $J(T^*(w)) = \phi_w$ in V^*, hence

$$(T(x), w)_W = \langle \phi_w, x\rangle = \langle J(T^*(w)), x\rangle = (x, T^*(w))_V.$$

This yields a well-defined linear map $T^* : W \to V$.

Once we know a map T^* satisfying (6.6) exists, it is easy to use the scalar identities (6.6) to verify that T^* is a linear operator and verify its important properties. For linearity we first observe that two vectors v_1, v_2 are equal in V if and only if $(v_1, x) = (v_2, x)$ for all $x \in V$ because the inner product is positive definite. Then $T^*(w_1 + w_2) = T^*(w_1) + T^*(w_2)$ in V follows: for all $v \in V$ we have

$$\begin{aligned}(T^*(w_1 + w_2), v)_V &= (w_1 + w_2, T(v))_W = (w_1, T(v))_W + (w_2, T(v))_W \\ &= (T^*(w_1), v)_V + (T^*(w_2), v)_V \quad \text{(definition of } T^*(w_k)) \\ &= (T^*(w_1) + T^*(w_2), v)_V \quad \text{(linearity of } (*,*) \text{ in first entry).}\end{aligned}$$

Similarly, $T^*(\lambda w) = \lambda T^*(w)$ for all $\lambda \in \mathbb{K}$, $w \in W$ (check that λ comes forward instead of $\overline{\lambda}$). □

NOTE: Regarding calculations with adjoints: Don't look at $T^*(v)$; instead, look at $(T^*(v), w)$ for all $v \in V$, $w \in W$.

LEMMA 6.30. *On an inner product space, $(T^*)^* = T$ as linear maps from $V \to W$.*

PROOF. It suffices to check the scalar identities $(T^{**}(v), w)_W = (T(v), w)_W$ for all $v \in V$, $w \in W$. But by definition,

$$(T^{**}(v), w)_W = (v, T^*(w))_V = (T(v), w)_W,$$

and we are done. □

The adjoint $T^* : W \to V$ of a linear operator $T : V \to W$ between inner products space is analogous to the transpose $T^{\mathsf{T}} : W^* \to V^*$. In fact, if V, W are inner product spaces and we identify $V = V^*$, $W = W^*$ via the maps $J_V : V \to V^*$, $J_W : W \to W^*$, then T^* *becomes* the transpose $T^{\mathsf{T}} : W^* \to V^*$ in the sense that the following diagram commutes:

$$\begin{array}{ccc} W & \xrightarrow{T^*} & V \\ {\scriptstyle J_W}\downarrow & & \downarrow{\scriptstyle J_V} \\ W^* & \xrightarrow{T^{\mathsf{T}}} & V^* \end{array}$$

so that

$$T^{\mathsf{T}} \circ J_W = J_V \circ T^* \quad (\text{or } T^* = J_V^{-1} \circ T^{\mathsf{T}} \circ J_W).$$

EXERCISE 6.31. Prove this last identity from the definitions.

Furthermore, as remarked earlier, when V is just a finite-dimensional vector space there is a natural identification of $V \cong V^{**}$:

$$j : V \to V^{**} \quad \langle j(v), \ell \rangle = \ell(v) \quad \text{for all } \ell \in V^*, v \in V.$$

Under this identification of $V^{**} \cong V$ we also have $T^{\mathrm{TT}} = T$ for any linear operator $T : V \to W$, because the following diagram commutes:

$$\begin{array}{ccc} V^{**} & \xrightarrow{T^{\mathrm{TT}}} & W^{**} \\ {\scriptstyle j_V}\uparrow & & \uparrow{\scriptstyle j_W} \\ V & \xrightarrow{T} & W \end{array}$$

If V, W are inner product spaces, we may actually identify $V \cong V^*$ (something that cannot be done in any natural way in the absence of the extra structure that an inner product provides). Then we may identify $V \cong V^* \cong V^{**} \cong V^{***} \cong \cdots$ and $W \cong W^* \cong W^{**} \cong W^{***} \cong \cdots$; when we do, T^{T} becomes T^* and T^{TT} becomes $T^{**} = T$.ii

EXERCISE 6.32 (Basic Properties of Adjoints). Use (6.6) to prove:

(a) $I^* = I$ and $(\lambda I)^* = \overline{\lambda} I$
(b) $(T_1 + T_2)^* = T_1^* + T_2^*$
(c) $(\lambda T)^* = \overline{\lambda} T^*$ (conjugate linearity)

EXERCISE 6.33 (Contravariance of Adjoint Operation). Given linear operators $V \xrightarrow{S} W \xrightarrow{T} Z$ between finite-dimensional inner product spaces, prove that

$$(T \circ S)^* = S^* \circ T^* : Z \to V.$$

Note the reversal of order when we take adjoints.

EXERCISE 6.34. If $A \in \mathrm{M}(n, \mathbb{C})$ and $(A^*)_{ij} = \overline{A_{ji}}$ is the usual adjoint matrix, consider the operator $L_A : \mathbb{C}^n \to \mathbb{C}^n$ such that $L_A(\mathbf{z}) = A \cdot \mathbf{z}$. If $\mathbb{C}^n$ is given the standard inner product, prove that

(a) If $\mathfrak{X} = \{\mathbf{e}_1, \dots, \mathbf{e}_n\}$ is the standard orthonormal basis, then $[L_A]_{\mathfrak{X}\mathfrak{X}} = A$.
(b) $(L_A)^* = L_{A^*}$ as operators on $\mathbb{C}^n$.

NOTE: This is not true unless $\mathfrak{X}$ is orthonormal.

Linear Projections vs. Orthogonal Projections.

EXAMPLE 6.35 (Self-Adjointness of Orthogonal Projections). Given an unadorned vector space V the "idempotent" relation $P^2 = P$ identifies the linear operators that are projections associated with an ordinary direct sum decomposition $V = M \oplus N$. The same is true of an inner product space, but if we only know $P = P^2$, the subspaces M, N are not necessarily orthogonal. We now show that an idempotent operator P on an inner product space corresponds to an *orthogonal* direct sum decomposition $V = M \oplus N$ if and only if it is *self-adjoint* ($P^* = P$), so that

$$P^2 = P = P^*. \tag{6.7}$$

DISCUSSION: If $M \perp N$, it is fairly easy to verify (Exercise 6.36) that the associated projection P_M of V onto $M = \text{range}(P_M)$ along $N = \ker(P_M)$ is self-adjoint. If $v, w \in V$, let us indicate the components by writing $v = v_M + v_N$, $w = w_M + w_N$. With (6.7) in mind, suppose $M \perp N$. Self-adjointness of P_M emerges from the following calculation:

$$\begin{aligned}(v, P_M^*(w)) &= (P_M(v), w) = (v_M, w_M + w_N) \quad \text{(definition of } P_M(v) = v_M) \\ &= (v_M, w_M) \quad (\text{since } w_N \perp w_M) \\ &= (v_M + v_N, w_M) = (v, w_M) = (v, P_M(w)) .\end{aligned}$$

Since this true for all $v \in V$, we get $P_M^*(w) = P_M(w)$ for all w, whence $P_M^* = P_M$ as operators.

For the converse we must prove: If the projection P_M associated with an ordinary direct sum decomposition $V = M \oplus N$ is self-adjoint, so that $P_M^* = P_M$, then the subspaces must be orthogonal. We leave this proof as an exercise.

EXERCISE 6.36. If $P : V \to V$ is a linear operator on a vector space such that $P^2 = P$, it is the projection operator associated with the decomposition

$$V = R \oplus K \quad \text{where } R = \text{range}(P),\ K = \ker(P).$$

If V is an inner product space, prove that the subspaces must be orthogonal ($R \perp K$) if the projection is self-adjoint, so $P^2 = P = P^*$.

Matrix realizations of adjoints are easily computed, provided we restrict attention to *orthonormal* bases in both V and W. With respect to arbitrary bases the computation of $[T^*]_{\mathfrak{X}\mathfrak{Y}}$ can be quite a mess.

PROPOSITION 6.37. *Let $T : V \to W$ be a linear operator between finite-dimensional inner product spaces and let $\mathfrak{X} = \{e_i\}$, $\mathfrak{Y} = \{f_j\}$ be orthonormal bases in V, W. Then*

$$(6.8) \qquad [T^*]_{\mathfrak{X}\mathfrak{Y}} = \left([T]_{\mathfrak{Y}\mathfrak{X}}\right)^* \quad \textit{(taking the matrix adjoint on the right)}$$

where A^* is the usual $m \times n$ "matrix-adjoint" of $A \in \mathrm{M}(n \times m, \mathbb{K})$, the conjugate-transpose of A such that $(A^*)_{ij} = \overline{A_{ji}}$. (This is just the transpose A^{T} when $\mathbb{K} = \mathbb{R}$.)

PROOF. By definition, the entries T_{ij} of $[T]_{\mathfrak{Y}\mathfrak{X}}$ are determined by the vector identities

$$T(e_i) = \sum_{k=1}^{n} T_{ki}\, f_k \quad \text{which imply} \quad \left(T(e_i), f_j\right)_W = \sum_{k=1}^{n} T_{ki}\,(f_k, f_j)_W = T_{ji}$$

for all i, j. Since $(e_k, e_j) = \delta_{kj}$ (Kronecker delta), we also get

$$T^*(f_i) = \sum_{k=1}^{n} [T^*]_{ki}\, e_k \Rightarrow (T^*(f_i), e_j) = [T^*]_{ji},$$

from which we see that

$$\begin{aligned}[T^*]_{ij} &= \left(T^*(f_j), e_i\right)_V \\ &= (f_j, T(e_i))_W = \overline{(T(e_i), f_j)_W} = \overline{[T]_{ji}} = ([T]^*)_{ij} .\end{aligned}$$

□

Adjoint T^* vs. Transpose T^{T}.

REMARK 6.38. For a vector space over $\mathbb{K} = \mathbb{C}$ or $\mathbb{R}$, the operator $T^* : V \to V$ can only be defined when V is equipped with an inner product. The transpose T^{T} makes sense for any vector space but is not an operator on V; instead $T^{\mathsf{T}} : V^* \to V^*$ acts on the dual space V^*, which is not at all the same thing as the original space V where T was defined.

Further confusion can arise when operators T are represented by matrices because there are obvious adjoint and transpose operations on the matrices that represent T. These send $A \in \mathrm{M}(n, \mathbb{K})$ to the *adjoint matrix* A^* or the *transpose matrix* A^{T} with

$$(A^*)_{ij} = \overline{A_{ji}} \quad \text{or} \quad (A^{\mathsf{T}})_{ij} = A_{ji}.$$

So, if $T : V \to V$ why can't we just take a basis $\mathfrak{X}$ and define new operators T^* and T^{T} on V such that

$$[T^*]_{\mathfrak{X}\mathfrak{X}} = [T]^*_{\mathfrak{X}\mathfrak{X}} \quad \text{(the adjoint matrix of } [T]_{\mathfrak{X}\mathfrak{X}})$$

or

$$[T^{\mathsf{T}}]_{\mathfrak{X}\mathfrak{X}} = [T]^{\mathsf{T}}_{\mathfrak{X}\mathfrak{X}} \quad \text{(the transpose of matrix } [T]_{\mathfrak{X}\mathfrak{X}})?$$

The problem is that the operator versions of adjoint and transpose match up with the matrix versions only when certain conditions are met, because vital information has been lost in passing from $T \to [T]_{\mathfrak{X}\mathfrak{X}}$.

DISCUSSION: To see what can go wrong, consider the identity

$$[T^*]_{\mathfrak{X}\mathfrak{X}} = ([T]_{\mathfrak{X}\mathfrak{X}})^* \quad (\mathfrak{X} \text{ a basis in } V)$$

proved in Proposition 6.37. For this to be true the basis $\mathfrak{X}$ has to be *orthonormal*, a condition that makes no sense in an ordinary vector space; and even if V has an inner product, the identity fails to be true for bases that aren't orthonormal. For instance, if $\mathfrak{X}$ and $\mathfrak{Y}$ are bases in an inner product space such that $\mathfrak{X}$ is orthonormal but $\mathfrak{Y}$ is not, we always have $[T^*]_{\mathfrak{X}\mathfrak{X}} = [T]^*_{\mathfrak{X}\mathfrak{X}}$ (matrix adjoint), but we could have $[T^*]_{\mathfrak{Y}\mathfrak{Y}} \neq [T]^*_{\mathfrak{Y}\mathfrak{Y}}$. (See Additional Exercise 3 for Section 6.3 at the end of this chapter.)

Computing an Operator Adjoint.

EXERCISE 6.39. Let V_N be the restrictions to $[0, 1]$ of polynomials $f \in \mathbb{C}[x]$ having degree $\leq N$. Give this $(N+1)$-dimensional subspace of $\mathcal{C}[0, 1]$ the usual L^2 inner product $(f, h)_2 = \int_0^1 f(t)\overline{h(t)}dt$ inherited from the larger space of continuous functions. Let $D : V_N \to V_N$ be the differentiation operator

$$D\left(a_0 + a_1 t + a_2 t^2 + \cdots + a_N t^N\right) = a_1 + 2a_2 t + 3a_3 t^2 + \cdots + N a_n t^{N-1}.$$

(a) Is D one-to-one? Onto? What are $R(D) = \text{range}(D)$ and $K(D) = \ker(D)$?
(b) With respect to the vector basis $\mathfrak{X} = \{1, x, x^2, \dots, x^N\}$, determine the matrix $[D]_{\mathfrak{X}\mathfrak{X}}$.
(c) Determine the eigenvalues of $D : V_N \to V_N$ and their multiplicities.

(d) Compute the L^2-inner product $(f,h)_2$ in terms of the coefficients a_k, b_k that determine $f = \sum_{k=0} a_k x^k$ and $h = \sum_{k=0} b_k x^k$.
(e) Is D a self-adjoint operator? Skew-adjoint?

EXERCISE 6.40. Assume that D^* is the adjoint of the differentiation operator $D : V_N \to V_N$. The entries D^*_{ij} in its matrix $[D^*]_{\mathfrak{X}\mathfrak{X}}$ with respect to the basis $\mathfrak{X} = \{1, x, x^2, \dots, x^N\}$ are determined by the vector identities $D^*(x^i) = \sum_{k=0}^N D^*_{ki}\, x^k$. By definition of the adjoint operator D^* we have

$$\left(x^i, D(x^j)\right)_2 = \left(D^*(x^i), x^j\right)_2 = \sum_{k=0}^{N} D^*_{ki}(x^k, x^j)_2 \quad \text{for } 0 \le i, j \le N,$$

and since $\mathfrak{X}$ is a basis, these identities implicitly determine the D^*_{ij}. As in the previous problem, $D(x^k) = k \cdot x^{k-1}$, $D(1) = 0$, and inner products in V_N are integrals

$$(f,h)_2 = \int_0^1 f(x) \cdot \overline{h(x)}dx$$

for polynomials $f, h \in V_N$. Find D^*_{ij} for $0 \le i, j \le 2$.

HINT: The powers x^i are NOT an orthonormal basis, so you will have to use some algebraic brute force instead of (6.8). This could get complicated. For something more modest, just compute the action of D^* on $V = \mathbb{C}\text{-span}\{1, x, x^2\}$.

EXERCISE 6.41. Let $V = \mathcal{C}_c^\infty(\mathbb{R})$ be the space of real-valued functions $f(t)$ on the real line that have continuous derivatives $D^k f$ of all orders, and have "bounded support"—each f is zero off of some bounded interval (which is allowed to vary with f). Because all such functions are "zero near ∞," there is a well-defined inner product

$$(f,h)_2 = \int_{-\infty}^{\infty} f(t)\overline{h(t)}dt.$$

The derivative $Df = df/dt$ is a linear operator on this infinite-dimensional space.

(a) Prove that D is *skew-adjoint* with $D^* = -D$.
(b) Prove that the second derivative $D^2 = d^2/dt^2$ is *self-adjoint*.

HINT: Integration by parts.

Self-Adjoint, Unitary, and Normal Operators. Various classes of linear operators $T : V \to V$ can be defined on a finite-dimensional inner product space.

1. SELF-ADJOINT: $T^* = T$
2. SKEW-ADJOINT: $T^* = -T$
3. UNITARY: $T^*T = I$
   This implies $TT^* = I$ because $T : V \to V$ is one-to-one $\Leftrightarrow$ onto $\Leftrightarrow$ bijective. Thus "T is unitary" is equivalent to saying that $T^* = T^{-1}$, at

least when V is finite dimensional. (In the infinite-dimensional case we need both identities $TT^* = T^*T = I$ to get $T^* = T^{-1}$.)

4. NORMAL: $T^*T = TT^*$ (T commutes with T^*)

These definitions apply even when V is an inner product space over $\mathbb{R}$, but then "unitary" becomes "real-orthogonal," which means $T^{\mathsf{T}} = T^{-1}$.

The *spectrum* $\mathrm{sp}_{\mathbb{K}}(T) = \{\lambda \in \mathbb{K} : E_\lambda(T) \neq (0)\}$ of T is closely related to that of T^*.

LEMMA 6.42. *If T is a linear operator on a finite-dimensional inner product space, then λ is an eigenvalue for $T \Leftrightarrow \overline{\lambda}$ is an eigenvalue for T^*, so that*

$$\mathrm{sp}(T^*) = \overline{\mathrm{sp}(T)} = \{\overline{\lambda} : \lambda \in \mathrm{sp}(T)\}.$$

PROOF. Eigenvalues of T are the roots in $\mathbb{K}$ of the characteristic polynomial, so

$$0 = \det(T - \lambda I) = \det([T]_{\mathfrak{X}\mathfrak{X}} - \lambda I_{n\times n})$$

for any basis $\mathfrak{X}$ in V. If $\mathfrak{X}$ is an orthonormal basis, we have $[T^*]_{\mathfrak{X}\mathfrak{X}} = [T]^*_{\mathfrak{X}\mathfrak{X}} = \overline{[T]^{\mathsf{T}}_{\mathfrak{X}\mathfrak{X}}}$ and then

$$\begin{aligned}\det([T^*]_{\mathfrak{X}\mathfrak{X}} - \overline{\lambda} I_{n\times n}) &= \overline{\det\left([T]^{\mathsf{T}}_{\mathfrak{X}\mathfrak{X}} - \lambda I_{n\times n}\right)} = \overline{\det\left([T]_{\mathfrak{X}\mathfrak{X}} - \lambda I_{n\times n}\right)^{\mathsf{T}}} \\ &= \overline{\det\left([T]_{\mathfrak{X}\mathfrak{X}} - \lambda I_{n\times n}\right)} = 0\end{aligned}$$

because

$$\det(A^{\mathsf{T}}) = \det(A) \quad \text{and} \quad \det(\overline{A}) = \overline{\det(A)}.$$

Hence $\overline{\lambda} \in \mathrm{sp}(T^*)$ and $\overline{\mathrm{sp}(T)} \subseteq \mathrm{sp}(T^*)$. The reverse inclusion follows by the same argument since $T^{**} = T$. □

If $A \in \mathrm{M}(n, \mathbb{K})$, its matrix adjoint $(A^*)_{ij} = \overline{A_{ji}}$ obviously has determinant

$$\det(A^*) = \overline{\det(A^{\mathsf{T}})} = \overline{\det(A)}.$$

PROPOSITION 6.43. *If $T : V \to V$ is a linear map on an inner product space, the characteristic polynomial satisfies*

$$p_{T^*}(\lambda) = \overline{p_T(\overline{\lambda})} \quad \textit{or equivalently} \quad p_T(\overline{\lambda}) = \overline{p_T(\lambda)}$$

for all $\lambda \in \mathbb{K}$.

PROOF. Since $I^* = I$ and $(\lambda I)^* = \overline{\lambda} I$ we get

$$\begin{aligned}p_{T^*}(\lambda) &= \det(T^* - \lambda I) = \det(T^* - (\overline{\lambda} I)^*) \\ &= \det((T - \overline{\lambda} I)^*) = \overline{\det(T - \overline{\lambda} I)} = \overline{p_T(\overline{\lambda})}.\end{aligned}$$

□

6.4. Diagonalization in Inner Product Spaces

If M is a T-invariant subspace of inner product space V, it does not follow that $T^*(M) \subseteq M$. The true relationship between invariance under T and under T^* is:

LEMMA 6.44. *If V is any inner product space and $T : V \to V$ a linear map, then:*

1. *A subspace $M \subseteq V$ is T-invariant (so $T(M) \subseteq M$) $\Rightarrow M^\perp$ is T^*-invariant.*
2. *If $\dim_{\mathbb{K}}(V) < \infty$ (so $M^{\perp\perp} = M$), then $T(M) \subseteq M \Leftrightarrow T^*(M^\perp) \subseteq M^\perp$.*

PROOF. If $y \in M^\perp$, then $(T^*(y), M) = (y, T(M)) = (0)$ and $T^*(M^\perp) \subseteq M^\perp$. For the reverse inclusion,

$$T^*(M^\perp) \subseteq M^\perp \Rightarrow T^{**}\left(M^{\perp\perp}\right) \subseteq M^{\perp\perp}.$$

But $T^{**} = T$ and by Proposition 6.16 we have $M^{\perp\perp} = M$ when $|V| < \infty$, which proves (2.). □

PROPOSITION 6.45. *If $T : V \to W$ is a linear map between finite-dimensional inner product spaces, let $R(T) = \text{range}(T)$, $K(T) = \ker(T)$. Then $T^* : W \to V$ and*

$$K(T^*) = R(T)^\perp \text{ in } W \quad \text{and} \quad R(T^*) = K(T)^\perp \text{ in } V.$$

In particular, if T is self-adjoint and $W = V$, then $\ker(T) \perp \text{range}(T)$, and we have an orthogonal direct sum decomposition $V = K(T) \dot{\oplus} R(T)$.

PROOF. If $w \in W$ then

$$\begin{aligned} T^*(w) = 0 &\Leftrightarrow (v, T^*(w))_V = 0 \quad \text{for all } v \in V \\ &\Leftrightarrow 0 = (v, T^*(w))_V = (T(v), w)_W \quad \text{for all } v \in V \\ &\Leftrightarrow w \perp R(T). \end{aligned}$$

Hence $w \in K(T^*)$ if and only if $w \perp R(T)$. The second part follows because $T^{**} = T$ and $M^{\perp\perp} = M$ for any subspace. □

We will often invoke this result.

Orthogonal Diagonalization. Not all linear operators $T : V \to V$ are diagonalizable, let alone orthogonally diagonalizable, but if V is an inner product space we can always find a basis that at least puts $[T]_{\mathfrak{X}\mathfrak{X}}$ into *upper-triangular* form, which can be helpful. In fact, this can be achieved via an othonormal basis provided the characteristic polynomial splits into linear factors over $\mathbb{K}$ (always true if $\mathbb{K} = \mathbb{C}$).

Schur Normal Form.

THEOREM 6.46 (Schur Normal Form). *Let $T : V \to V$ be a linear operator on a finite-dimensional inner product space over $\mathbb{R}$ or $\mathbb{C}$. If $p_T(x) = \det(T - xI)$*

splits over $\mathbb{K}$, then there are scalars $\lambda_1, \dots, \lambda_n$ and an orthonormal basis $\mathfrak{X}$ in V such that

$$[T]_{\mathfrak{X}\mathfrak{X}} = \begin{pmatrix} \lambda_1 & & & * \\ & \lambda_2 & & \\ & & \ddots & \\ 0 & & & \lambda_n \end{pmatrix}.$$

PROOF. Work by induction on $n = \dim_{\mathbb{K}}(V)$; the case $n = 1$ is trivial. For $n > 1$, since p_T splits there is an eigenvalue λ in $\mathbb{K}$ and a vector $v_0 \neq 0$ such that $T(v_0) = \lambda v_0$. Then $\overline{\lambda}$ is an eigenvalue for T^*, so there is some $w_0 \neq 0$ such that $T^*(w_0) = \overline{\lambda} w_0$.

Let $M = \mathbb{K}w_0$; this one-dimensional space is T^*-invariant, so $M^\perp$ is invariant under $(T^*)^* = T$ and has dimension $n - 1$. Scale w_0 if necessary to make $\|w_0\| = 1$. By the induction hypothesis there is an orthonormal basis $\mathfrak{X}_0 = \{e_1, \dots, e_{n-1}\}$ in $M^\perp$ such that

$$[T|_{M^\perp}]_{\mathfrak{X}_0} = \begin{pmatrix} \lambda_1 & & & * \\ & \lambda_2 & & \\ & & \ddots & \\ 0 & & & \lambda_{n-1} \end{pmatrix}.$$

Then letting $e_n = w_0$ (norm = 1) we get an orthonormal basis for V such that $[T]_{\mathfrak{X}\mathfrak{X}}$ has the form:

$$[T]_{\mathfrak{X}\mathfrak{X}} = \left(\begin{array}{ccc|c} \lambda_1 & & * & c_1 \\ & \ddots & & \vdots \\ 0 & & \lambda_{n-1} & c_{n-1} \\ \hline 0 & & 0 & \lambda_n \end{array}\right)$$

where

$$T(e_n) = T(w_0) = \lambda_n e_n + \sum_{j=1}^{n-1} c_j e_j$$

(Remember: $M = \mathbb{K}w_0$ need not be invariant under T.) □

EXERCISE 6.47. Explain why the diagonal entries in the Schur normal form must be the roots in $\mathbb{K}$ of the characteristic polynomial $p_T(x) = \det(T - xI)$, each counted according to its algebraic multiplicity.

NOTE: Nevertheless, it might not be possible to find an orthonormal basis such that all occurrences of a particular eigenvalue $\lambda \in \mathrm{sp}_{\mathbb{K}}(T)$ appear in a consecutive string $\lambda, \dots, \lambda$ on the diagonal of the Schur normal form. A linear operator $T : V \to V$ on an inner product space is *normal* if it commutes with its adjoint, so that $T^*T = TT^*$. We will eventually show that when $\mathbb{K} = \mathbb{C}$ (or when $\mathbb{K} = \mathbb{R}$ and the characteristic polynomial of T splits into linear factors: $p_T(x) = \prod_{i=1}^{n}(x - \alpha_i)$ with $\alpha_i \in \mathbb{R}$), then T is orthogonally diagonalizable *if and only if T is normal*. Note carefully what this does not say: T might be

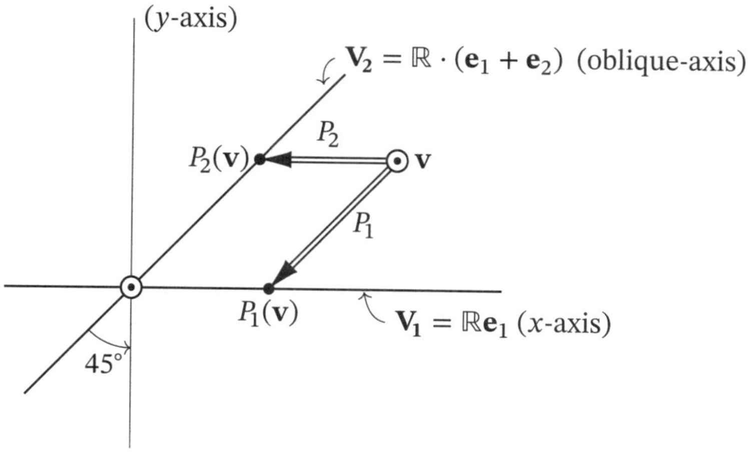

FIGURE 6.7. The (nonorthogonal) basis vectors $\mathbf{f}_1 = \mathbf{e}_1$ and $\mathbf{f}_2 = \mathbf{e}_1 + \mathbf{e}_2$ in Exercise 6.48.

(nonorthogonally) diagonalizable over $\mathbb{K} = \mathbb{C}$ even if T is not normal. This latter issue can only be resolved by determining the pattern of eigenspaces $E_\lambda(T)$ and demonstrating that they span all of V.

EXERCISE 6.48. Let $\{\mathbf{e}_1, \mathbf{e}_2\}$ be the standard orthonormal basis vectors in $V = \mathbb{K}^2$, and consider the ordinary direct sum decomposition

$$V = V_1 \oplus V_2 = \mathbb{K}\mathbf{f}_1 \oplus \mathbb{K}\mathbf{f}_2 \quad \text{where } \mathbf{f}_1 = \mathbf{e}_1,\ \mathbf{f}_2 = \mathbf{e}_1 + \mathbf{e}_2$$

whose subspaces are not orthogonal with respect to the standard euclidean inner product

$$(x_1\mathbf{e}_1 + x_2\mathbf{e}_1, y_1\mathbf{e}_1 + y_2\mathbf{e}_2) = x_1\overline{y_1} + x_2\overline{y_2}$$

Define a $\mathbb{K}$-linear map $T : V \to V$, letting

$$T(\mathbf{e}_1) = 2\mathbf{e}_1, \quad T(\mathbf{e}_1 + \mathbf{e}_2) = \frac{1}{2}(\mathbf{e}_1 + \mathbf{e}_2).$$

(see Figure 6.7). Then T is diagonalized by the basis $\mathfrak{Y} = \{\mathbf{f}_1, \mathbf{f}_2\}$ (which is obviously not orthonormal),

$$[T]_{\mathfrak{Y}\mathfrak{Y}} = \begin{pmatrix} 2 & 0 \\ 0 & \frac{1}{2} \end{pmatrix}.$$

(a) Determine the action of T on the standard orthonormal basis vectors $\mathfrak{X} = \{\mathbf{e}_1, \mathbf{e}_2\}$ in $\mathbb{R}^2$ and find $[T]_{\mathfrak{X}\mathfrak{X}}$.
(b) Describe the adjoint operator T^* by determining its action on the standard orthonormal basis $\mathfrak{X}$ and find $[T^*]_{\mathfrak{X}\mathfrak{X}}$.
(c) Explain why T is not a normal operator on V. Explain why no orthonormal basis $\{\mathbf{f}_1, \mathbf{f}_2\}$ in V can possibly diagonalize T.

HINT: The discussion is exactly the same for $\mathbb{K} = \mathbb{R}$ and $\mathbb{C}$, so assume $\mathbb{K} = \mathbb{R}$ if that makes you more comfortable.

Diagonalizing Self-Adjoint and Normal Operators. We now show that a linear operator $T : V \to V$ on a finite-dimensional inner product space is orthogonally diagonalizable if and only if T is normal. First, we analyze the special case of self-adjoint operators ($T^* = T$), which motivates the more subtle proof needed for normal operators.

THEOREM 6.49 (Diagonalizing Self-Adjoint Operators). *Any self-adjoint linear operator $T : V \to V$ on a finite-dimensional inner product space is orthogonally diagonalizable.*

PROOF. If $\mu, \lambda \in \text{sp}_{\mathbb{K}}(T)$, we first observe that:

1. If $T = T^*$ all eigenvalues are real, so $\text{sp}_{\mathbb{K}}(T) \subseteq \mathbb{R} + i0$ (the real axis in $\mathbb{C}$).
   PROOF: If $v \in E_\lambda(T)$, $v \neq 0$, we have
   $$\lambda\|v\|^2 = (Tv, v) = (v, T^*v) = (v, Tv) = (v, \lambda v) = \overline{\lambda}\|v\|^2,$$
   which implies $\lambda = \overline{\lambda}$.
2. If $\lambda \neq \mu$ in sp(T), the eigenspaces $E_\lambda(T)$ and $E_\mu(T)$ must be orthogonal.
   PROOF: If $v \in E_\lambda(T)$, $w \in E_\mu(T)$ then
   $$\lambda(v, w) = (Tv, w) = (v, T^*w) = (v, \mu w) = \overline{\mu}(v, w) = \mu(v, w)$$
   since eigenvalues are real when $T^* = T$. But $\mu \neq \lambda$, hence $(v, w) = 0$ and we have $E_\lambda(T) \perp E_\mu(T)$. Thus the linear span $E = \sum_\lambda E_\lambda(T)$ (which is always a direct sum) is actually an orthogonal sum $E = \dot{\bigoplus}_{\lambda \in \text{sp}(T)} E_\lambda(T)$.
3. If $T^* = T$ the span of the eigenspaces is all of V, hence T is orthogonally diagonalizable.
   PROOF: $E_\lambda(T) \neq (0)$ if $\lambda \in \text{sp}_{\mathbb{K}}(T)$ and $\dim(M) < \dim(V)$ for $M = E_\lambda(T)^\perp$. By Lemma 6.44 the orthogonal complement is T^*-invariant, hence T-invariant because $T^* = T$. It is easy (see 6.50 below) to check that if $W \subseteq V$ is T-invariant and $T^* = T$ on V, then the restriction $T|_W : W \to W$ is self-adjoint on W if one equips W with the restricted inner product from V.

EXERCISE 6.50. If $T : V \to V$ is linear and $T^* = T$, prove that
$$(T|_W)^* = (T^*|_W)$$
for any T-invariant subspace $W \subseteq V$ equipped with the restricted inner product. In particular, $T|_M$ is self-adjoint $\Leftrightarrow T^* = T$ on W.

To complete our discussion we show by induction on $n = \dim(V)$ that adjoint operators are orthogonally diagonalizable. This is clear if $\dim(V) = 1$, so assume it true whenever $\dim(V) \leq n$ and consider a space of dimension $n + 1$. Since all roots in $\mathbb{C}$ of the characteristic polynomial are real, there is a nontrivial eigenspace $M = E_\lambda(T)$ even if $\mathbb{K} = \mathbb{R}$. If E_λ is all of V we're done: $T = \lambda I$.

Otherwise, M has lower dimension, and by Exercise 6.50 it has an orthonormal basis that diagonalizes $T|_M$. But $V = M \oplus M^\perp$ (an orthogonal direct sum), and $M = E_\lambda$ has an orthonormal basis of eigenvectors because all vectors in E_λ are eigenvectors. Combining these bases we get an orthonormal diagonalizing basis for all of V. That completes the proof of Theorem 6.49. □

We now elaborate the basic properties of normal operators on an inner product space.

PROPOSITION 6.51. *A normal linear operator $T : V \to V$ on a finite-dimensional inner product space has the following properties:*

1. *If $T : V \to V$ is normal, then $\|T(v)\| = \|T^*(v)\|$ for all $v \in V$, even if $T(v) \neq T^*(v)$.*

   PROOF: *We have*

   $$\|T(v)\|^2 = (Tv, Tv) = (T^*T(v), v) = (TT^*(v), v)$$
   $$= (T^*v, T^*v) = \|T^*(v)\|^2.$$

2. *For any $c \in \mathbb{K}$, $T - cI$ is also normal because $(T - cI)^* = T^* - \bar{c}I$ and cI commutes with all operators.*
3. *If $T(v) = \lambda v$, then for the same vector v we have $T^*(v) = \bar{\lambda} v$. In particular $E_{\bar{\lambda}}(T^*) = E_\lambda(T)$ when T is normal. (This is a much stronger statement than our earlier observation that $\mathrm{sp}_{\mathbb{K}}(T^*) = \overline{\mathrm{sp}_{\mathbb{K}}(T)} = \{\bar{\lambda} : \lambda \in \mathrm{sp}_{\mathbb{K}}(T)\}$).*

   PROOF: *$(T - \lambda)$ is also normal. Therefore if $v \in V$ and $T(v) = \lambda v$, we have*

   $$T(v) = \lambda v \Rightarrow ((T - \lambda)^*(T - \lambda)\, v, v) = \|(T - \lambda)\, v\|^2 = 0.$$

   Since T is normal so is $(T - \lambda I)$, so we get

   $$0 = ((T - \lambda)(T - \lambda)^*\, v, v) = \|(T^* - \bar{\lambda} I)\, v\|^2 \Rightarrow T^*(v) = \bar{\lambda} v.$$

   NOTE: *Although $\mathrm{sp}(T^*) = \overline{\mathrm{sp}(T)}$, when T is not normal, the spaces $E_{\bar{\lambda}}(T^*)$ and $E_\lambda(T)$ are generally not the same.*
4. *If $\lambda \neq \mu$ in $\mathrm{sp}_{\mathbb{K}}(T)$, then $E_\lambda \perp E_\mu$.*

   PROOF: *If v, w are in E_λ, E_μ then by (3.)*

   $$\lambda(v, w) = (\lambda v, w) = (Tv, w) = (v, T^*w) = \big(v, \bar{\mu}\, w\big) = \mu(v, w).$$

   Therefore $(v, w) = 0$ if $\mu \neq \lambda$.

COROLLARY 6.52. *If $T : V \to V$ is normal and $\mathbb{K} = \mathbb{C}$ (or if $\mathbb{K} = \mathbb{R}$ and the characteristic polynomial p_T splits over $\mathbb{R}$), there is a diagonalizing orthonormal basis $\{e_i\}$ and V is an orthogonal direct sum $\dot{\bigoplus}_{\lambda \in \mathrm{sp}(T)} E_\lambda(T)$.*

PROOF. When $\mathbb{K} = \mathbb{C}$ the characteristic polynomial $p_T(x) = \det(T - xI)$ splits in $\mathbb{C}[x]$, so there is an eigenvalue λ_0 and $T(v_0) = \lambda_0 v_0$ for some $v_0 \neq 0$.

The one-dimensional space $M = \mathbb{C}v_0$ is T-invariant, but is also T^*-invariant since $T^*(v_0) = \overline{\lambda}_0 v_0$ by (3.). Then

$$T^*(M) \subseteq M \Rightarrow T^{**}(M^\perp) = T(M^\perp) \subseteq M^\perp \quad \text{(by Lemma 6.44)}.$$

We also have $T^*(M^\perp) \subseteq M^\perp$ because $T(M) \subseteq M \Leftrightarrow T^*(M^\perp) \subseteq M^\perp$.

When $\mathbb{K} = \mathbb{R}$ the argument is the same except that p_T splits in $\mathbb{R}[x]$ and $\lambda_0 \in \mathbb{R}$.

EXERCISE 6.53. If N is a subspace in an inner product space that is invariant under both T and T^*, prove that:

(a) $(T|_N)^* = (T^*|_N)$.
(b) $T|_N$ is a normal operator when N is given the inner product it inherits from V.

NOTE: Here we do not assume $T^* = T$ as we did in Exercise 6.50.

Taking $N = M^\perp$ in this exercise, we see that $T|_{M^\perp}$ is again a normal operator with respect to the inner product $M^\perp$ inherits from V. Since $\dim(M^\perp) < \dim(V)$ we may argue by induction to get an orthonormal basis of eigenvectors, completing the proof of Corollary 6.52. □

THEOREM 6.54 (Orthogonal Diagonalization). *Let $T : V \to V$ be a linear operator on a finite-dimensional inner product space. Assume that the characteristic polynomial $p_T(x)$ splits over $\mathbb{K}$ (always true when $\mathbb{K} = \mathbb{C}$). Then there is an orthonormal basis that diagonalizes $T \Leftrightarrow T$ is normal: $T^*T = TT^*$.*

NOTE: *It follows that $V = \dot{\bigoplus}_{\lambda \in \mathrm{sp}_{\mathbb{K}}(T)} E_\lambda(T)$ when T is normal, and in particular the eigenspaces are pairwise orthogonal. Once the eigenspaces are determined it is easy to construct the diagonalizing orthonormal basis for T.*

PROOF OF ($\Rightarrow$). This has just been done. □

PROOF OF ($\Leftarrow$). If there is an orthonormal basis $\mathfrak{X} = \{e_i\}$ that diagonalizes T, then

$$[T]_{\mathfrak{X}\mathfrak{X}} = \begin{pmatrix} \lambda_1 & & & 0 \\ & \lambda_2 & & \\ & & \ddots & \\ 0 & & & \lambda_n \end{pmatrix}.$$

But $[T^*]_{\mathfrak{X}\mathfrak{X}}$ is the adjoint of the matrix $[T]_{\mathfrak{X}\mathfrak{X}}$,

$$[T^*]_{\mathfrak{X}\mathfrak{X}} = \overline{[T]^{\mathrm{T}}_{\mathfrak{X}\mathfrak{X}}} = \begin{pmatrix} \overline{\lambda_1} & & & 0 \\ & \overline{\lambda_2} & & \\ & & \ddots & \\ 0 & & & \overline{\lambda_n} \end{pmatrix}.$$

Obviously these diagonal matrices commute (all diagonal matrices do), so

$$[T^*T]_{\mathfrak{X}\mathfrak{X}} = [T^*]_{\mathfrak{X}\mathfrak{X}}[T]_{\mathfrak{X}\mathfrak{X}} = [T]_{\mathfrak{X}\mathfrak{X}}[T^*]_{\mathfrak{X}\mathfrak{X}} = [TT^*]_{\mathfrak{X}\mathfrak{X}},$$

which implies $T^*T = TT^*$ as operators on V. □

EXAMPLE 6.55. Let $L_A : \mathbb{C}^2 \rightarrow \mathbb{C}^2$ be the multiplication operator determined by

$$A = \begin{pmatrix} 1 & 1 \\ 0 & 2 \end{pmatrix}$$

so that $L_A(\mathbf{e}_1) = \mathbf{e}_1$ and $L_A(\mathbf{e}_1 + \mathbf{e}_2) = 2 \cdot (\mathbf{e}_1 + \mathbf{e}_2)$, where $\mathfrak{X} = \{\mathbf{e}_1, \mathbf{e}_2\}$ is the standard orthonormal basis. As we saw in Chapter 2, $[L_A]_{\mathfrak{X}\mathfrak{X}} = A$. But L_A is obviously diagonalizable with respect to the *nonorthonormal* basis $\mathfrak{Y} = \{\mathbf{f}_1, \mathbf{f}_2\}$ where $\mathbf{f}_1 = \mathbf{e}_1$, $\mathbf{f}_2 = \mathbf{e}_1 + \mathbf{e}_2$. The $\mathbf{f}_i$ are basis vectors for the (one-dimensional) eigenspaces of L_A, which are uniquely determined without any reference to the inner product in $V = \mathbb{C}^2$; if there were an orthonormal basis that diagonalized L_A, the eigenspaces would be orthogonal, which they are not. This operator cannot be orthogonally diagonalized with respect to the standard inner product in $\mathbb{C}^2$.

EXERCISE 6.56. Let $T : \mathbb{C}^2 \rightarrow \mathbb{C}^2$ be L_A for a matrix A in $\mathrm{M}(2, \mathbb{C})$. Determine the eigenvalues in $\mathbb{C}$ and eigenspaces, and exhibit an orthonormal basis $\mathfrak{Y} = \{\mathbf{f}_1, \mathbf{f}_2\}$ that diagonalizes T for each of the following matrices:

$$(a)\ A = A^* = \begin{pmatrix} 1 & -1 \\ -1 & 2 \end{pmatrix} \qquad (b)\ A = A^* = \begin{pmatrix} 1 & -i \\ i & 3 \end{pmatrix}$$

EXERCISE 6.57. Prove that $|\lambda| = 1$ for all eigenvalues of a unitary operator, so λ lies on the unit circle if $\mathbb{K} = \mathbb{C}$ (or $\lambda = \pm 1$ if $\mathbb{K} = \mathbb{R}$).

EXERCISE 6.58. Let P be a projection on a finite-dimensional vector space (so $P^2 = P$).

(a) Explain why P is diagonalizable over any field $\mathbb{K}$. What are the and eigenspaces?
(b) Give an explicit example of a projection operator on a finite-dimensional inner product space that is not *orthogonally diagonalizable.*

EXERCISE 6.59. If P is a projection operator (so $P^2 = P$) on a finite-dimensional inner product space, prove that P is a normal operator $\Leftrightarrow$ $K(P) = \ker(P)$ and $R(P) = \mathrm{range}(P)$ are orthogonal subspaces.

NOTE: ($\Rightarrow$) is trivial since $K(P) = E_{\lambda=0}(P)$ and $R(P) = E_{\lambda=1}(P)$.

EXERCISE 6.60. A projection operator P (with $P^2 = P$) on an inner product space is fully determined once we know its kernel $K(P)$ and range $R(P)$, since $V = R(P) \oplus K(P)$. The adjoint P^* is also a projection operator because $(P^*P^*) = (PP)^* = P^*$.

(a) In an inner product space, show that $K(P) = R(P^*)^\perp$ and $K(P^*) = R(P)^\perp$.
(b) For the nonorthogonal direct sum decomposition of Exercise 6.48 describe the subspaces $K(P^*)$ and $R(P^*)$. (Find bases for each.)

If $T : V \rightarrow V$ is an arbitrary linear operator on an inner product space we showed in Lemma 6.42 that $\mathrm{sp}(T^*)$ is equal to $\overline{\mathrm{sp}(T)}$, and in part (3.) of

Proposition 6.51 we showed that

$$E_{\overline{\lambda}}(T^*) = E_\lambda(T) \quad (\lambda \in \text{sp}(T))$$

for *normal* operators. Unfortunately this is not true in general.

EXERCISE 6.61. If $\lambda \in \text{sp}(T)$ and $T : V \to V$ is a linear operator on an inner product space, use Proposition 6.45 to prove the following:

(a) $E_{\overline{\lambda}}(T^*) = K(T^* - \overline{\lambda}I)$ is equal to $R(T - \lambda I)^\perp$.
(b) $\dim E_{\overline{\lambda}}(T^*) = \dim E_\lambda(T)$, although the eigenspaces themselves need not be equal.
(c) T diagonalizable $\Leftrightarrow$ T^* is diagonalizable.
(d) T orthogonally diagonalizable $\Leftrightarrow$ T^* is orthogonally diagonalizable.

As the next examples show, $E_{\overline{\lambda}}(T^*) = K(T^* - \overline{\lambda}I)$ is not always equal to $E_\lambda(T)$ unless T is normal.

EXERCISE 6.62. If $P : V \to V$ is an idempotent operator on a finite-dimensional vector space (so $P^2 = P$), explain why P must be diagonalizable over any field. If $P \neq 0$ and $P \neq I$, what are its eigenvalues and its eigenspaces?

EXERCISE 6.63. Let $V = R(P) \oplus K(P)$ be a *nonorthogonal* direct sum decomposition of an inner product space. If P is the projection of V onto $R(P)$ along $K(P)$, then the adjoint P^* is a projection onto $R(P^*)$ along $K(P^*)$.

(a) What are the eigenvalues and eigenspaces for P and P^*?
(b) For $\lambda = 1$, is $E_{\overline{\lambda}}(T^*) = K(T^* - \overline{\lambda}I)$ equal to $E_\lambda(T)$?

HINT: See Exercises 6.60 and 6.61.

Unitary Equivalence of Operators vs. Similarity. Two operators T, T' on a vector space V are *similar*, indicated by writing $T' \sim T$, if there is an invertible linear operator S such that $T' = SAS^{-1}$; that means they are represented by the same matrix $[T']_{\mathfrak{Y}\mathfrak{Y}} = [T]_{\mathfrak{X}\mathfrak{X}}$ with respect to suitably chosen bases in V. We say T' is *unitarily equivalent* to T if there is a unitary operator U such that $T' = UTU^* (= UTU^{-1})$. This relation, indicated by writing $T' \cong T$, is an RST equivalence relation between operators on an inner product space V, but is more stringent than mere similarity. We now show $T' \cong T$ if and only if there are *orthonormal* bases $\mathfrak{X}, \mathfrak{Y}$ such that $[T']_{\mathfrak{Y}\mathfrak{Y}} = [T]_{\mathfrak{X}\mathfrak{X}}$.

DEFINITION 6.64. A *linear isometry* is a linear operator $U : V \to W$ between inner product spaces that preserve distances in mapping points from V into W,

$$\|Uv - Uv'\|_W = \|U(v - v')\|_W = \|v - v'\|_V \; ; \tag{6.9}$$

in particular, $\|U(v)\|_W = \|v\|_V$ for all $v \in V$. Isometries are clearly one-to-one but need not be bijections unless $\dim V = \dim W$ (see exercises below). A linear map $U : V \to W$ is *unitary* if $U^*U = \text{id}_V$ and $UU^* = \text{id}_W$, which means U is invertible with $U^{-1} = U^*$ (hence $\dim V = \dim W$). Obviously the inverse map $U^{-1} : W \to V$ is also unitary. Unitary operators $U : V \to W$ are also isometries since

$$\|Ux\|_W^2 = (Ux, Ux)_W = (x, U^*Ux)_V = \|x\|_V^2.$$

Thus unitary maps are precisely the bijective linear isometries from V to W. If V, W are inner product spaces over $\mathbb{R}$, unitary operators are referred to as *real-orthogonal* operators.

If V is finite dimensional and we restrict attention to the case $V = W$, either of the conditions $UU^* = \mathrm{id}_V$ or $U^*U = \mathrm{id}_V$ implies U is invertible with $U^{-1} = U^*$ because

$$U \text{ one-to-one} \quad \Leftrightarrow \quad U \text{ is surjective} \quad \Leftrightarrow \quad U \text{ is bijective}$$

for any linear operator $U : V \to V$ when $\dim(V) < \infty$.

EXERCISE 6.65. If V, W are inner product spaces of the same *finite* dimension, explain why there must exist a bijective linear isometry $T : V \to W$. Is T unique? Is the adjoint $T^* : W \to V$ also an isometry?

EXERCISE 6.66. Let $V = \mathbb{C}^m$, $W = \mathbb{C}^n$ with the usual inner products. Exhibit examples of linear operators $U : V \to W$ such that

(a) $UU^* = \mathrm{id}_W$ but $U^*U \neq \mathrm{id}_V$,
(b) $U^*U = \mathrm{id}_V$ but $UU^* \neq \mathrm{id}_W$.

NOTE: This might not be possible for all choices of m, n (for instance, $m = n$).

EXERCISE 6.67. If $m < n$ and the coordinate spaces $\mathbb{K}^m$, $\mathbb{K}^n$ are equipped with the standard inner products, consider the linear operator

$$T : \mathbb{K}^m \to \mathbb{K}^n, \quad T(x_1, \dots, x_m) = (x_1, \dots, x_m, 0, \dots, 0).$$

This is an isometry that embeds $\mathbb{K}^m$ into $\mathbb{K}^n$, with trivial kernel $K(T) = (0)$ and range $R(T) = \mathbb{K}^m \times (0)$ in $\mathbb{K}^n = \mathbb{K}^m \oplus \mathbb{K}^{n-m}$.

(a) Provide an explicit description of the adjoint operator $T^* : \mathbb{K}^n \to \mathbb{K}^m$ and determine $K(T^*), R(T^*)$.
(b) Compute the matrices of $[T]$ and $[T^*]$ with respect to the standard orthonormal bases $\mathfrak{X}, \mathfrak{Y}$ in $\mathbb{K}^m, \mathbb{K}^n$.
(c) How is the action of T^* related to the subspaces $K(T), R(T^*)$ in $\mathbb{K}^m$ and $R(T), K(T^*)$ in $\mathbb{K}^n$? Can you give a description of this action?

Unitary operators can be described in several different ways, each with its own advantages in applications.

THEOREM 6.68. *The statements below are equivalent for a linear operator $U : V \to W$ between finite-dimensional inner product spaces.*

(a) $UU^* = \mathrm{id}_W$ *and* $U^*U = \mathrm{id}_V$ (*so* $U^* = U^{-1}$ *and* $\dim V = \dim W$).
(b) *U maps some orthonormal basis $\{e_i\}$ in V to an orthonormal basis $\{f_i = U(e_i)\}$ in W.*
(c) *U maps* EVERY *orthonormal basis $\{e_i\}$ in V to an orthonormal basis $\{f_i = U(e_i)\}$ in W.*

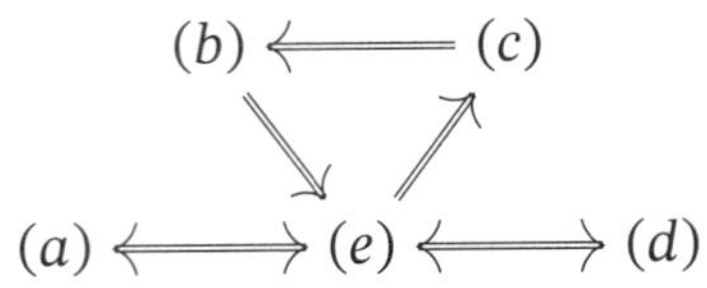

FIGURE 6.8. The pattern of implications in proving Theorem 6.68.

(d) *U is a surjective isometry, so distances are preserved:*

$$\|U(x) - U(y)\|_W = \|x - y\|_V \quad \textit{for } x, y \in V.$$

(Then U is invertible and U^{-1} is also an isometry).

(e) *U is a bijective map that preserves inner products, so that*

$$(U(x), U(y))_W = (x, y)_V \quad \textit{for all } x, y \in V.$$

This remains valid for inner product spaces over $\mathbb{R}$.

PROOF. We prove the implications shown in Figure 6.8. □

PROOF OF (d) ⇔ (e). Clearly (e) ⇒ (d). For the converse, (d) implies that U preserves lengths of vectors, with $\|Ux\|_W = \|x\|_V$ for all x. Then by the polarization identity for inner products

$$(x, y) = \frac{1}{4} \sum_{k=0}^{3} \frac{1}{i^k} \|x + i^k y\|^2,$$

so inner products are preserved, proving (d) ⇒ (e) when $\mathbb{K} = \mathbb{C}$; the same argument applies but with only two terms if $\mathbb{K} = \mathbb{R}$. □

PROOF OF (e) ⇒ (c) ⇒ (b). These are obvious since "orthonormal basis" is defined in terms of the inner product. For instance, if (e) holds and $\mathfrak{X} = \{e_i\}$ is an orthonormal basis in V, then $\mathfrak{Y} = \{f_i = U(e_i)\}$ is an orthonormal family in W because

$$(f_i, f_j)_W = \big(U(e_i), U(e_j)\big)_W = (e_i, U^*Ue_j)_V = (e_i, e_j)_V = \delta_{ij},$$

where δ_{ij} is the Kronecker delta symbol. But

$$\mathbb{K}\text{-span}\{f_j = U(e_j)\} = U\left(\mathbb{K}\text{-span}\{e_j\}\right) = U(V) = W,$$

so $\mathfrak{Y}$ spans W and therefore is a basis. □

PROOF OF (a) ⇔ (e). We have

$$U^*U = \mathrm{id}_V \Leftrightarrow U^*Ux = x \text{ for all } x \Leftrightarrow (Ux, Uy)_W = (x, U^*Uy)_V = (x, y)_V$$

for all $x, y \in V$. □

PROOF OF (b)$\Rightarrow$(e). Given an orthonormal basis $\mathfrak{X} = \{e_i\}$ in V such that the vectors $\mathfrak{Y} = \{f_i = U(e_i)\}$ are an orthonormal basis in W, we may write $x, y \in V$ as $x = \sum_i (x, e_i)\, e_i$, etc. Then

$$U(x) = \sum_i (x, e_i)_V\, U(e_i) = \sum_i (x, e_i)_V\, f_i, \text{ etc.}$$

Hence by orthonormality

$$\begin{aligned}(Ux, Uy)_W &= \Big(\sum_i (x, e_i)_V f_i, \sum_j (y, e_j)_V f_j\Big)_W = \sum_{i,j} (x, e_i)_V \overline{(y, e_j)_V} (f_i, f_j)_W \\ &= \sum_k (x, e_k)_V (e_k, y)_V = (x, y)_V.\end{aligned}$$

□

Here we applied a formula worth remembering.

LEMMA 6.69 (Parseval's Identity). *If* $x = \sum_i a_i e_i$, $y = \sum_j b_j e_j$ *with respect to an orthonormal basis in a finite-dimensional inner product space, then* $(x, y) = \sum_{k=1}^n a_k \overline{b_k}$. *Equivalently, since* $a_i = (x, e_i)$ *and* $\overline{b_k} = \overline{(y, e_k)} = (e_k, y)$, *we have*

$$(x, y) = \sum_{k=1}^{n} (x, e_k)(e_k, y) \quad \textit{for all } x, y$$

in any finite-dimensional inner product space.

The Matrix Groups U(n), SU(n), O(n), SO(n). A *matrix group.* is a set of matrices $G \subseteq \mathrm{M}(n, \mathbb{K})$ that satisfies the following *group axioms*:

(i) $I \in G$ (ii) $A, B \in G \Rightarrow AB \in G$ (iii) $A \in G \Rightarrow$ inverse matrix $A^{-1} \in G$.

DEFINITION 6.70. A matrix $A \in \mathrm{M}(n, \mathbb{K})$ is *unitary* if $AA^* = I$ (which holds $\Leftrightarrow AA^* = I \Leftrightarrow A^* = A^{-1}$), where A^* is the *adjoint matrix* such that $(A^*)_{ij} = \overline{A_{ji}}$. The set of all unitary matrices is a group since products and inverses of such matrices are again unitary. When $\mathbb{K} = \mathbb{C}$ this is the *unitary group*:

$$\text{UNITARY GROUP:} \quad \begin{aligned} \mathrm{U}(n) &= \{A \in \mathrm{M}(n, \mathbb{C}) : A^*A = I\} \\ &= \{A \in \mathrm{M}(n, \mathbb{C}) : A^* = A^{-1}\}. \end{aligned}$$

But when $\mathbb{K} = \mathbb{R}$ and $A^* = A^{\mathrm{T}}$ (the transpose matrix), it goes by another name and is called the *(real) orthogonal group*,

$$\mathrm{O}(n) = \{A \in \mathrm{M}(n, \mathbb{R}) : A^{\mathrm{T}}A = I\} = \{A \in \mathrm{M}(n, \mathbb{R}) : A^{\mathrm{T}} = A^{-1}\}.$$

Both groups lie within the *general linear group* of nonsingular matrices

$$\text{GENERAL LINEAR GROUP:} \quad \mathrm{GL}(n, \mathbb{K}) = \{A \in \mathrm{M}(n, \mathbb{K}) : \det(A) \neq 0\},$$

and both contain noteworthy subgroups

$$\text{SPECIAL UNITARY GROUP:} \quad \mathrm{SU}(n) = \{A : A^*A = I \text{ and } \det(A) = +1\}$$

$$\text{SPECIAL ORTHOGONAL GROUP:} \quad \mathrm{SO}(n) = \{A : A^{\mathrm{T}}A = I \text{ and } \det(A) = +1\}$$

PROPOSITION 6.71. *If $A \in \mathrm{M}(n,\mathbb{K})$ with $\mathbb{K} = \mathbb{R}$ or $\mathbb{C}$, the following conditions are equivalent:*

1. $L_A : \mathbb{K}^n \to \mathbb{K}^n$ *is unitary.*
2. *A is a unitary matrix, so $A^*A = AA^* = I$ in $\mathrm{M}(n,\mathbb{K})$.*
3. *The rows in A form an orthonormal basis in $\mathbb{K}^n$.*
4. *The columns in A form an orthonormal basis in $\mathbb{K}^n$.*

PROOF. With respect to the standard basis $\mathfrak{X} = \{e_1, \dots, e_n\}$ in $\mathbb{K}^n$, we know that $[L_A]_{\mathfrak{X}\mathfrak{X}} = A$, but since $\mathfrak{X}$ is an orthonormal basis, we also have $[(L_A)^*]_{\mathfrak{X}\mathfrak{X}} = [L_A]^*_{\mathfrak{X}\mathfrak{X}} = A^*$ (the adjoint matrix), by Proposition 6.37. Next observe that

$$L_{A^*} = (L_A)^* \quad \text{as operators on } \mathbb{K}^n.$$

(This may sound obvious, but it actually needs to be proved if we keep in mind how the various adjoints are defined—see Proposition 2.35 and Exercise 6.73 below.) Then

$$\begin{aligned} A^*A = I \Leftrightarrow \mathrm{id}_{\mathbb{K}^n} &= [L_{A^*A}]_{\mathfrak{X}} = [L_{A^*}]_{\mathfrak{X}} \cdot [L_A]_{\mathfrak{X}} = [(L_A)^*]_{\mathfrak{X}} \cdot [L_A]_{\mathfrak{X}} \\ &\Leftrightarrow (L_A)^* L_A = \mathrm{id}_{\mathbb{K}^n} \Leftrightarrow (L_A \text{ is a unitary operator}), \end{aligned}$$

proving (1.) $\Leftrightarrow$ (2.).

By definition of row-column matrix multiplication, we have

$$\delta_{ij} = (AA^*)_{ij} = \sum_k A_{ik}(A^*)_{kj} = \sum_k A_{ik}\overline{A_{jk}} = (\mathrm{Row}_i(A), \mathrm{Row}_j(A))_{\mathbb{K}^n}.$$

This says precisely that the rows are an orthonormal basis with respect to the standard inner product in $\mathbb{K}^n$. Thus (2.) $\Leftrightarrow$ (3.), and similarly $A^*A = I \Leftrightarrow$ the columns form an orthonormal basis in $\mathbb{K}^n$. $\square$

Similar criteria allow us to decide when linear operators are unitary.

PROPOSITION 6.72. *A linear operator $T : V \to V$ on a finite-dimensional inner product space is unitary $\Leftrightarrow$ its matrix $A = [T]_{\mathfrak{X}\mathfrak{X}}$ with respect to any orthonormal basis is a unitary matrix (so $AA^* = A^*A = I$).*

PROOF. If $\mathfrak{X}$ is orthonormal we have (by Proposition 6.37)

$$I = [\mathrm{id}_V]_{\mathfrak{X}\mathfrak{X}} = [T^*T]_{\mathfrak{X}\mathfrak{X}} = [T^*]_{\mathfrak{X}\mathfrak{X}}\,[T]_{\mathfrak{X}\mathfrak{X}} = ([T]_{\mathfrak{X}\mathfrak{X}})^*\,[T]_{\mathfrak{X}\mathfrak{X}} = A^*A,$$

and similarly $AA^* = I$, so A is a unitary matrix.

Conversely, if $A = [T]_{\mathfrak{X}\mathfrak{X}}$ is a unitary matrix,

$$\begin{aligned} (Te_i, Te_j) &= \Big(\sum_k A_{ki}\,e_k, \sum_\ell A_{\ell j}\,e_\ell\Big) = \sum_{k,\ell} A_{ki}\overline{A_{\ell j}}\,\delta_{k\ell} \\ &= \sum_k A_{ki}(A^*)_{jk} = (AA^*)_{ji} = \delta_{ji} = (e_i, e_j). \end{aligned}$$

By 6.68(c) T maps orthonormal basis $\mathfrak{X}$ to a new orthonormal basis $\mathfrak{Y} = \{T(e_i)\}$, and T is unitary. $\square$

EXERCISE 6.73. If $A \in \mathrm{M}(n,\mathbb{K})$, give a careful proof that $A^*A = I \Leftrightarrow AA^* = I$.

EXERCISE 6.74. Given two orthonormal bases $\{e_i\}, \{f_j\}$ in finite-dimensional inner product spaces V, W of the same dimension, construct a unitary operator $U : V \to W$ such that $U(e_i) = f_i$ for all i.

Change of Orthonormal Basis. If $T : V \to V$ is a linear operator on a finite-dimensional inner product space, and we know its matrix $[T]_{\mathfrak{X}\mathfrak{X}}$ with respect to one orthonormal basis, what is its matrix realization with respect to a different orthonormal basis $\mathfrak{Y}$?

DEFINITION 6.75. Matrices $A, B \in \mathrm{M}(n, \mathbb{K})$ are *unitarily equivalent*, indicated by writing $A \cong B$, if there is some unitary matrix $S \in \mathrm{M}(n, \mathbb{K})$ such that $B = SAS^* = SAS^{-1}$.

THEOREM 6.76 (Change of Orthonormal Basis). *If $\mathfrak{X} = \{e_i\}$ and $\mathfrak{Y} = \{f_j\}$ are orthonormal bases in a finite-dimensional inner product space and $T : V \to V$ is any linear operator, the corresponding matrices $A = [T]_{\mathfrak{X}\mathfrak{X}}$ and $B = [T]_{\mathfrak{Y}\mathfrak{Y}}$ are unitarily equivalent: there is some unitary matrix S such that*

$$[T]_{\mathfrak{Y}\mathfrak{Y}} = S\,[T]_{\mathfrak{X}\mathfrak{X}}S^* = S\,[T]_{\mathfrak{X}\mathfrak{X}}S^{-1} \tag{6.10}$$

where $S = [\mathrm{id}_V]_{\mathfrak{Y}\mathfrak{X}} = [\mathrm{id}_V]^{-1}_{\mathfrak{X}\mathfrak{Y}} = [\mathrm{id}_V]^*_{\mathfrak{X}\mathfrak{Y}}$. The identity (6.10) *remains true if the transition matrix S is multiplied by any scalar such that $|\lambda|^2 = \lambda\overline{\lambda} = 1$.*

PROOF. For *arbitrary* vector bases $\mathfrak{X}, \mathfrak{Y}$ in V, we have $[\mathrm{id}]_{\mathfrak{X}\mathfrak{Y}} = [\mathrm{id}]^{-1}_{\mathfrak{Y}\mathfrak{X}}$ and

$$[T]_{\mathfrak{Y}\mathfrak{Y}} = [\mathrm{id}]_{\mathfrak{Y}\mathfrak{X}} \cdot [T]_{\mathfrak{X}\mathfrak{X}} \cdot [\mathrm{id}]_{\mathfrak{X}\mathfrak{Y}} = S\,[T]_{\mathfrak{X}\mathfrak{X}}S^{-1} \tag{6.11}$$

where $S = [\mathrm{id}]_{\mathfrak{Y}\mathfrak{X}}$ is given by the vector identities $e_i = \mathrm{id}(e_i) = \sum_j S_{ji}f_j$. But we also have $e_i = \sum_j (e_i, f_j)\, f_j$, so $S_{ij} = (e_j, f_i)$ for $1 \le i, j \le n$.

In fact, the transition matrix S in (6.11) is unitary because $S_{ij} = (e_j, f_i) \Rightarrow$

$$\begin{aligned}\big(\mathrm{Row}_i(S), \mathrm{Row}_j(S)\big)_{\mathbb{K}^n} &= \sum_k S_{ik}\overline{S_{jk}} = \sum_k (e_k, f_i)\overline{(e_k, f_j)} \\ &= \sum_k (f_j, e_k)\,(e_k, f_i) = (f_j, f_i) = \delta_{ij}\end{aligned}$$

by Parseval's identity. Then $S^* = S^{-1} = [\mathrm{id}]^{-1}_{\mathfrak{Y}\mathfrak{X}} = [\mathrm{id}]_{\mathfrak{X}\mathfrak{Y}}$ and

$$[T]_{\mathfrak{Y}\mathfrak{Y}} = S\,[T]_{\mathfrak{X}\mathfrak{X}}S^* = S\,[T]_{\mathfrak{X}\mathfrak{X}}S^{-1}.$$ □

Conclusion: The various matrix realizations of T with respect to orthonormal bases in V are related by unitary equivalence (similarity modulo a unitary matrix) rather than similarity modulo a matrix that is merely invertible. Unitary equivalence is a more stringent condition on two matrices than similarity (as defined in Chapter 5).

Elements U in the unitary group $\mathrm{U}(n)$ act on matrix space $X = \mathrm{M}(n, \mathbb{K})$ by conjugation, sending

$$A \mapsto UAU^{-1} = UAU^*.$$

This group action $\mathrm{U}(n) \times X \to X$ partitions X into disjoint orbits

$$\mathcal{O}_A = \mathrm{U}(n) \cdot A = \{UAU^* : U \in \mathrm{U}(n)\},$$

which are the *unitary equivalence classes* in matrix space. When $\mathbb{K} = \mathbb{R}$, there is a similar group action $\mathrm{O}(n) \times \mathrm{M}(n, \mathbb{R}) \to \mathrm{M}(n, \mathbb{R})$ of the real-orthogonal group on real matrices. Recall that the *similarity class* of an $n \times n$ matrix A is its orbit $\mathrm{GL}(n, \mathbb{K}) \cdot A = \{EAE^{-1} : E \text{ in } \mathrm{GL}(n, \mathbb{K})\}$ under the action of the general linear group $\mathrm{GL}(n, \mathbb{K}) = \{A : \det(A) \neq 0\}$, which is considerably larger than $\mathrm{U}(n)$ or $\mathrm{O}(n)$ and has larger orbits.

Diagonalization over $\mathbb{K} = \mathbb{C}$: A Summary. We recall the properties of spectra $\mathrm{sp}_{\mathbb{C}}(T)$ of operators $T : V \to V$ over $\mathbb{C}$ and their adjoints:

1. For any T, $\mathrm{sp}(T^*) = \overline{\mathrm{sp}(T)}$ and $\dim E_{\overline{\lambda}}(T^*) = \dim E_\lambda(T)$. But as we saw in Exercises 6.62 and 6.63, the $\overline{\lambda}$ eigenspace $E_{\overline{\lambda}}(T^*)$ need not be equal to $E_\lambda(T)$ unless T is normal.
2. If $T = T^*$ then T is orthogonally diagonalizable, and all eigenvalues are *real* because $T(v) = \lambda v$ implies
$$\lambda\|v\|^2 = (T(v), v) = (v, T^*(v)) = (v, \lambda v) = \overline{\lambda}\|v\|^2.$$
3. If T is unitary, then all eigenvalues satisfy $|\lambda| = 1$ and lie on the unit circle in $\mathbb{C}$, because if $v \neq 0$ we have
$$\begin{aligned} T(v) = \lambda \cdot v \Rightarrow \|v\|^2 &= (T^*Tv, v) = (Tv, Tv) = (\lambda v, \lambda v) = |\lambda|^2 \cdot \|v\|^2 \\ &\Rightarrow |\lambda|^2 = 1. \end{aligned}$$
4. If T is skew-adjoint, so $T^* = -T$, then all eigenvalues are pure imaginary because
$$\lambda\|v\|^2 = (Tv, v) = (v, T^*v) = (v, -T(v)) = (v, -\lambda v) = -\overline{\lambda}\|v\|^2.$$
Consequently, $\overline{\lambda} = -\lambda$ and $\lambda \in 0 + i\mathbb{R}$ in $\mathbb{C}$.
5. A general normal operator is orthogonally diagonalizable, but there are no restrictions on the pattern of eigenvalues.

In Theorem 6.54 we proved the following necessary and sufficient condition for a linear operator on a complex inner product space to be diagonalizable.

THEOREM 6.77 (Orthogonal Diagonalization). *A linear operator $T : V \to V$ on a finite-dimensional complex inner product space is orthogonally diagonalizable if and only if T is normal (so $T^*T = TT^*$).*

6.5. Reflections, Rotations, and Rigid Motions on $\mathbb{R}^n$

In this section we take $\mathbb{K} = \mathbb{R}$, but in this context unitary operators are referred to as *(real)-orthogonal transformations*. The corresponding matrices $A = [T]_{\mathfrak{X}\mathfrak{X}}$ with respect to orthonormal bases satisfy $A^{\mathrm{T}}A = I = AA^{\mathrm{T}}$, so $A^{\mathrm{T}} = A^{-1}$ in $\mathrm{M}(n, \mathbb{R})$. An orthogonal transformation might not have enough real eigenvalues to be diagonalizable, which happens if and only if the eigenspaces $E_\lambda(T)$ $(\lambda \in \mathbb{R})$ fail to span V. In fact, there might not be any real eigenvalues at all. For example, if $R_\theta =$ (counterclockwise rotation about origin by θ radians) in $\mathbb{R}^2$, and if θ is not an integer multiple of π, then with respect to the standard $\mathbb{R}$-basis $\mathfrak{X} = \{e_1, e_2\}$ we have

$$[R_\theta]_{\mathfrak{X}\mathfrak{X}} = \begin{pmatrix} \cos\theta & -\sin\theta \\ \sin\theta & \cos\theta \end{pmatrix}$$

whose complex eigenvalues are $e^{i\theta}$ and $e^{-i\theta}$. There are no real eigenvalues if $\theta \neq n\pi$, even though R_θ is a normal operator. (A rotation by $\theta \neq n\pi$ radians cannot send a vector $v \neq 0$ to a scalar multiple of itself.)

The Group of Rigid Motions M(n). *Rigid motions* on $\mathbb{R}^n$ are the bijective maps $\rho : \mathbb{R}^n \to \mathbb{R}^n$ that preserve distances between points,

$$||\rho(x) - \rho(y)|| = ||x - y|| \quad \text{for all } x, y.$$

We *do not* assume ρ is linear. The rigid motions form a group $\mathrm{M}(n)$ under composition; it includes two important subgroups:

1. TRANSLATIONS: Operators $T = \{t_b : b \in \mathbb{R}^n\}$ where

   $$t_b(x) = x + b \quad \text{for all } x \in \mathbb{R}^n \ (b \in \mathbb{R}^n \text{fixed})$$

   Under the bijective map $\phi : \mathbb{R}^n \to T$ with $\phi(t) = t_b$ we have

   $$\phi(s+t) = \phi(s) \circ \phi(t) \quad \text{and} \quad \phi(0) = \mathrm{id}_{\mathbb{R}^n}.$$

   Obviously translations are isometric mappings since

   $$||\, t_b(x) - t_b(y)|| = ||(x+b) - (y-b)|| = ||x - y|| \quad \text{for all } b \text{ and } x, y,$$

   but they are *not* linear operators on $\mathbb{R}^n$ unless $b = 0$; the zero element does not remain fixed because $t_b(0) = b$.
2. LINEAR ISOMETRIES: Operators $H = \{L_A : A \in \mathrm{O}(n)\}$ where $L_A(x) = A \cdot x$ and A is any orthogonal real $n \times n$ matrix (so A is invertible with $A^{\mathrm{T}} = A^{-1}$).

Although rigid motions need not be linear operators, it is remarkable that they are nevertheless simple combinations of a linear isometry (an orthogonal linear mapping on $\mathbb{R}^n$) and a translation operator:

$$\rho(x) = (t_b \circ L_A)(x) = A \cdot x + b \quad (b \in \mathbb{R}^n, A \in \mathrm{O}(n)) \tag{6.12}$$

for all $x \in \mathbb{R}^n$. In particular, any rigid motion $\rho : \mathbb{R}^n \to \mathbb{R}^n$ that leaves the origin fixed is *automatically linear.*

PROPOSITION 6.78. *If $\rho : \mathbb{R}^n \to \mathbb{R}^n$ is a rigid motion that fixes the origin (so $\rho(\mathbf{0}) = \mathbf{0}$), then ρ is in fact a* LINEAR *operator on $\mathbb{R}^n$, with $\rho = L_A$ for some $A \in \mathrm{O}(n)$. In general, every rigid motion is a composite of the form* (6.12).

PROOF. The second statement is immediate from the first, for if ρ moves the origin to $b = \rho(\mathbf{0})$, the operation $t_{-b} \circ \rho$ is a rigid motion that fixes the origin, and $\rho = t_b \circ (t_{-b} \circ \rho)$.

To prove the first assertion, let $\{\mathbf{e}_j\}$ be the standard orthonormal basis in $\mathbb{R}^n$ written as column vectors and let $\mathbf{e}'_j = \rho(\mathbf{e}_j)$. Since $\rho(\mathbf{0}) = \mathbf{0}$ lengths are preserved because $||\rho(\mathbf{x})|| = ||\rho(\mathbf{x})-\rho(\mathbf{0})|| = ||\mathbf{x}-\mathbf{0}|| = ||\mathbf{x}||$; then inner products are also preserved because

$$\begin{aligned} -2\,(\rho(\mathbf{x}),\rho(\mathbf{y})) &= ||\rho(\mathbf{x})-\rho(\mathbf{y})||^2 - ||\rho(\mathbf{x})||^2 - ||\rho(\mathbf{y})||^2 \\ &= ||\mathbf{x}-\mathbf{y}||^2 - ||\mathbf{x}||^2 - ||\mathbf{y}||^2 = -2\,(\mathbf{x},\mathbf{y}). \end{aligned}$$

Hence the images $\mathbf{e}_i' = \rho(\mathbf{e}_i)$ of the standard basis vectors are also an orthonormal family of vectors.

Now let A be the matrix whose i^{th} column is $\mathbf{e}_i'$, so $L_A(\mathbf{e}_i) = A\cdot\mathbf{e}_i = \mathbf{e}_i'$. Then by Proposition 6.71 A is in $\mathrm{O}(n)$, L_A and $(L_A)^{-1} = L_{A^{-1}}$ are both *linear* orthogonal transformations on $\mathbb{R}^n$, and the product $L_A^{-1}\circ\rho$ is a rigid motion that fixes each $\mathbf{e}_i$ as well as the zero vector. But any such motion must be the identity map. In fact, if $\mathbf{x}\in\mathbb{R}^n$ then $(\mathbf{x},\mathbf{e}_i) = (\rho(\mathbf{x}),\rho(\mathbf{e}_i)) = (\rho(\mathbf{x}),\mathbf{e}_i)$ since $\rho(\mathbf{e}_i) = \mathbf{e}_i$ by hypothesis. Thus

$$\rho(\mathbf{x}) = \sum_i(\rho(\mathbf{x}),\mathbf{e}_i)\,\mathbf{e}_i = \sum_i(\mathbf{x},\mathbf{e}_i)\mathbf{e}_i = \mathbf{x},$$

as claimed. □

Reflections on Inner Product Spaces. Every rigid motion on $\mathbb{R}^n$,

$$T(x) = A\cdot x + t_b = (t_b\circ L_A)x \quad \text{with } A\in\mathrm{O}(n) \text{ and } b\in\mathbb{R}^n,$$

has two components, an orthogonal *linear* map L_A and a translation t_b. Rigid motions are of two types, *orientation preserving* and *orientation reversing reversing*. Translations always preserve orientation of geometric figures, so the nature of a rigid motion T is determined by its linear component L_A, which preserves orientation if $\det(A) > 0$ and reverses it if $\det(A) < 0$. As a simple illustration, consider the matrices (with respect to the standard basis $\mathfrak{X}$ in $\mathbb{R}^2$) of a rotation about the origin R_θ (orientation preserving), and a reflection r_y across the y-axis (orientation reversing).

$$[R_\theta]_{\mathfrak{X}\mathfrak{X}} = \begin{pmatrix}\cos\theta & -\sin\theta\\ \sin\theta & \cos\theta\end{pmatrix} \qquad [r_y]_{\mathfrak{X}\mathfrak{X}} = \begin{pmatrix}-1 & 0\\ 0 & 1\end{pmatrix}$$

$$\textit{Rotation: } \det[R_\theta] = +1 \qquad \textit{Reflection: } \det[r_y] = -1$$

Reflections can be described in terms of the standard inner product in $\mathbb{R}^n$.

EXAMPLE 6.79 (Reflections in Inner Product Spaces). If V is a finite-dimensional inner product space over $\mathbb{R}$, a *hyperplane* in V is any vector subspace M with $\dim(M) = n-1$, so M has codimension 1 in V. This determines a reflection of vectors across M.

DISCUSSION: Since $V = M\dot{\oplus}M^\perp$ (orthogonal direct sum), every vector v splits uniquely as $v = v_{\|} + v_\perp$ (with parallel component $v_{\|}\in M$, and $v_\perp\in M^\perp$). By definition, reflection r_M across M is the (linear) operator that reverses the perpendicular component $v_\perp$, so that

$$r_M(v_{\|} + v_\perp) = v_{\|} - v_\perp = v - 2\cdot v_\perp \tag{6.13}$$

as shown in Figure 6.9.

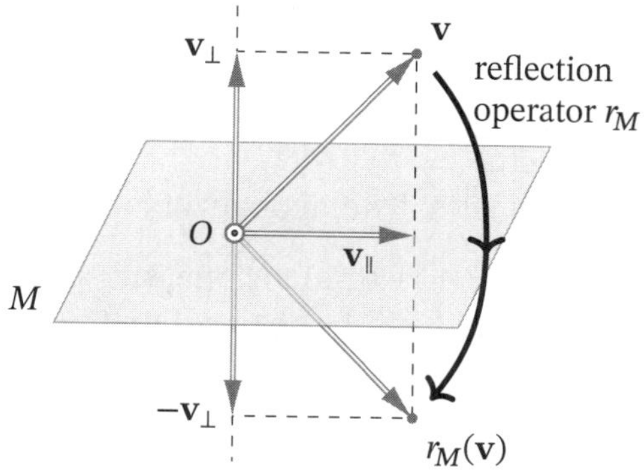

FIGURE 6.9. Geometric meaning of the reflection r_M across an $(n-1)$-dimensional hyperplane in an n-dimensional inner product space over $\mathbb{R}$.

Now let $\{e_1, \dots, e_{n-1}\}$ be an orthonormal basis in the subspace M and let e_n be $v_\perp$ renormalized to make $\|e_n\| = 1$, so $M^\perp = \mathbb{R}e_n$. We have seen that

$$v_\| = \sum_{k=1}^{n-1} (v, e_k)\, e_k,$$

so $v_\perp = v - v_\| = c \cdot e_n$ for some $c \in \mathbb{R}$. But in fact $c = (v, e_n)$ because

$$c = (c \cdot e_n, e_n) = (v - v_\|, e_n) = (v, e_n) + 0.$$

This yields an important formula for reflections that involves only the inner product:

$$r_M(v) = v_\| - v_\perp = (v_\| + v_\perp) - 2 \cdot v_\perp = v - 2(v, e_n) \cdot e_n. \tag{6.14}$$

Note that we need $\|e_n\| = 1$ to make this work.

EXERCISE 6.80. Show that (6.14) implies the following properties for any reflection.

(a) $r_M \circ r_M = \mathrm{id}_V$, so r_M is invertible and is its own inverse.
(b) $\det(r_M) = -1$, so all reflections are orientation-reversing.
(c) M is the set of fixed points $\mathrm{Fix}(r_M) = \{x : r_M(x) = x\}$.

EXERCISE 6.81. Prove that every reflection r_M on an inner product space preserves distances,

$$\|r_M(x) - r_M(y)\| = \|x - y\|$$

for all $x, y \in V$.

EXERCISE 6.82. In $V = \mathbb{R}^3$ equipped with the usual euclidean inner product, let $M = (\mathbb{R}f_3)^\perp$ where $f_3 = (1, -2, 3)$.

(a) Find an orthonormal basis $\{e_1, e_2\}$ for M.
(b) If $R : \mathbb{R}^3 \to \mathbb{R}^3$ is the linear operator that reflects vectors across the hyperplane M, give an explicit formula

$$(y_1, y_2, y_3) = R(x_1, x_2, x_3)$$

that describes this operation in terms of inner products. Find the image of the particular vector $\mathbf{x} = (1, 1, -3)$.

(c) What is the matrix $[R]_{\mathfrak{Y}\mathfrak{Y}}$ with respect to the orthonormal basis $\mathfrak{Y} = \{e_1, e_2, e_3\}$ in which $e_3 = f_3/\|f_3\|$?

(d) Is $R : V \to V$ orthogonally diagonalizable? Explain.

EXERCISE 6.83. If M is an $(n-1)$-dimensional subspace in an n-dimensional real inner product space V and $b \notin M$, the translate $E = b + M$ a coset in V/M) is a typical hyperplane parallel to M but is not a vector subspace because $0 \notin E$.

(a) Verify that the operator $T = t_b \circ r_M \circ t_{-b}$ on V has the algebraic property (i) $T^2 = \text{id}_V$ and (ii) that its set of *fixed points* $\text{Fix}(T) = \{x \in V : T(x) = x\}$ is precisely the hyperplane E.

By definition, $R_E =$ (*reflection across* E) sends a point $x \in V$ to a point $R(x)$ on the other side of E that has the same perpendicular distance from E so that $\text{dist}(Rx, E)$ is equal to $\text{dist}(x, E)$.

(b) In your own words, describe the geometric action of each successive factor in the product

$$T(x) = t_b \circ r_M \circ t_{-b} \quad \text{for all } x \in V$$

and give an algebraic formula for $T(x)$.

(c) Give a geometric explanation why T and the reflection R_E are the same operator on $\mathbb{R}^3$.

Euler's Theorem: Rotations on $\mathbb{R}^3$. In another direction, we prove Euler's famed geometric characterization of orientation-preserving real-orthogonal transformations $L_A : \mathbb{R}^3 \to \mathbb{R}^3$, with $A^{\mathsf{T}}A = I$ in $\text{M}(3, \mathbb{R})$ and $\det(A) > 0$. In fact, $\det(A) = +1$ since $A^{\mathsf{T}}A = I$ implies $(\det(A))^2 = 1$, so $\det(A) = \pm 1$ for $A \in \text{O}(n)$.

THEOREM 6.84 (Euler's Theorem). *Let A be a matrix in*

$$\text{SO}(3) = \{A \in \text{M}(3, \mathbb{R}) : A^{\mathsf{T}}A = I \text{ and } \det(A) = 1\}.$$

If $A \neq I$ then $\lambda = 1$ is an eigenvalue for L_A such that $\dim_{\mathbb{R}}(E_{\lambda=1}) = 1$. If $v_0 \neq 0$ in $E_{\lambda=1}$ and $\ell = \mathbb{R}v_0$, there is some angle $\theta \notin 2\pi\mathbb{Z}$ such that L_A is equal to

$$R_{\ell,\theta} = (\text{rotation by } \theta \text{ radians about the oriented line } \ell \text{ through the origin}).$$

Rotations by a positive angle are determined by the usual "right-hand rule": when you wrap the fingers of your right hand around around ℓ in the direction of the rotation, your thumb should be pointing in the direction of v_0).

PROOF. The characteristic polynomial $p_T(x)$ for $T = L_A$ has real coefficients. Regarded as a polynomial $p_T \in \mathbb{R}[x] \subseteq \mathbb{C}[x]$, its complex roots are either real or occur in conjugate pairs $z = x + iy$, $\overline{z} = x - iy$ with $y \neq 0$. Since $\text{degree}(p_T) = 3$ there must be at least one real root λ. But because $T = L_A$ is real-orthogonal its complex eigenvalues have $|\lambda| = 1$, for if $v \neq 0$ in E_λ we have

$$\|v\|^2 = (T(v), T(v)) = (\lambda v, \lambda v) = |\lambda|^2\,\|v\|^2 \Rightarrow |\lambda|^2 = 1.$$

If λ is real the only possibilities are $\lambda = \pm 1$. The real roots cannot all be -1, for then $\det(T) = (-1)^3 = -1$ and we require $\det(T) = +1$. Thus $\lambda = 1$ is an eigenvalue, and we will see below that $\dim_{\mathbb{R}}(E_{\lambda=1}) = 1$ unless $T = I$.

If $v_0 \neq 0$ in $E_{\lambda=1}$, let $M = \mathbb{R}v_0$. Then $M^{\perp}$ is 2-dimensional and is invariant under both T and $T^* = T^{-1}$. Furthermore (see Exercise 6.85 below) the restriction $T|_{M^{\perp}}$ is a real-orthogonal transformation on the 2-dimensional space $M^{\perp}$ equipped with the inner product it inherits from $\mathbb{R}^3$. If we fix an orthonormal basis $\mathfrak{X}' = \{f_1, f_2\}$ in $M^{\perp}$ and let $f_0 = v_0/\|v_0\|$, we obtain an orthonormal basis for $\mathbb{R}^3$. First observe that the matrix A of $T|_{M^{\perp}}$ with respect to $\mathfrak{X}'$ is in

$$\mathrm{SO}(2) = \{A \in \mathrm{M}(2,\mathbb{R}) : A^{\mathrm{T}}A = I \text{ and } \det(A) = 1\}$$

because the matrix of T with respect to the orthonormal basis $\mathfrak{X} = \{f_0, f_1, f_2\}$ has the form

$$[T]_{\mathfrak{X}\mathfrak{X}} = \begin{pmatrix} 1 & 0 \\ 0 & A \end{pmatrix} \quad \text{for some } 2 \times 2 \text{ real matrix } A.$$

Furthermore, $[T]_{\mathfrak{X}\mathfrak{X}}$ is orthogonal by hypothesis, so $[T]_{\mathfrak{X}\mathfrak{X}}^{-1} = [T]_{\mathfrak{X}\mathfrak{X}}^{*}$. Therefore $A^* = A^{-1}$ and $A \in \mathrm{SO}(2)$ as claimed. If $A \in \mathrm{M}(2,\mathbb{R})$ is a real-orthogonal matrix ($A^{\mathrm{T}}A = I$), it follows from Theorem 6.68 and Proposition 6.71 that its rows are an orthonormal basis in $\mathbb{R}^2$ (likewise for its columns). Hence there exist $a, b \in \mathbb{R}$ such that

$$a^2 + b^2 = 1 \quad \text{and} \quad A = \begin{pmatrix} a & -b \\ b & a \end{pmatrix}.$$

If we also have $\det A = +1$, so $A \in \mathrm{SO}(2)$, it follows that (see Exercise 6.86 below) there is some $\theta \in \mathbb{R}$ such that

$$A = \begin{pmatrix} \cos\theta & -\sin\theta \\ \sin\theta & \cos\theta \end{pmatrix}.$$

This is the matrix $A = [R_\theta]_{\mathfrak{X}'\mathfrak{X}'}$ of a rotation by θ radians about the origin in $M^{\perp}$, so $T : \mathbb{R}^3 \to \mathbb{R}^3$ is a rotation $R_{\ell,\theta}$ by θ radians about the axis $\ell = \mathbb{R}v_0$. □

We cannot have $\theta \in 2\pi\mathbb{Z}$ because then $T = \mathrm{id}$ is not really a rotation about any well-defined axis, which is why we required $A \neq I$ in the theorem.

EXERCISE 6.85. Let $T : V \to V$ be a unitary (or real-orthogonal) linear operator on a finite-dimensional inner product space, and M a subspace that is invariant under both T and T^*. Prove that the restriction $(T|_M) : M \to M$ is unitary (real-orthogonal) with respect to the inner product that M inherits from V.
HINT: Recall Exercise 6.53.

EXERCISE 6.86. If $A = [a, b; c, d] \in \mathrm{M}(2,\mathbb{R})$, verify that $A^{\mathrm{T}}A = I \Leftrightarrow$ rows of A are an orthonormal basis in $\mathbb{R}^2$, so that

$$a^2 + b^2 = 1, \quad c^2 + d^2 = 1, \quad ac + bd = 0.$$

If we also have

$$\det(A) = ad - bc = +1,$$

prove that $c = -b$, $d = a$, and $a^2 + b^2 = 1$, and then explain why there is some $\theta \in \mathbb{R}$ such that $a = \cos(\theta)$ and $b = -\sin(\theta)$.
NOTE: Thus $L_A : \mathbb{R}^2 \to \mathbb{R}^2$ is a counterclockwise rotation about the origin by θ radians.

HINT: For the last step, think of $a^2 + b^2 = 1$ in terms of a right triangle whose hypoteneuse has length $= 1$.

EXERCISE 6.87. Consider the linear map $L_A : \mathbb{R}^2 \to \mathbb{R}^2$ for the matrix

$$A = \frac{1}{\sqrt{2}}\begin{pmatrix} -1 & 1 \\ 1 & 1 \end{pmatrix} \quad \text{in } \mathrm{O}(2).$$

What is the geometric action of L_A? If a rotation, find the rotation angle θ; if not, show that the set of fixed points for L_A is a line L through the origin and that $L_A =$ *(reflection across L)*.

EXERCISE 6.88. Let $A = \begin{pmatrix} a & b \\ c & d \end{pmatrix}$ be in $\mathrm{O}(2)$ with $\det(A) = -1$,

(a) Prove that $L_A : \mathbb{R}^2 \to \mathbb{R}^2$ is reflection across some line ℓ through the origin.
(b) Explain why

$$a^2 + b^2 = 1, \quad c^2 + d^2 = 1, \quad ac + bd = 0, \quad \det(A) = ad - bc = -1,$$

and then show there is some θ such that $A = \begin{pmatrix} \cos\theta & \sin\theta \\ \sin\theta & -\cos\theta \end{pmatrix}$.

NOTE: The preceding matrix is *not* a rotation matrix since $\det(A) = -1$. The parameter θ is now related to the angle between the line of reflection ℓ and the $+x$-axis.

HINTS: The map $L_A : \mathbb{C}^2 \to \mathbb{C}^2$ is unitary, and in particular is orthogonally diagonalizable. What are the possible patterns of complex eigenvalues (counted according to multiplicity), and how do they relate to the requirement that $\det(A) = -1$?

Further Comments on Euler. We have observed in Chapter 2, and in this section, that the linear operators L_A corresponding to matrices in the full orthogonal group $\mathrm{O}(n)$ are precisely the rigid motions on $\mathbb{R}^n$ that happen to be linear operators. Furthermore, algebraic operations in $\mathrm{O}(n)$ and in the space of linear operators on $\mathbb{R}^n$ match up under the bijective correspondence $A \mapsto L_A$, with

$$L_{AB} = L_A \circ L_B \quad \text{for all } A, B \in \mathrm{O}(n).$$

This correspondence intertwines the group operations in matrix space (matrix multiply) and those in the space of linear operators on $\mathbb{R}^n$ (compositon of operators).

Euler's theorem establishes a bijection $\phi : A \to L_A$ between the matrix group $\mathrm{SO}(3)$ and the family of rotations $R_{\ell,\theta}$ by any angle θ about any oriented axis ℓ through the origin in $\mathbb{R}^3$.

EXERCISE 6.89. Use Euler's theorem to prove the following:

> THEOREM 6.90. *The composition product* $R_{\ell_1,\theta_1} \circ R_{\ell_2,\theta_2}$ *of two rotations about axes through the origin in* $\mathbb{R}^3$ *is either the identity* I *or is a new rotation* R_{ℓ_3,θ_3}.

In other words, the family of rotations in $\mathbb{R}^3$ is closed under formation of composition products and is a *group of linear operators* on $\mathbb{R}^3$.

This is not easy to prove without benefit of Euler. To understand the issues involved you might consider axes $\ell_i = \mathbb{R} \cdot \mathbf{u}_i$ where $\mathbf{u}_i$ are unit vectors, and rotation angles θ_i about each. Determine the new axis ℓ_3 and angle of rotation θ_3 about it for the product $R_{\ell_1,\theta_1} \circ R_{\ell_2,\theta_2}$.
HINT: The new axis is determined by the calculus cross product $\mathbf{u}_1 \times \mathbf{u}_2$, a fact of some importance in angular momentum calculations in classical mechanics and quantum mechanics.

6.6. Spectral Theorem for Vector and Inner Product Spaces

If V is a vector space over a field $\mathbb{K}$ (not necessarily an inner product space), and if $T : V \to V$ is diagonalizable over $\mathbb{K}$, there is an ordinary direct sum decomposiiton $V = \bigoplus_{\lambda \in \mathrm{sp}(T)} E_\lambda(T)$; see Proposition 2.29. This determines projection operators $P_\lambda = P_\lambda^2$ of V onto $E_\lambda(T)$ along the complementary subspaces $\bigoplus_{\mu \neq \lambda} E_\mu(T)$. The projections $P_\lambda = P_\lambda(T)$ have the following easily verified properties:

1. $P_\lambda^2 = P_\lambda$
2. $P_\lambda P_\mu = P_\mu P_\lambda = 0$ if $\lambda \neq \mu$ in $\mathrm{sp}(T)$
3. $I = \sum_\lambda P_\lambda$

Condition (1.) simply reflects the fact that P_λ is a projection operator. Each v in V has a unique decomposition $v = \sum_\lambda v_\lambda$ with $v_\lambda \in E_\lambda(T)$, and (by definition) $P_\lambda(v) = v_\lambda$. Property (3.) follows from this. For (2.) write $v = \sum_\lambda v_\lambda$ and consider distinct $\alpha \neq \beta$ in $\mathrm{sp}(T)$. Then

$$P_\alpha P_\beta(v) = P_\alpha P_\beta\Big(\sum_\lambda v_\lambda\Big) = P_\alpha(v_\beta) = 0 \quad (\text{since } \alpha \neq \beta)$$

and similarly for $P_\beta P_\alpha$. The operators $\{P_\lambda : \lambda \in \mathrm{sp}_{\mathbb{K}}(T)\}$ are the *spectral projections* associated with the diagonalizable operator T.

Now let V be an inner product space. If T is orthogonally diagonalizable, we have additional information regarding the spectral projections $P_\lambda(T)$:

4. The eigenspaces $E_\lambda(T)$ are orthogonal, $E_\lambda \perp E_\mu$ if $\lambda \neq \mu$, and $V = \dot{\bigoplus}_\lambda E_\lambda(T)$ is an orthogonal direct sum decomposition.
5. The P_λ are *orthogonal* projections, hence they are *self-adjoint* in addition to having the preceding properties, so that $P_\lambda^2 = P_\lambda = P_\lambda^*$.

In this setting we can prove useful facts relating diagonalizability and eigenspaces of an operator $T : V \to V$ and its adjoint T^*. These follow by recalling that there is a natural isomorphism between any finite-dimensional inner product space V and its dual space V^*, as explained in Lemma 6.27. Therefore given any basis $\mathfrak{X} = \{e_1, \dots, e_n\}$ in V, there exists WITHIN V a matching basis $\mathfrak{X}' = \{f_1, \dots, f_n\}$ that is "dual to" $\mathfrak{X}$ in the sense that

$$(e_i, f_j) = \delta_{ij} \quad \text{(Kronecker delta)}$$

These paired bases in V can be extremely useful in comparing properties of T with those of its adjoint T^*.

EXERCISE 6.91. Let $\mathfrak{X} = \{e_1, \dots, e_n\}$ be an arbitrary basis (not necessarily orthonormal) in a finite-dimensional inner product space V.

(a) Use induction on n to prove that there exist vectors $\mathfrak{Y} = \{f_1, \dots, f_n\}$ such that $(e_i, f_j) = \delta_{ij}$.
(b) Explain why the f_j are uniquely determined and a basis for V.

NOTE: If the initial basis $\mathfrak{X}$ is orthonormal, then $f_i = e_i$ and the result is trivial. Here we are interested in *arbitrary* bases in an inner product space.

EXERCISE 6.92. Let V be an inner product space and T a linear operator that is diagonalizable in the ordinary sense, but not necessarily orthogonally diagonalizable. Prove the following:

(a) The adjoint operator T^* is diagonalizable. What can you say about its eigenvalues and eigenspaces?
(b) If T is *orthogonally* diagonalizable so is T^*.

HINT: If $\{e_i\}$ diagonalizes T, what does the "dual basis" $\mathfrak{Y} = \{f_j\}$ defined in Exercise 6.91 do for T^*?

EXERCISE 6.93. If V is a finite-dimensional inner product space and $T : V \to V$ is diagonalizable in the ordinary sense, prove that the spectral projections for T^* are adjoints of those for T, with

$$P_{\bar{\lambda}}(T^*) = (P_\lambda(T))^* \quad \text{for all } \lambda \in \text{sp}(T).$$

HINT: Use Exercise 6.92 and dual diagonalizing bases; we know $\text{sp}(T^*) = \overline{\text{sp}(T)}$ already.

NOTE: $(P_\lambda(T))^*$ might differ from $P_{\bar{\lambda}}(T)$.

We now prove the spectral theorem and examine its many applications.

Spectral Theorem.

THEOREM 6.94. *If a linear operator $T : V \to V$ is diagonalizable on a finite-dimensional vector space V over a field $\mathbb{K}$, and if $\{P_\lambda : \lambda \in \text{sp}_{\mathbb{K}}(T)\}$ are the spectral rems for projections, $P_\lambda : V \to E_\lambda(T)$, then T has the following* SPECTRAL DECOMPOSITION *in terms of those projections:*

$$T = \sum_{\lambda \in \text{sp}(T)} \lambda \cdot P_\lambda. \tag{6.15}$$

If $f(x) = \sum_{k=0} c_k x^k \in \mathbb{K}[x]$ is any polynomial, the operator $f(T) = \sum_{k=0} c_k T^k$ is also diagonalizable and takes the form

$$f(T) = \sum_{\lambda \in \mathrm{sp}(T)} f(\lambda) \cdot P_\lambda. \tag{6.16}$$

In particular, all powers T^k are diagonalizable, with $T^k = \sum_{\lambda \in \mathrm{sp}(T)} \lambda^k \cdot P_\lambda$.

If we define the map $\Phi : \mathbb{K}[x] \to \mathrm{Hom}_{\mathbb{K}}(V, V)$ from polynomials to linear operators on V, letting $\Phi(1) = I$ and

$$\Phi(f) = \sum_{k=0} c_k T^k \quad \text{for } f(x) = \sum_{k=0} c_k x^k,$$

then Φ is linear and a homomorphism of associative algebras over $\mathbb{K}$ so that

$$\Phi(1) = I \quad \text{and} \quad \Phi(fg) = \Phi(f) \circ \Phi(g) \quad \text{for } f, g \in \mathbb{K}[x]. \tag{6.17}$$

Finally, $\Phi(f) = \mathbf{0}$ (the zero operator on V) if and only if $f(\lambda) = 0$ for each $\lambda \in \mathrm{sp}_{\mathbb{K}}(T)$. Thus $\Phi(f) = \Phi(g)$ if and only if f and g take the same values on the spectrum $\mathrm{sp}(T)$. In particular, various polynomials $f \in \mathbb{K}[x]$ yield the same operator $f(T)$.

NOTE: This all remains true for orthogonally diagonalizable operators on an inner product space, but in that case we have the additional property

$$\Phi(\overline{f}) = \Phi(f)^* \quad \text{(adjoint operator)} \tag{6.18}$$

where $\overline{f}(x) = \sum_{k=0} \overline{c_k} x^k$ and $\bar{c}$ is the complex conjugate of c.

PROOF. If $v = \sum_\lambda v_\lambda$ in $V = \bigoplus_{\lambda \in \mathrm{sp}(T)} E_\lambda(T)$, then

$$\begin{aligned} T(v) = T\Big(\sum_\lambda v_\lambda\Big) &= \sum_\lambda \lambda \cdot v_\lambda = \sum_\lambda \lambda \cdot P_\lambda(v) \\ &= \Big(\sum_{\lambda \in \mathrm{sp}(T)} \lambda \cdot P_\lambda\Big) v \end{aligned}$$

for all $v \in V$, proving (6.15). Then $T^k = \sum_\lambda \lambda^k P_\lambda$ becomes

$$T^k(v) = T^k\Big(\sum_\lambda v_\lambda\Big) = \sum_\lambda T^k v_\lambda = \sum_\lambda \lambda^k v_\lambda$$

because $T(v_\lambda) = \lambda v_\lambda$ implies

$$T^2(v_\lambda) = T(\lambda \cdot v_\lambda) = \lambda^2 v_\lambda, \quad T^3(v_\lambda) = \lambda^3 v_\lambda, \quad \text{etc.}$$

If $v = \sum_\lambda v_\lambda$ we get

$$T^k(v) = \sum_\lambda \lambda^k v_\lambda = \sum_\lambda \lambda^k P_\lambda(v) = \Big(\sum_\lambda \lambda^k P_\lambda\Big) v$$

for all $v \in V$. Then (6.16) follows because the powers T^k and the sum $f(T)$ are linear operators. Finally, for any $f(x) = \sum_k c_k x^k$ we have

$$\begin{aligned}
f(T)(v) &= f(T)\Big(\sum_\lambda v_\lambda\Big) = \sum_\lambda f(T)(v_\lambda) \\
&= \sum_\lambda \Big(\sum_k c_k T^k\Big)(v_\lambda) = \sum_\lambda \sum_k c_k T^k(v_\lambda) \\
&= \sum_\lambda \sum_k c_k \lambda^k v_\lambda = \sum_\lambda \Big(\sum_k c_k \lambda^k\Big) v_\lambda \\
&= \sum_\lambda f(\lambda)\, v_\lambda = \sum_\lambda f(\lambda) P_\lambda(v) \\
&= \Big(\sum_\lambda f(\lambda) P_\lambda\Big) v \quad \text{for all } v \in V.
\end{aligned}$$

Thus $f(T) = \sum_\lambda f(\lambda) P_\lambda$ as operators on V.

When $f(x)$ is the constant polynomial $f(x) = 1$ we get

$$\sum_{\lambda \in \mathrm{sp}(T)} f(\lambda) P_\lambda = \sum_\lambda P_\lambda = I$$

as expected. Linearity of Φ is easily checked by applying the operators on either side to a typical vector. As for the multiplicative property, let $f = \sum_{k=0} a_k x^k$ and $g = \sum_{\ell \geq 0} b_\ell x^\ell$, so $fg = \sum_{k,\ell=0} a_k b_\ell\, x^{k+\ell}$. First notice that the multiplicative property holds for monomials $f = x^k$, $g = x^\ell$ because

$$\begin{aligned}
\Phi(x^k)\Phi(x^\ell) &= \Big(\sum_{\lambda \in \mathrm{sp}(T)} \lambda^k P_\lambda\Big) \cdot \Big(\sum_{\mu \in \mathrm{sp}(T)} \mu^\ell P_\mu\Big) \\
&= \sum_{\lambda,\mu} \lambda^k \mu^\ell\, P_\lambda P_\mu = \sum_\lambda \lambda^{k+\ell} P_\lambda = \Phi(x^{k+\ell})
\end{aligned}$$

($P_\lambda P_\mu = 0$ if $\lambda \neq \mu$, and $P_\lambda^2 = P_\lambda$). Then use linearity of Φ to get

$$\begin{aligned}
\Phi(fg) &= \Phi\Big(\sum_{k,\ell=0} a_k b_\ell\, x^{k+\ell}\Big) = \sum_{k,\ell=0} a_k b_\ell \Phi(x^{k+\ell}) \\
&= \sum_{k,\ell=0} a_k b_\ell \Phi(x^k)\Phi(x^\ell) = \Big(\sum_{k=0} a_k \Phi(x^k)\Big) \cdot \Big(\sum_{\ell=0} b_\ell \Phi(x^\ell)\Big) \\
&= \Phi(f) \circ \Phi(g).
\end{aligned}$$

That completes the proof of Theorem 6.6. □

Although the operator $\Phi(f) = \sum_\lambda f(\lambda) P_\lambda$ was defined for polynomials $f \in \mathbb{K}[x]$, this sum involves only the values of f on the finite subset $\mathrm{sp}(T) \subseteq \mathbb{K}$, so it makes sense for all functions $h : \mathrm{sp}(T) \to \mathbb{K}$ whether or not they are defined off of the spectrum or related in any way to polynomials. Thus the spectral decomposition of T determines a linear map

(6.19) $$\Phi : \mathcal{E} \to \mathrm{Hom}_{\mathbb{K}}(V, V), \quad \Phi(h) = \sum_{\lambda \in \mathrm{sp}(T)} h(\lambda) P_\lambda,$$

defined on the larger algebra $\mathcal{E} \supseteq \mathbb{F}[x]$ whose elements are arbitrary functions h from $\text{sp}_{\mathbb{K}}(T) \to \mathbb{K}$. The same argument used for polynomials shows that the extended version of Φ is again a homomorphism between associative algebras, as in (6.17). Incidentally, the Lagrange interpolation formula tells us that any $h(x)$ in $\mathcal{E}$ is equal to the restriction of some (nonunique) polynomial $f(x)$, so that

$$\Phi(h) = \Phi(f|_{\text{sp}(T)}) = \Phi(f).$$

All this applies to matrices as well as operators since a matrix is diagonalizable if and only if the left multiplication operator $L_A : \mathbb{K}^m \to \mathbb{K}^m$ on coordinate space is diagonalizable.

We can now define functions $h(T)$ of an operator for a much broader class of functions than polynomials, as in the next examples.

EXAMPLE 6.95. If a diagonalizable linear operator $T : V \to V$ over $\mathbb{C}$ has spectral decomposition $T = \sum_\lambda \lambda \cdot E_\lambda$, we can define operators $h(T)$ such as the following:

1. $|T| = \sum_\lambda |\lambda| P_\lambda$ where we take $h(z) = |z|$
2. $e^T = \sum_\lambda e^\lambda P_\lambda$ where we take $h(z) = e^z = \sum_{n=0}^{\infty} z^n/n!$
3. $\sqrt{T} = \sum_\lambda \lambda^{1/2} P_\lambda$ where we assign any (complex) determination of $h(z) = \sqrt{z}$ at each point in the spectrum. Thus there are 2^r possible operator square roots if T has r distinct nonzero eigenvalues. Every such square root has the property $h(T)^2 = T$.
4. The *indicator function* of a finite subset $E \subseteq \mathbb{C}$ is defined to be

$$1_E(z) = \begin{cases} 1 & \text{if } z \in E, \\ 0 & \text{otherwise.} \end{cases}$$

By the property (6.17), $1_E(T)$ is a projection operator with $1_E(T)^2 = 1_E(T)$ since $1_E \cdot 1_E = 1_E$, and in particular if $E = \{\lambda_1, \dots, \lambda_s\} \subseteq \text{sp}(T)$ we have

$$1_E(T) = \sum_{\lambda \in E} P_\lambda = \bigoplus_{i=1}^{s} P_{\lambda_i} \quad \left(\text{projection onto } \bigoplus_{i=1}^{s} E_{\lambda_i}(T) \right).$$

We get $1_E(T) = I$ if $E \supseteq \text{sp}(T)$, and if $E = \{\lambda_0\}$ a single eigenvalue, we recover the individual spectral projections: $1_E(T) = P_{\lambda_0}$.

EXERCISE 6.96. Let $T : V \to V$ be a diagonalizable linear operator over any ground field. If T is invertible ($\lambda = 0$ not in the spectrum), explain why

$$h(T) = \sum_{\lambda \in \text{sp}(T)} \frac{1}{\lambda} \cdot P_\lambda \quad \text{with } h(x) = \frac{1}{x} \text{ for } x \neq 0$$

is the usual inverse T^{-1} of T.

HINT: Show $T \circ h(T) = h(T) \circ T = I$.

Similarly, we have $T^k = \sum_\lambda \lambda^k P_\lambda$ for $k \geq 1$, $T^0 = I$, and

$$T^{-k} = \left(T^{-1}\right)^k = \sum_\lambda \frac{1}{\lambda^k} P_\lambda$$

for negative exponents.

EXERCISE 6.97. Prove (6.18) when V is an inner product space over $\mathbb{C}$. (There is nothing to prove when $\mathbb{K} = \mathbb{R}$.)

EXERCISE 6.98. Prove that a normal operator $T : V \to V$ on a finite-dimensional inner product space over $\mathbb{C}$ is self-adjoint if and only if its spectrum is real: $\mathrm{sp}_{\mathbb{C}}(T) \subseteq \mathbb{R} + i0$.

NOTE: We already proved ($\Rightarrow$).

EXERCISE 6.99. Let $S \in \mathrm{M}(2, \mathbb{C})$ be a *symmetric* matrix, so $A^{\mathrm{T}} = A$.

(a) Is $L_A : \mathbb{C}^2 \to \mathbb{C}^2$ diagonalizable in the ordinary sense?
(b) Is $L_A : \mathbb{C}^2 \to \mathbb{C}^2$ orthogonally diagonalizable when $\mathbb{C}^2$ is given the usual inner product?

Prove or provide a counterexample.

NOTE: If we take $\mathbb{R}$ instead of $\mathbb{C}$ the answer is "yes" for both (a) and (b) because $A^* = A^{\mathrm{T}}$ when $\mathbb{K} = \mathbb{R}$. Recall that $(L_A)^* = L_{A^*}$ for the standard inner product on $\mathbb{C}^2$—see Exercise 6.34. Self-adjoint matrices are diagonalizable over both $\mathbb{R}$ and $\mathbb{C}$, but we are not assuming $A = A^*$ here, only $A = A^{\mathrm{T}}$.

Functions of Operators: e^T and $\sqrt{T}$ Revisited. If T is a linear operator on a finite-dimensional space over $\mathbb{R}$ or $\mathbb{C}$, the exponential matrix e^T has been defined in Chapter 2 as the sum of the exponential series

$$e^T = \sum_{k=0}^{\infty} \frac{1}{k!} T^k .$$

When T is diagonalizable the spectral decomposition $T = \sum_\lambda \lambda P_\lambda$ provides an alternative definition

$$e^T = \sum_\lambda e^\lambda P_\lambda .$$

But do these constructs yield the same operator?

PROPOSITION 6.100. *If T is diagonalizable over $\mathbb{R}$ or $\mathbb{C}$ then*

$$e^T = \sum_{k=0}^{\infty} \frac{1}{k!} T^k = \sum_{\lambda \in \mathrm{sp}(T)} e^\lambda P_\lambda ,$$

where $T = \sum_\lambda \lambda P_\lambda$ is the spectral decomposition.

PROOF. To discuss convergence of the matrix-valued exponential series we fix a basis $\mathfrak{X}$ in V. Then a sequence $[T_n]$ converges, with $T_n \to T$ as $n \to \infty$,

$\Leftrightarrow$ if the corresponding matrices converge entry-by-entry, $[T_n]_{\mathfrak{X}\mathfrak{X}} \to [T]_{\mathfrak{X}\mathfrak{X}}$ as $n \to \infty$ as described in Section 5.3. The partial sums

$$S_n = I + T + \frac{1}{2!}T^2 + \cdots + \frac{1}{n!}T^n$$

of the exponential series for T have a limit e^T $\Leftrightarrow$ the matrices $[S_n]_{\mathfrak{X}\mathfrak{X}}$ converge to a limit

$$[S_n]_{\mathfrak{X}\mathfrak{X}} \to e^{[T]_{\mathfrak{X}\mathfrak{X}}} = [e^T]_{\mathfrak{X}\mathfrak{X}}$$

in matrix space as $n \to \infty$.

The spectral decomposition of T^k is $T^k = \sum_\lambda \lambda^k P_\lambda$, so the partial sums of the exponential series decompose as

$$\begin{aligned} S_n = \sum_{k=0}^{n} \frac{1}{k!}T^k &= \sum_{k=0}^{n} \frac{1}{k!}\Big(\sum_\lambda \lambda^k P_\lambda\Big) \\ &= \sum_\lambda \Big(\sum_{k=0}^{n} \frac{1}{k!}\lambda^k\Big) \cdot P_\lambda \longrightarrow \sum_\lambda e^\lambda P_\lambda \end{aligned}$$

as $n \to \infty$. Convergence to $\sum_\lambda e^\lambda P_\lambda$ in the last line holds because the sum over eigenvalues is finite. □

EXERCISE 6.101 (Uniqueness of Spectral Decompositions). Suppose $T : V \to V$ is diagonalizable on an arbitrary vector space (not necessarily an inner product space), so $T = \sum_{i=1}^{r} \lambda_i P_{\lambda_i}$ where $\mathrm{sp}(T) = \{\lambda_1, \dots, \lambda_r\}$ and P_{λ_i} is the projection onto the λ_i-eigenspace. Now suppose $T = \sum_{j=1}^{s} \mu_j Q_j$ is some other decomposition such that

$$Q_j^2 = Q_j \neq 0, \quad Q_j Q_k = Q_k Q_j = 0 \ \text{ if } j \neq k, \quad \sum_{j=1}^{s} Q_j = I,$$

and $\{\mu_1, \dots, \mu_r\}$ are distinct. Prove that

(a) $r = s$ and if the μ_j are suitably relabeled we have $\mu_i = \lambda_i$ for $1 \le i \le r$.
(b) $Q_i = P_{\lambda_i}$ for $1 \le i \le r$.

HINT: First show $\{\mu_1, \dots, \mu_s\} \subseteq \{\lambda_1, \dots, \lambda_r\} = \mathrm{sp}(T)$; then relabel.

Here is another useful observation about spectra of diagonalizable operators.

LEMMA 6.102 (Spectral Mapping Theorem). *If $T : V \to V$ is a diagonalizable operator on a finite-dimensional vector space, and $f(x)$ is any function $f : \mathrm{sp}(T) \to \mathbb{C}$, then $f(T)$ is diagonalizable and*

$$\mathrm{sp}(f(T)) = f(\mathrm{sp}(T)) = \{f(\lambda) : \lambda \in \mathrm{sp}(T)\}.$$

PROOF. We have shown that $T = \sum_{\lambda \in \mathrm{sp}(T)} \lambda P_\lambda$ where the P_λ are the spectral projections determined by the direct sum decomposition $V = \bigoplus_\lambda E_\lambda(T)$. Then $f(T) = \sum_\lambda f(\lambda) P_\lambda$, from which it is obvious that $f(T)v = f(\lambda)v$ for $v \in E_\lambda(T)$; hence $f(T)$ is diagonalizable. The eigenvalues are the values $f(\lambda)$ for $\lambda \in \mathrm{sp}(T)$, but notice that we might have $f(\lambda) = f(\mu)$ for different eigenvalues of T. To get the eigenspace $E_\alpha(f(T))$ we must add together all these spaces

$$E_\alpha(f(T)) = \bigoplus_{\{\lambda : f(\lambda)=\alpha\}} E_\lambda(T) \quad \text{for every } \alpha \in f(\text{sp}(T)).$$

The identity is now clear. □

As an extreme illustration, if $f(z) \equiv 1$ then $f(T) = I$ and $\text{sp}(T) = \{1\}$.

Computing a Spectral Decomposition.

EXAMPLE 6.103. We shall compute the spectral decomposition of the multiplication operator $T = L_A$ where

$$A = [T]_{\mathfrak{X}\mathfrak{X}} = \begin{pmatrix} 7 & \sqrt{3} \\ \sqrt{3} & 5 \end{pmatrix}$$

with respect to the standard orthonormal basis $\mathfrak{X} = \{\mathbf{e}_1, \mathbf{e}_2\}$ in $V = \mathbb{C}^2$.

DISCUSSION: Since $A = A^*$ and $\mathfrak{X}$ is an *orthonormal* basis, Proposition 6.37 implies

$$[T^*]_{\mathfrak{X}\mathfrak{X}} = [L_{A^*}]_{\mathfrak{X}\mathfrak{X}} = A^* = A = [L_A]_{\mathfrak{X}\mathfrak{X}} = [T]_{\mathfrak{X}\mathfrak{X}}.$$

Thus $T^* = T$ and T is self-adjoint.

THE SPECTRUM $\text{sp}(T)$. By equation (5.2), the characteristic polynomial can be written as

$$p_A(x) = \det(A - xI) = x^2 - \text{Tr}(A)\,x + \det(A) = x^2 - 12x + 32.$$

Both roots in $\mathbb{C}$ are real,

$$\lambda_\pm = \frac{12 \pm \sqrt{144 - 4(32)}}{2} = 8 \text{ or } 4,$$

so $\text{sp}(T) = \{4, 8\}$.

THE EIGENSPACES: We compute $E_{\lambda_\pm} = K(A - \lambda_\pm I)$ using symbolic row operations to solve the matrix equations $[A - \lambda I]\,\mathbf{x} = \mathbf{0}$. We have

$$[A - \lambda_\pm I] = \begin{pmatrix} 7 - \lambda_\pm & \sqrt{3} \\ \sqrt{3} & 5 - \lambda_\pm \end{pmatrix} \xrightarrow{\text{Row Ops}} \begin{pmatrix} 1 & \sqrt{3}/(7 - \lambda_\pm) \\ 0 & 0 \end{pmatrix}.$$

Thus x_2 is a free variable and

$$\mathbf{x} \in E_{\lambda_\pm}(A) \Leftrightarrow x_1 = x_2 \left(\frac{\lambda_\pm - 7}{\sqrt{3}} \right),$$

so $\dim E_{\lambda_\pm}(A) = 1$ and the eigenspaces are

$$E_{\lambda=4} = \mathbb{C} \cdot \mathbf{f}_1 = \mathbb{C} \cdot \begin{pmatrix} -1/\sqrt{3} \\ 1 \end{pmatrix}$$

when $\lambda = \lambda_1 = 4$ and

$$E_{\lambda=8} = \mathbb{C} \cdot \mathbf{f}_2 = \mathbb{C} \cdot \begin{pmatrix} \sqrt{3} \\ 1 \end{pmatrix}$$

when $\lambda = \lambda_2 = 8$. Since $\mathbf{f}_1 \perp \mathbf{f}_2$ we may scale both vectors to get an orthonormal basis of eigenvectors $\mathfrak{Y} = \{\mathbf{h}_1, \mathbf{h}_2\}$,

$$\mathbf{h}_1 = \begin{pmatrix} -1/2 \\ \sqrt{3}/2 \end{pmatrix} \quad \text{and} \quad \mathbf{h}_2 = \begin{pmatrix} \sqrt{3}/2 \\ 1/2 \end{pmatrix}$$

Thus T is orthogonally diagonalizable.

Determining the Spectral Projections P_λ. With respect to the diagonalizing basis $\mathfrak{Y} = \{\mathbf{h}_1, \mathbf{h}_2\}$, the matrix of T is diagonal, $[T]_{\mathfrak{Y}\mathfrak{Y}} = D = \text{diag}(4, 8)$. Therefore the matrices of the spectral projections P_{λ_k} with respect to this basis can be written down immediately.

$$[P_{\lambda_1}]_{\mathfrak{Y}\mathfrak{Y}} = \begin{pmatrix} 1 & 0 \\ 0 & 0 \end{pmatrix} \quad [P_{\lambda_2}]_{\mathfrak{Y}\mathfrak{Y}} = \begin{pmatrix} 0 & 0 \\ 0 & 1 \end{pmatrix}$$

and the spectral decomposition takes the form

$$[T]_{\mathfrak{Y}\mathfrak{Y}} = \text{diag}(\lambda_1, \lambda_2) = 4 \cdot [P_{\lambda_1}]_{\mathfrak{Y}\mathfrak{Y}} + 8 \cdot [P_{\lambda_2}]_{\mathfrak{Y}\mathfrak{Y}}.$$

To describe the projection matrices $[P_{\lambda_i}]_{\mathfrak{X}\mathfrak{X}}$ with respect to the standard orthonormal basis $\mathfrak{X} = \{\mathbf{e}_1, \mathbf{e}_2\}$ in $\mathbb{C}^2$, we must compute the transition matrix S that diagonalizes $A = [T]_{\mathfrak{X}\mathfrak{X}}$, so that $SAS^{-1} = D = \text{diag}(\lambda_1, \lambda_2)$. Recall that

$$D = [T]_{\mathfrak{Y}\mathfrak{Y}} = [\text{id}]_{\mathfrak{Y}\mathfrak{X}} \cdot [T]_{\mathfrak{X}\mathfrak{X}} \cdot [\text{id}]_{\mathfrak{X}\mathfrak{Y}} = [\text{id}]_{\mathfrak{Y}\mathfrak{X}} \cdot A \cdot [\text{id}]_{\mathfrak{X}\mathfrak{Y}},$$

hence the transition map we seek is $S = [\text{id}]_{\mathfrak{Y}\mathfrak{X}} = [\text{id}]_{\mathfrak{Y}\mathfrak{X}}^{-1}$. But the inverse $S^{-1} = [\text{id}]_{\mathfrak{X}\mathfrak{Y}}$ of this matrix is easier to compute: it is the transpose of the array of coefficients appearing in the system of vector identities relating $\mathfrak{X}$ and $\mathfrak{Y}$,

$$\mathbf{h}_1 = (-1/2)\mathbf{e}_1 + (\sqrt{3}/2)\mathbf{e}_2, \quad \mathbf{h}_2 = (\sqrt{3}/2)\mathbf{e}_1 + (1/2)\mathbf{e}_2.$$

We conclude that

$$S^{-1} = [\text{id}]_{\mathfrak{Y}\mathfrak{X}} = \frac{1}{2} \cdot \begin{pmatrix} -1 & \sqrt{3} \\ \sqrt{3} & 1 \end{pmatrix},$$

with $\det(S^{-1}) = -1$. By Cramer's Rule the matrix $S = (S^{-1})^{-1}$ is

$$S = \frac{1}{\det(S^{-1})} \cdot \text{Cof}(S^{-1})^{\mathrm{T}} = \frac{1}{2} \begin{pmatrix} 1 & \sqrt{3} \\ \sqrt{3} & -1 \end{pmatrix}.$$

Rewriting the similarity transform $D = SAS^{-1}$ as $A = S^{-1}DS$ we see that

$$[T]_{\mathfrak{X}\mathfrak{X}} = A = S^{-1}D\,S = S^{-1}[T]_{\mathfrak{Y}\mathfrak{Y}}\,S.$$

Using the Spectral Theorem we can now show that *every* polynomial function $f(T)$ of T transforms the same way in passing from $\mathfrak{X}$ basis to $\mathfrak{Y}$ basis, so that

$$(6.20) \qquad [f(T)]_{\mathfrak{X}\mathfrak{X}} = S^{-1}[f(T)]_{\mathfrak{Y}\mathfrak{Y}}S \quad \text{for all polynomials } f \text{ in } \mathbb{C}[x].$$

By Lagrange interpolation (Example 1.66), if $\lambda_1, \dots, \lambda_r$ are the distinct eigenvalue of T, then for each $1 \le k \le r$ there is a polynomial $f_k(x) \in \mathbb{C}[x]$ such that

$$f_k(\lambda_k) = 1 \quad \text{and} \quad f_k(\lambda_j) = 0 \quad \text{for all } j \neq k.$$

By the Spectral Theorem, $h(T) = \sum_i h(\lambda_i) P_{\lambda_i}$ for every polynomial $h(x)$ if $T = \sum_i \lambda_i P_{\lambda_i}$. When $h = f_k$ as above it is immediate from these formulas that $f_k(T) = P_{\lambda_k}$, and then by (6.20) we get

$$[P_{\lambda_h}]_{\mathfrak{X}\mathfrak{X}} = S^{-1}[P_{\lambda_k}]_{\mathfrak{Y}\mathfrak{Y}}\, S \quad \text{for } 1 \le k \le r.$$

Since the matrices $[P_{\lambda_k}]_{\mathfrak{Y}\mathfrak{Y}}$ have already been computed, each of the desired matrices $[P_{\lambda_k}]_{\mathfrak{X}\mathfrak{X}}$ stand revealed as a product of three known matrices

$$[P_{\lambda_1}]_{\mathfrak{X}\mathfrak{X}} = S^{-1}\begin{pmatrix} 1 & 0 \\ 0 & 0 \end{pmatrix} \cdot S = \frac{1}{4} \cdot \begin{pmatrix} -1 & -\sqrt{3} \\ \sqrt{3} & 3 \end{pmatrix}$$

and

$$[P_{\lambda_2}]_{\mathfrak{X}\mathfrak{X}} = S^{-1}\begin{pmatrix} 0 & 0 \\ 0 & 1 \end{pmatrix} \cdot S = \frac{1}{4} \cdot \begin{pmatrix} 3 & -\sqrt{3} \\ \sqrt{3} & -1 \end{pmatrix}.$$

To complete our discussion it remains only to verify (6.20).

Proof of Equation (6.20). If T has spectral decomposition $T = \sum_{\lambda_i} \lambda_i P_{\lambda_i}$ where $\text{sp}(T) = \{\lambda_1, \dots, \lambda_r\}$, consider bases $\mathfrak{X}, \mathfrak{Y}$ and let S be an invertible matrix that effects the transition $[T]_{\mathfrak{X}\mathfrak{X}} = S^{-1}([T]_{\mathfrak{Y}\mathfrak{Y}})S$. Direct calculations yield

$$\begin{aligned} [T^k]_{\mathfrak{X}\mathfrak{X}} = [T]^k_{\mathfrak{X}\mathfrak{X}} &= S^{-1}([T]_{\mathfrak{Y}\mathfrak{Y}})S \cdot \dots \cdot S^{-1}([T]_{\mathfrak{Y}\mathfrak{Y}})S \\ &= S^{-1}([T]_{\mathfrak{Y}\mathfrak{Y}})^k S \quad \text{(cancellation)}. \end{aligned}$$

Taking finite linear combinations with complex coefficients, we have $f(T) = \sum_k c_k T^k$ and

$$\begin{aligned} [f(T)]_{\mathfrak{X}\mathfrak{X}} = \sum_{k=1}^{r} c_k [T^k]_{\mathfrak{X}\mathfrak{X}} &= \sum_{k=1}^{r} c_k S^{-1}([T]_{\mathfrak{Y}\mathfrak{Y}})^k S \\ &= S^{-1}\Big(\sum_k c_k [T]^k_{\mathfrak{Y}\mathfrak{Y}}\Big)S = S^{-1}[f(T)]^k_{\mathfrak{Y}\mathfrak{Y}} S. \end{aligned}$$

In particular, if $f(x)$ is a polynomial such that $f(\lambda_j) = \delta_{ij}\lambda_i$, the spectral decomposition of T reduces to $f(T) = P_{\lambda_i}$, so we get

$$[P_{\lambda_i}]_{\mathfrak{X}\mathfrak{X}} = S^{-1}[P_{\lambda_i}]_{\mathfrak{Y}\mathfrak{Y}} S \quad \text{for } 1 \le k \le r$$

:2 and for arbitrary f

$$[f(T)]_{\mathfrak{X}\mathfrak{X}} = \sum_{k=1}^{r} f(\lambda_k)\,[P_{\lambda_k}]_{\mathfrak{X}\mathfrak{X}},$$

which completes our discussion of this example.

□

Exercise 6.104 (Computing $\sqrt{T}$). Use the results in Exercise 6.103 for the following:

(a) Give an explicit formula for the spectral decomposition $\sum_k \lambda_i P_{\lambda_k}$ of the operator T in Exercise 6.103 with respect to the standard basis $\mathfrak{X}$ in $\mathbb{C}^2$.

(b) Compute the matrix of the square root operator

$$\sqrt{T} = \sum_{\lambda \in \mathrm{sp}(T)} \sqrt{\lambda}\, P_\lambda \quad \text{(taking positive roots)}$$

with respect to the basis $\mathfrak{X}$.

EXERCISE 6.105. Let $T : \mathbb{C}^2 \to \mathbb{C}^2$ be the multiplication operator $T = L_A$ for

$$A = \begin{pmatrix} 1 & 1 \\ -2 & 1 \end{pmatrix}$$

on the inner product space $V = \mathbb{C}^2$ equipped with the standard inner product $(z, w) = z_1\overline{w_1} + z_2\overline{w_2}$. Then

(a) Determine the spectrum $\mathrm{sp}_{\mathbb{C}}(T) = \{\lambda_1, \lambda_2\}$.
(b) Find the eigenspaces $E_\lambda(T)$ in $\mathbb{C}^2$ and a vector basis $\{f_1, f_2\}$ that diagonalizes T. Is T diagonalizable? Is T orthogonally diagonalizable?
(c) Find an invertible matrix $S \in \mathrm{GL}(2, \mathbb{C})$ such that

$$SAS^{-1} = \begin{pmatrix} \lambda_1 & 0 \\ 0 & \lambda_2 \end{pmatrix}$$

REMARK: This problem involves complex eigenvalues, and eigenvectors with complex entries.

6.7. Positive Operators and Polar Decomposition

If $T : V \to W$ but $V \neq W$ one cannot speak of "diagonalizing T." (What would "eigenvector" and "eigenvalue" mean in that context?) But we can still seek other decompositions of T as a product of particularly simple, easily understood operators. Even when $V = W$ one might profitably explore such options if T fails to be diagonalizable—diagonalization is not the only useful decomposition of a linear operator.

When $V = W$ and V is an inner product space over $\mathbb{R}$ or $\mathbb{C}$, all self-adjoint (or normal) operators are orthogonally diagonalizable, and among them the positive operators are particularly simple.

DEFINITION 6.106. A linear operator $T : V \to V$ on an inner product space is *positive* if

$$\text{(i)}\ T^* = T \qquad \text{(ii)}\ (Tv, v) \geq 0 \text{ for all } v \in V.$$

It is *positive definite* if $(Tv, v) = 0$ only when $v = 0$. We write $T \geq 0$ or $T > 0$, respectively, to indicate these possibilities. A matrix $A \in \mathrm{M}(n, \mathbb{C})$ is said to be positive (or positive definite) if the multiplication operator $L_A : \mathbb{C}^n \to \mathbb{C}^n$ is positive (positive definite) with respect to the usual inner product, so that $(Av, v) \geq 0$ for all v.

Note that self-adjoint projections $P^2 = P^* = P$ are examples of positive operators, and sums of positive operators are again positive (but not linear combinations unless the coefficients are positive).

A self-adjoint operator T with $\mathrm{sp}(T) = \{\lambda_1, \ldots, \lambda_r\}$, and spectral decomposition $T = \sum_i \lambda_i P_{\lambda_i}$ is positive $\Leftrightarrow \lambda_i \geq 0$ for all i, so $\mathrm{sp}(T)$ is contained in the interval $[0, +\infty) + i0$. In fact, if $T \geq 0$ and $v_i \in E_{\lambda_i}$, we have $(Tv_i, v_i) = \lambda_i \|v_i\|^2 \geq 0$. Conversely, if all $\lambda_i \geq 0$ and $v = \sum_{i=1}^r v_i$, we get $(Tv, v) = \sum_{i,j} (Tv_i, v_j)$, and since $E_{\lambda_i} \perp E_{\lambda_j}$ for $i \neq j$ this reduces to $\sum_i (Tv_i, v_i) = \sum_i \lambda_i \|v_i\|^2 \geq 0$.

If T is positive *definite* then $\lambda_i = 0$ cannot occur in $\mathrm{sp}(T)$ and T is invertible, with

$$T^{-1} = \sum_i \frac{1}{\lambda_i} P_{\lambda_i} \quad \text{(also a positive definite operator).}$$

Positive Square Roots. If $T \geq 0$ there is a positive square root (a self-adjoint positive operator $S \geq 0$ such that $S^2 = T$), namely

$$\sqrt{T} = \sum_i \sqrt{\lambda_i}\, P_{\lambda_i} \quad \left(\sqrt{\lambda_i} = \text{the nonnegative square root of } \lambda_i \geq 0\right),$$

which is also denoted by $T^{1/2}$. This is a square root because

$$S^2 = \sum_{i,j} \sqrt{\lambda_i}\sqrt{\lambda_j}\, P_{\lambda_i} P_{\lambda_j} = \sum_i \lambda_i P_{\lambda_i} = T\,,$$

since $P_{\lambda_i} P_{\lambda_j} = \delta_{ij} \cdot P_{\lambda_i}$. Notice that the spectral decompositions of T and $\sqrt{T}$ involve the same spectral projections P_{λ_i}, and the eigenspaces match up too because $E_{\lambda_i}(T) = E_{\sqrt{\lambda_i}}(\sqrt{T})$ for all i.

Subject to the requirement that $S \geq 0$, this square root is *unique*, as a consequence of uniqueness of the spectral decomposition on any vector space (see Exercise 6.101).

EXERCISE 6.107. Use uniqueness of spectral decompositions to show that the positive square root operator $\sqrt{T} = \sum_i \sqrt{\lambda_i} P_{\lambda_i}$ defined above is *unique*; i.e., if $A \geq 0$ and $B \geq 0$ and $A^2 = B^2 = T$ for some $T \geq 0$, then $A = B$.

Positivity of $T : V \to V$ has an interesting connection with the exponential map on matrices $\mathrm{Exp} : \mathrm{M}(n, \mathbb{C}) \to \mathrm{M}(n, \mathbb{C})$,

$$\mathrm{Exp}(A) = e^A = \sum_{n=0}^{\infty} \frac{1}{n!} A^n.$$

We indicated in Section 5.3 that commuting matrices A, B satisfy the exponent law $e^{A+B} = e^A \cdot e^B$, with $e^0 = I$. In particular, all matrices in the range of Exp are invertible, with $(e^A)^{-1} = e^{-A}$.

EXERCISE 6.108. Let $\mathcal{P}$ be the set of *positive definite* matrices A in $\mathrm{M}(n, \mathbb{C})$, which are all self-adjoint by definition of $A > 0$. Let $\mathcal{H}$ be the set of all self-adjoint matrices in $\mathrm{M}(n, \mathbb{C})$, which is a vector subspace *over* $\mathbb{R}$ but not over $\mathbb{C}$ since iA is skew-adjoint if A is self-adjoint. Prove the following:

(a) The exponential matrix e^H is positive and invertible for self-adjoint matrices H.

(b) The exponential map $\text{Exp}(H) = e^H$ is a *bijection* from $\mathcal{H} \to \mathcal{P}$.

HINT: Explain why $(e^A)^* = e^{A^*}$ and then use the exponent law applied to matrices e^{tA}, $t \in \mathbb{R}$ (you could also invoke the Spectral Theorem).

It follows that every positive definite matrix $A > 0$ has a unique self-adjoint *logarithm* $\text{Log}(A)$ such that

$$\begin{aligned} \text{Exp}(\text{Log}(A)) &= A \quad \text{for } A \in \mathcal{P}, \\ \text{Log}(\text{Exp}(H)) = \text{Log}(e^H) &= H \quad \text{for } H \in \mathcal{H}, \end{aligned}$$

so Log is the inverse of the bijection $\text{Exp} : \mathcal{H} \to \mathcal{P}$. In terms of spectral decompositions,

$$\text{Log}(T) \text{ of a positive definite } T \text{ is } \text{Log}(T) = \sum_i \text{Log}(\lambda_i) P_{\lambda_i} \quad \text{if } T = \sum_i \lambda_i P_{\lambda_i},$$

$$\text{Exp}(H) \text{ of a self-adjoint matrix } H \text{ is } e^H = \sum_i e^{\mu_i} Q_{\lambda_i} \qquad \text{if } H = \sum_i \mu_i Q_{\mu_i}.$$

Polar Decompositions $T = UP$ (Invertible T). When $V = W$ *unitary* operators $U : V \to V$ are another well-understood family of (diagonalizable) operators on an inner product space. They are particularly interesting and easy to understand because they correspond to the possible choices of orthonormal bases in V. Every unitary U is obtained by specifying a pair of orthonormal bases $\mathfrak{X} = \{e_i\}$ and $\mathfrak{Y} = \{f_j\}$ and defining U to be the unique linear map such that

$$U\Big(\sum_{i=1}^{n} c_i e_i\Big) = \sum_{j=1}^{n} c_j f_j \quad (\text{arbitrary } c_i \in \mathbb{C}).$$

The positive operators $P \geq 0$ and unitary operators U on an inner product space provide a natural *polar decomposition* $T = U \cdot P$ of any linear operator $T : V \to V$. In its simplest form (when T is invertible) it asserts that any *invertible* map T has a *unique* factorization

$$T = U \cdot P \quad \begin{cases} U : V \to V & \text{unitary (a bijective isometry of } V), \\ P : V \to V & \text{positive definite, invertible} = e^H \text{ with } H \text{ self-adjoint.} \end{cases}$$

Both factors are orthogonally diagonalizable (U because it is normal and P because it is self-adjoint), but the original operator T need not itself be diagonalizable over $\mathbb{C}$, let alone orthogonally diagonalizable.

We will develop the polar decomposition, first for an invertible operator $T : V \to V$ since that proof is particularly transparent, and then address the general result (often referred to as the *singular value decomposition* when it is stated for matrices). This involves operators that are not necessarily invertible and may be maps $T : V \to W$ between inner product spaces of different dimensions. The positive component $P : V \to V$ is still unique but the unitary component U may be nonunique (in a harmless sort of way). The *singular values* of T are the eigenvalues $\lambda_i \geq 0$ of the positive component P.

THEOREM 6.109 (Polar Decomposition I). *Let V be a finite-dimensional inner product space over $\mathbb{C}$. Every invertible operator $T : V \to V$ has a unique decomposition $T = U \cdot P$ where*

$$U \in \mathrm{U}(n) = (\textit{the group of unitary operators } U^* = U^{-1}),$$
$$P \in \mathcal{P} = (\textit{invertible positive definite operators } P > 0).$$

By Exercise 6.108 *we can also write $T = U \cdot e^H$ for a unique self-adjoint operator $H \in \mathcal{H}$.*

This is the linear operator (or matrix) analogue of the decomposition

$$z = |z|e^{i\theta} = r \cdot e^{i\theta} \quad \text{with } r > 0 \text{ and } \theta \text{ real (so } |e^{i\theta}| = 1)$$

for nonzero complex numbers. If we think of "positive definite" = "positive," "self-adjoint" = "real," and "unitary" = "absolute value 1," the analogy with the polar decomposition $z = re^{i\theta}$ of a nonzero complex number z is clear.

SOME PRELIMINARY REMARKS. If $T : V \to W$ is a linear map between two inner product spaces, its *absolute value* is the linear map $|T| : V \to V$ determined in the following way:

The product $T^*T : V \to V$ is a positive operator because

$$(T^*T)^* = T^*T^{**} = T^*T \quad (T^* \text{ maps } W \to V \text{ and } T^{**} = T \text{ on } V),$$
$$(T^*Tv, v) = (Tv, Tv) = ||Tv||^2 \geq 0 \quad \text{for } v \in V.$$

Thus T^*T is self-adjoint and has a spectral decomposition $T^*T = \sum_i \lambda_i P_{\lambda_i}$, with eigenvalues $\lambda_i \geq 0$ and self-adjoint projections $P_{\lambda_i} : V \to E_{\lambda_i}(T^*T)$ onto orthogonal subspaces. The *absolute value* $|T| : V \to V$ is then defined as the unique positive square root

$$|T| = (T^*T)^{1/2} = \sum_i \sqrt{\lambda_i}\, P_{\lambda_i},$$

whose spectral decomposition involves the same projections that appeared in T^*T. For any linear operator $T : V \to W$ we have $T^*T = |T|^2$ and hence

$$\begin{aligned} ||Tv||_W^2 &= (Tv, Tv)_W = (T^*Tv, v)_V \\ &= (|T|^2 v, v)_V = ||\,|T|(v)||_V^2 \quad \text{for } v \in V. \end{aligned} \tag{6.21}$$

Thus $|T|(v) \in V$ and $Tv \in W$ have the same norm for every $v \in V$. It follows from (6.21) that T, T^*T, and $|T|$ have the same kernel because

$$\begin{aligned} Tv = 0 \Rightarrow\ T^*T(v) = 0 &\Rightarrow (T^*Tv, v) = (|T|^2(v), v)) = ||\,|T|(v)||^2 = 0 \\ &\Rightarrow |T|(v) = 0 \Rightarrow Tv = 0 \quad (\text{by (6.21)}). \end{aligned}$$

Thus the kernels coincide

$$K(T) = K(T^*T) = K(|T|) \tag{6.22}$$

even if the ranges differ. Since T^*T and $|T|$ map $V \to V$, these operators are invertible in V (hence strictly positive definite) $\Leftrightarrow T : V \to W$ is one-to-one but perhaps not an onto map.

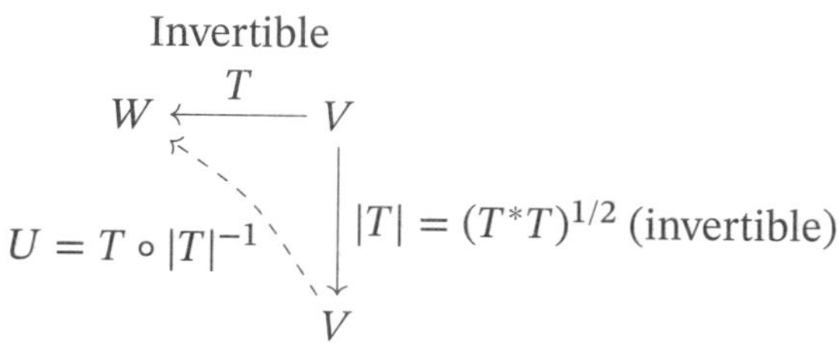

FIGURE 6.10. The maps involved in defining $U : V \to W$ in the polar decomposition $T = U \cdot |T|$ of an invertible map $T : V \to W$ between inner product spaces. In our discussion we show that the positive operator $|T| = (T^*T)^{1/2}$ is invertible and $R(T^*T) = R(|T|) = R(T) = W$ when T is invertible. The induced bijection $U = T \circ |T|^{-1} : V \to W$ is a bijective linear isometry (a unitary map of $V \to W$).

(Comparisons between T and $|T|$ do not follow from spectral theory because T itself need not be diagonalizable, even when $V = W$.)

PROOF OF THEOREM 6.109 FOR INVERTIBLE T. The proof in the invertible case is simple. For any linear operator $T : V \to W$ we have $T^*T = |T|^2 : V \to V$ and have seen in (6.21) that $|T|(v)$ and Tv always have the same norm. When T is invertible, $K(T)$ and $K(|T|)$ are trivial, so $|T| : V \to V$ is invertible, and then $R(T) = W$ and $R(|T|) = V$. The identities (6.21) determine a bijective isometry $U : V \to W$ that sends $T(v) \in W$ to $|T|(v) \in V$ for all $v \in V$, as indicated in Figure 6.10. This map is also linear because $U = T \circ |T|^{-1}$ is a composite of linear operators from $V \to V \to W$. Thus when T is invertible the desired polar decomposition is

$$U \circ |T| = \left(T \circ |T|^{-1}\right) \circ |T| = T$$

As for uniqueness (valid only in the invertible case), suppose $T = UP = U_0P_0$ with U, U_0 unitary and P, P_0 positive definite. Then $P^* = P, P_0^* = P_0$ since the positive components are self-adjoint, and $T^* = P^*U^* = PU^* = P_0^*U_0^* = P_0U_0^*$, hence

$$P^2 = PU^*UP = PU^*\left(PU^*\right)^* = P_0U_0^*U_0P_0 = P_0^2.$$

Now $P^2 = P^*P$ is a positive operator that has a unique positive square root, namely P; likewise for P_0^2. By uniqueness we get $P_0 = P$, from which $U_0 = U$ follows. □

Computing the Polar Decomposition for Invertible $T : V \to W$. Even if we happen to have $W = V$ we shall deliberately write $T : V \to W$ to emphasize the different roles played by the initial space V and range $W = R(T)$; this will motivate our discussion of the general polar decomposition.

Determining the positive part $P = |T|$ is straightforward: $P^2 = T^*T$ is self-adjoint from $V \to V$, and its spectral decomposition can be computed in the usual way. If $\{e_i\}$ is an orthonormal basis of eigenvectors for T^*T, which are also eigenvectors for $P = |T|$, we have

$$T^*T(e_i) = \lambda_i e_i \quad \text{and} \quad |T|(e_i) = \sqrt{\lambda_i}\, e_i,$$

with all $\lambda_i > 0$ when $|T|$ is invertible. From this we get

$$|T|^{-1}(e_i) = |T|^{-1}\left(\frac{1}{\sqrt{\lambda_i}}\,|T|(e_i)\right) = \frac{1}{\sqrt{\lambda_i}}\, e_i$$

$$\Downarrow$$

$$U(e_i) = T(|T|^{-1}e_i) = \frac{1}{\sqrt{\lambda_i}} T(e_i). \tag{6.23}$$

By its construction U is unitary on V, so the vectors

$$f_i = \frac{1}{\sqrt{\lambda_i}}\, T(e_i)$$

are a new orthonormal basis in $R(T) = W$. This completely determines U. Note that

$$\|U(e_i)\| = \frac{1}{\sqrt{\lambda_i}} \|T(e_i)\| = \frac{1}{\sqrt{\lambda_i}} \|\,|T|(e_i)\| = \frac{\sqrt{\lambda_i}}{\sqrt{\lambda_i}} = 1$$

and

$$\begin{aligned}(U(e_i), U(e_j)) &= \frac{(T(e_i), T(e_j))}{\sqrt{\lambda_i \lambda_j}} = \frac{(T^*T(e_i), e_j)}{\sqrt{\lambda_i \lambda_j}} \\ &= \frac{(|T|^2(e_i), e_j)}{\sqrt{\lambda_i \lambda_j}} = \frac{(|T|(e_i), |T|(e_j))}{\sqrt{\lambda_i \lambda_j}} = (e_i, e_j) = \delta_{ij}\end{aligned}$$

as expected.

The Singular Value Decomposition. The polar decomposition of an operator $T : V \to W$ is often rewritten in terms of bases and eigenvectors in V and W, a form that lends itself to numerical calculations. This variant is called the *singular value decomposition*. We state the general form of the theorem, but for now will only prove it for invertible operators, for which the proof takes its most transparent form.

For any linear map $T : V \to W$ between inner product spaces, the operator $T^*T : V \to V$ is self-adjoint and has a unique positive square root $|T| = \sqrt{T^*T}$, whose nonzero eigenvalues $\sigma_1 \geq \cdots \geq \sigma_r > 0$ are the called the *singular values* of T. Obviously these are the (positive) square roots $\sigma_k = \sqrt{\lambda_k}$ of the nonzero eigenvalues of T^*T. Their significance is revealed by recasting the polar decomposition in terms of orthonormal bases in V and W.

THEOREM 6.110 (Singular Value Decomposition). *Let $T : V \to W$ be a linear operator with* rank$(T) = r$ *between finite-dimensional inner product spaces.*

Then there are orthonormal bases $\mathfrak{X} = \{e_1, \dots, e_m\}$ in V and $\mathfrak{Y} = \{f_1, \dots, f_n\}$ in W such that

$$T(e_i) = \begin{cases} \sigma_i \cdot f_i & \text{for } 1 \le i \le r, \\ 0 & \text{for } r+1 \le i \le m, \end{cases} \tag{6.24}$$

where σ_i are the nonzero singluar values of T.

PROOF FOR INVERTIBLE T. We have $m = n = \text{rank}(T) = \text{rank}(|T|)$ when T is invertible, and (6.24) simplifies accordingly. In the polar decomposition $T = U \cdot |T|$, the positive part is an invertible map from $V \to V$ and $U : V \to W$ is a unitary operator (bijective isometry) given by the identities (6.23); it maps orthonormal bases in V to orthonormal bases in W. If $\{e_i\}$ is an orthonormal basis in V that diagonalizes $|T| : V \to V$, the vectors $f_j = U(e_j)$ form an orthonormal basis in W such that

$$T(e_i) = (T \circ |T|^{-1})(|T|(e_i)) = U(\sigma_i \cdot e_i) = \sigma_i \cdot f_i \quad \text{in } W$$

as desired. □

Computing a Singular Value Decomposition.

EXAMPLE 6.111 (Singular Value Decomposition). Consider the operator $T = L_A$ on $\mathbb{C}^2$ equipped with the usual inner product and the standard orthonormal basis $\mathfrak{X} = \{e_i\}$, where

$$A = \begin{pmatrix} 11 & -5 \\ -2 & 10 \end{pmatrix}.$$

We will show that T is diagonalizable, find its eigenspaces, and describe its polar decomposition by calculating $|T| = \sqrt{T^*T}$ and $U = T \circ |T|^{-1}$.
DISCUSSION: For any orthonormal basis $\mathfrak{X}$ we have

$$[T^*T]_{\mathfrak{X}\mathfrak{X}} = [T^*]_{\mathfrak{X}\mathfrak{X}}[T]_{\mathfrak{X}\mathfrak{X}} = [T]^*_{\mathfrak{X}\mathfrak{X}}[T]_{\mathfrak{X}\mathfrak{X}} = A^*A \quad \text{(by Proposition 6.37)},$$

and by direct calculation

$$A^*A = \begin{pmatrix} 125 & -75 \\ -75 & 125 \end{pmatrix},$$

so the characteristic polynomial is $p(x) = x^2 - 250x + 10{,}000$. The eigenvalues of A^*A are $\lambda_1 = 200$, $\lambda_2 = 50$ and their square roots (the singular values of T) are $\sigma_1 = 10\sqrt{2}$, $\sigma_2 = 5\sqrt{2}$. Symbolic row operations send

$$[A^*A - 200\,I] \longrightarrow \begin{pmatrix} 1 & 1 \\ 0 & 0 \end{pmatrix}, \quad \text{so} \quad E_{\lambda_1} = \mathbb{C} \cdot h_1 = \mathbb{C}\begin{pmatrix} -1 \\ 1 \end{pmatrix}$$

and similarly for $\lambda_2 = 50$,

$$E_{\lambda_2} = \mathbb{C} \cdot h_2 = \mathbb{C}\begin{pmatrix} 1 \\ 1 \end{pmatrix}.$$

The basis vectors h_1, h_2 are orthogonal by construction and the rescaled vectors $e_i = h_i/\sqrt{2}$ are an orthonormal basis in the initial space of T.

Write V for the initial space of T and $W = R(T) = \text{range}(U)$. With respect to the diagonalizing basis $\mathfrak{Y} = \{e_1, e_2\}$ in V, the positive part in $T = U \cdot |T|$ has matrix $\text{diag}(\sigma_1, \sigma_2) = \text{diag}(10\sqrt{2}, 5\sqrt{2})$. If we define $f_i = (1/\sigma_i) \cdot T(e_i)$ as in (6.17), we get an orthonormal basis in the range space W, and U is the unique unitary map from $V \to W$ that sends $e_i \to f_i$. This yields a complete description of the factors in the polar decomposition of T.

The General Polar Decomposition. When $T : V \to W$ is not invertible the polar decomposition is somewhat more complicated. The positive component in $T = U \cdot P$ is still the unique positive square root $P = |T| = (T^*T)^{1/2}$. But the unitary part is a uniquely determined *isometry* $U_0 : R(|T|) \to R(T)$ between proper subspaces in V and W. The fact that U_0 is not defined on all of V causes no problem in the factorization $T = U \cdot P$ because behavior of U off of $R(|T|)$ is completely irrelevant.

THEOREM 6.112 (Polar Decomposition II). *Any linear operator $T : V \to W$ between finite-dimensional complex inner product space has a factorization $T = U \cdot |T|$ where*

1. *$|T|$ is the positive square root of T, and*
2. *U is a isometry from $R(|T|)$ in V to $R(T)$ in W.*

The factor U is uniquely determined on the range $R(|T|)$, which is all that matters in the decomposition $R = U|T|$.

PROOF. We take the case $V = W$ first. Start by noting that

$$R(|T|) = K(|T|)^\perp = K(T)^\perp = K(T^*T)^\perp = R(T^*T),$$
$$R(|T|)^\perp = K(|T|) = K(T) = K(T^*T) = R(T^*T)^\perp.$$

The subspaces in the first row are just the orthocomplements of those in the second, so the identities in row 1 follow from those in row 2. The first and last identities in row 2 hold because $|T|$ and T^*T are self-adjoint (Proposition 6.45); the rest have been proved in (6.22). We now observe that by (6.21),

$$\|Tv\|^2 = (T^*Tv, v) = (|T|^2 v, v) = \| |T|(v) \|^2 \quad \text{for all } v \in V,$$

so we get a norm-preserving bijective map $U_0 : R(|T|) \to R(T)$ if we let

$$U_0\left(|T|(v)\right) = T(v) \quad \text{for } v \in V. \tag{6.25}$$

This makes sense despite its seeming ambiguity: If an element $y \in R(|T|)$ has realizations $y = |T|(v') = |T|(v)$, we get $|T|(v' - v) = 0$, and then

$$0 = T(v' - v) = T(v') - T(v)$$

because $|T|(v' - v)$ and $T(v' - v)$ have equal norms. Thus $T(v') = T(v)$ and the operator (6.25) is in fact a well-defined bijective map from $R(|T|)$ into $R(T)$. It is linear because

$$\begin{aligned} U_0\left(|T|v_1 + |T|v_2\right) &= U_0\left(|T|(v_1 + v_2)\right) = T(v_1 + v_2) \\ &= Tv_1 + Tv_2 = U_0(|T|v_1) + U_0(|T|v_2), \end{aligned}$$

and similarly $U_0\left(\lambda|T|(v)\right) = \lambda U_0\left(|T|(v)\right)$.

By definition of U_0 we have $\|U_0(y)\| = \|y\|$ for all $y \in R(|T|)$, and $R(U_0) \subseteq R(T)$, so $\dim R(U_0) \leq \dim R(T)$. But $R(U_0) = R(|T|)$ by definition of U_0, and furthermore $K(T) = K(|T|) \Rightarrow \dim R(T) = \dim R(|T|)$. Putting these facts together we get

$$\dim R(|T|) = \dim R(U_0) \leq \dim R(T) = \dim R(|T|).$$

We conclude that $R(U_0) = R(T)$ and $U_0 : R(|T|) \to R(T)$ is a bijective *linear* isometry between subspaces of equal dimension. Hence

$$T(v) = (U_0 \cdot |T|)(v) \quad \text{for all } v \in V.$$

When $V = W$ we can extend U_0 to a globally defined unitary map $U : V \to V$. In this situation $K(T) = K(|T|) \Rightarrow \dim R(T) = \dim R(|T|)$ and $\dim R(T)^{\perp} = \dim R(|T|)^{\perp}$, and because these spaces have the same dimension there exist various isometries

$$U_1 : R(|T|)^{\perp} \to R(T)^{\perp}$$

corresponding to orthonormal bases in these subspaces. Using the orthogonal decompositions

$$V = R(|T|) \dot{\bigoplus} R(|T|)^{\perp} = R(T) \dot{\bigoplus} R(T)^{\perp},$$

we then obtain a (nonunique) bijective isometry U on this direct sum,

$$U(v, v') = \big(U_0(v), U_1(v')\big),$$

such that $U|T| = U_0|T| = T$ on all of V. □

There is a similar decomposition for operators $T : V \to W$ between different inner product spaces; we merely sketch the proof. Once again we define the positive component $|T| = (T^*T)^{1/2}$. The identity

$$\|\, |T|(v) \,\|_V^2 = \|T(v)\|_W^2 \quad \text{for all } v \in V$$

holds exactly as in (6.21), and this induces a linear isometry U_0 from $M = R(|T|) \subseteq V$ to $N = R(T) \subseteq W$ such that

$$T = U_0 \cdot |T| = [T \cdot (|T|_M)^{-1}] \cdot |T|$$

where $|T|_M$ is the restriction of $|T|$ to $M = K(T)^{\perp}$ (see Figure 6.11).

The fact that U_0 is only defined on $R(|T|)$ is irrelevant, as it was in Theorem 6.110, but now U_0 cannot be extended to a unitary map (bijective isometry) from V to W unless $\dim(V) = \dim(W)$. On the other hand since $|T|$ is self-adjoint, we have

$$R(|T|) = K(|T|)^{\perp} = K(T)^{\perp}.$$

and can define $U \equiv 0$ on $K(T)$ to get a globally defined "*partial isometry*" $U : V \to W$ such that $K(U) = K(T)$, $R(U) = R(U_0) = R(T)$, and

$$U|_{K(T)} = 0, \quad U|_{K(T)^{\perp}} = U|_{R(|T|)} = U_0.$$

The players involved are shown in the diagram of Figure 6.11. The *singular value decomposition* stated in Theorem 6.110 is equivalent to the general polar decomposition of $T : V \to W$.

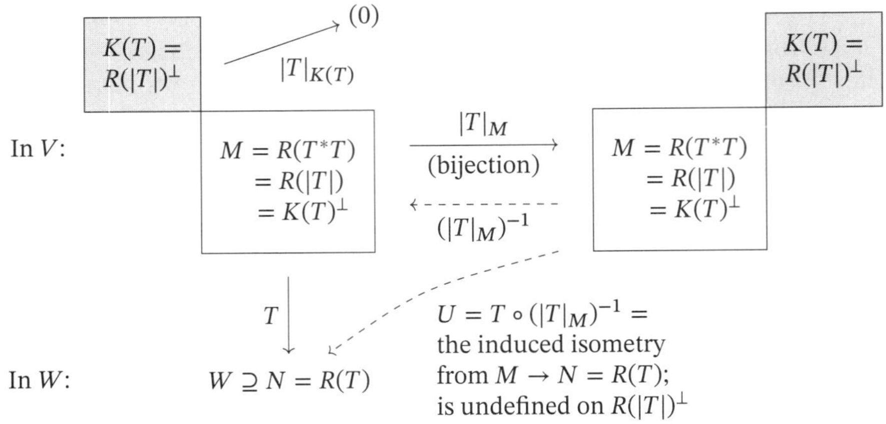

FIGURE 6.11. The maps involved in defining a polar decomposition $T = U_0 \cdot |T|$ for an arbitrary linear map $T : V \to W$ between different inner product spaces. We abbreviate $M = K(T)^\perp \subseteq V$ and $N = R(T) \subseteq W$. Here $U_0 : M \to N$ is a bijective isometry such that $T = U_0 \cdot |T|$.

Additional Exercises

Section 6.1. Basic Definitions and Examples

1. TRUE/FALSE QUESTIONS: ("True" if the statement is *always* true.)

(a) Inner products can only be defined on vector spaces over $\mathbb{R}$ or $\mathbb{C}$.

(b) The matrix-adjoint $A^* = \overline{A^\mathrm{T}}$ can only be defined for square matrices.

(c) If x, y, z are vectors in an inner product space and $(x, y) = (x, z)$ for all $x \in V$, then $y = z$.

(d) The Cauchy-Schwarz inequality $|(x, y)| \le \|x\| \cdot \|y\|$ remains valid for infinite-dimensional inner product spaces.

2. For the Hilbert-Schmidt inner product on matrix space $\mathrm{M}(2, \mathbb{C})$,

$$(A, B)_{\mathrm{HS}} = \sum_{i,j=1}^{2} A_{ij} \cdot \overline{B_{ij}},$$

and the matrices

$$A = \begin{pmatrix} 1 & 2+i \\ 3 & i \end{pmatrix}, \quad B = \begin{pmatrix} 1-i & 0 \\ i & -i \end{pmatrix},$$

compute:

(a) The Hilbert-Schmidt norms $\|A\|_{\mathrm{HS}}$, $\|B\|_{\mathrm{HS}}$, and $\|I\|_{\mathrm{HS}}$ where I is the 2×2 identity matrix.

(b) The HS-norm distances $\|A - B\|_{\mathrm{HS}}$. Which of the matrices A, B is closer to the identity matrix I in the HS-norm?

(c) Compare $\|A\|_{\mathrm{HS}} \cdot \|B\|_{\mathrm{HS}}$ with the HS-norm $\|AB\|_{\mathrm{HS}}$ of the product AB.

NOTE: The HS-norm on $\mathrm{M}(2, \mathbb{R})$ becomes the usual euclidean norm on $\mathbb{R}^4$ when we identify matrix A with the 4-tuple $\mathbf{a} = (A_{11}, A_{12}, A_{21}, A_{22})$.

3. Let A, B be arbitrary $n \times n$ matrices.

(a) Is there any universal inequality that relates the Hilbert-Schmidt norm $\|AB\|_{\text{HS}}$ to the product of norms $\|A\|_{\text{HS}} \cdot \|B\|_{\text{HS}}$? Explain your answer.

(b) Same question for the norms $\|A^k\|_{\text{HS}}$ vs. $\|A\|_{\text{HS}}^k$ for $k = 0, 1, 2, \ldots$.

REMARK: The relation in (a) is the analogue for matrix products of the triangle inequality $\|A - B\|_{\text{HS}}$ for sums.

HINT: Remember that $\|A\|_{\text{HS}}$ is an *inner product norm* on $\mathrm{M}(n, \mathbb{K})$.

4. Any 2×2 matrix A determines an operation $B : \mathbb{R}^2 \times \mathbb{R}^2 \to \mathbb{R}$ that sends ordered pairs of vectors in $\mathbb{R}^2$ to real scalars;

$$B(\mathbf{x}, \mathbf{y}) = \mathbf{x}^{\mathrm{T}} \cdot A \cdot \mathbf{y} \quad \text{(matrix product } (1 \times 2) \times (2 \times 2) \times (2 \times 1)),$$

where elements of $\mathbb{R}^2$ are regarded as 2×1 column vectors, and the $1{\times}1$ product as a scalar in $\mathbb{R}$. This operation is *bilinear*—linear in each input when the other is held fixed. What conditions on the entries in $A = [a, b; c, d]$ ensure that B is an inner product on $\mathbb{R}^2$?

NOTE: All inner products on $\mathbb{R}^2$ arise this way.

HINT: Start by explaining why $\det(A)$ must be nonzero, which means at most three of the scalars can be independent.

5. Prove the *parallelogram law*

$$\|x + y\|^2 + \|x - y\|^2 = 2\|x\|^2 + 2\|x\|^2 \quad \text{for all } x, y \in V$$

for vectors in any inner product space over $\mathbb{R}$.

NOTE: Whence the name? If $x, y \in \mathbb{R}^2$ are regarded as the edges of a parallelogram P with a vertex at the origin, then $x + y$ and $x - y$ correspond to the two diagonals of P. In $\mathbb{R}^n$ just consider what is going on in the plane spanned by x, y (assuming they are not collinear). All this is closely related to the *polarization identities* for inner products.

Section 6.2. Orthogonal Complements and Projections

1. TRUE/FALSE QUESTIONS: ("True" if the statement is *always* true.)

(a) Every finite-dimensional inner product space has an orthonormal basis.

(b) If M is a subspace of a finite-dimensional inner product space V, any orthonormal basis in M can be extended to an orthonormal basis in V by adjoining suitably chosen vectors from V.

(c) If $(x, y) = 0$ for all y in an inner product space, then $x = 0$.

(d) If M is a subspace of an inner product space V, and $y \in V$ is perpendicular to all vectors that are perpendicular to M, then y itself is in M.

(e) If S is a nonempty subset in an inner product space, then

$$S^{\perp} = \{x \in V : x \perp s \text{ for all } s \in S\}$$

is a vector subspace, and its intersection with S is trivial (either empty or consists only of the origin).

NOTE: An "AND" statement such as "P and Q are true" is true only when both statements P and Q are true.

2. A "square wave" of period $\Delta t = 1$ is the piecewise continuous function $f(t)$ on $\mathbb{R}$ whose restriction to the period interval $[0, 1)$ is

$$f(t) = \begin{cases} 1 & \text{for } 0 \leq t < \frac{1}{2}, \\ 0 & \text{for } \frac{1}{2} \leq t < 1; \end{cases}$$

see Figure 6.5(c).

(a) Compute the Fourier coefficients

$$c_k = \int_0^1 f(t) e^{-2\pi i k t} \, dt \quad \text{for } k \in \mathbb{Z}$$

in the Fourier series expansion $f(t) \sim \sum_{k \in \mathbb{Z}} c_k e^{2\pi i k t}$.

(b) How fast do the Fourier coefficients decay as $k \to \infty$? Is $|c_k| \approx 1/k$? Or $|c_k| \approx 1/k^2$? Or what?

3. Consider the following matrix A and subset S in $\mathrm{M}(2, \mathbb{R})$ equipped with the Hilbert-Schmidt inner product.

(a) Explain why the "matrix units" $E_{ij} = 1$ in the i, j spot and zero elsewhere, are an orthonormal basis.

(b) Taking

$$A = \begin{pmatrix} -1 & 27 \\ -4 & -8 \end{pmatrix} \quad S = \left\{ \begin{pmatrix} 3 & 5 \\ -1 & 1 \end{pmatrix}, \begin{pmatrix} -1 & 9 \\ 5 & 1 \end{pmatrix}, \begin{pmatrix} 7 & -17 \\ 2 & -6 \end{pmatrix} \right\},$$

use Gram-Schmidt to find an orthonormal basis $\{F_1, F_2, F_3\}$ for the subspace W spanned by S. What is $\dim(W)$?

(c) Evaluate the projection $P_W(A) = \sum_{k=1}^{3} (A, F_k)_{\mathrm{HS}} \cdot F_k$ of A onto W.

Section 6.3. Adjoints and Orthonormal Decompositions

1. TRUE/FALSE QUESTIONS: ("True" if the statement is *always* true.)

(a) In a finite-dimensional inner product space, the dimensions of the kernels $K(T)$ and $K(T^*)$ are the same, so $|K(T)| = |K(T^*)|$.

(b) In a finite-dimensional inner product space, the kernels $K(T)$ and $K(T^*)$ are identical.

(c) If we let matrices $A \in \mathrm{M}(n, \mathbb{C})$ act on $V = \mathbb{C}^n$ equipped with the standard inner product, then matrix adjoint and operator adjoint are related by

$$L_{A^*} = (L_A)^* \quad (\text{operator adjoint of } L_A : V \to V)$$

(d) If $T : V \to V$ acts on a finite-dimensional inner product space, its adjoint T^* is diagonalizable if T is.

(e) If $\mathbb{C}^n$ is equipped with the usual inner product and $A \in \mathrm{M}(n, \mathbb{C})$, the left multiplication operator L_A on $\mathbb{C}^n$ is self-adjoint if and only if $A = A^*$ (matrix adjoint).

2. If $A \in \mathrm{M}(n, \mathbb{C})$, let $L_A : \mathbb{C}^n \to \mathbb{C}^n$ be the multiplication operator $L_A(\mathbf{x}) = A \cdot \mathbf{x}$, and let $\mathfrak{X} = \{e_i\}$ be the usual orthonormal basis in $\mathbb{C}^n$ equipped with the standard inner product. Consider the operators $L_A, L_{A^*}, (L_A)^*$ and the associated matrices

$$[L_A]_{\mathfrak{X}\mathfrak{X}}, \quad [L_{A*}]_{\mathfrak{X}\mathfrak{X}}, \quad [(L_A)^*]_{\mathfrak{X}\mathfrak{X}}.$$

Which of the three possible pairings of these matrices yield matrices that are equal? Explain your reasoning.

HINT: For orthonormal bases $\mathfrak{X} \subseteq \mathbb{C}^n$ and $\mathfrak{Y} \subseteq \mathbb{C}^n$, and any linear operator $T : \mathbb{C}^n \to \mathbb{C}^m$, we have shown that $[T^*]_{\mathfrak{X}\mathfrak{Y}} = [T]^*_{\mathfrak{Y}\mathfrak{X}}$ (Proposition 6.37). We have also described how matrix entries can be computed in terms of inner products. In this question we are taking $\mathfrak{Y} = \mathfrak{X}$.

3. Let $\mathfrak{X} = \{\mathbf{e}_1, \mathbf{e}_2\}$ be the standard orthonormal basis in $V = \mathbb{R}^2$ and T a linear operator such that

$$[T]_{\mathfrak{X}\mathfrak{X}} = \begin{pmatrix} 0 & 1 \\ 1 & 0 \end{pmatrix}.$$

We know that $[T^*]_{\mathfrak{X}\mathfrak{X}} = [T]^*_{\mathfrak{X}\mathfrak{X}}$ (matrix adjoint) by Proposition 6.37. For the (nonorthonormal) basis $\mathfrak{Y} = \{\mathbf{f}_1, \mathbf{f}_2\}$ such that $\mathbf{f}_1 = \mathbf{e}_1 + \mathbf{e}_2$ and $\mathbf{f}_2 = \mathbf{e}_2$, verify that $[T^*]_{\mathfrak{Y}\mathfrak{Y}} \neq [T]^*_{\mathfrak{Y}\mathfrak{Y}}$.

Section 6.4. Orthogonal Diagonalization in Inner Product Spaces

1. TRUE/FALSE QUESTIONS: ("True" if the statement is *always* true.)

(a) The multiplication operator $L_A(\mathbf{x}) = A \cdot \mathbf{x}$ has a diagonalizing basis in $\mathbb{R}^n$ for any real $n \times n$ matrix such that $A^{\mathrm{T}} = A$.

(b) The multiplication operator $L_A(\mathbf{x}) = A \cdot \mathbf{x}$ has a diagonalizing basis in $\mathbb{C}^n$ for any complex $n \times n$ matrix such that $A^{\mathrm{T}} = A$.

(c) The multiplication operator $L_A(\mathbf{x}) = A \cdot \mathbf{x}$ has a diagonalizing basis in $\mathbb{C}^n$ for any complex $n \times n$ matrix such that $A^* = A$.

(d) On a finite-dimensional inner product space over $\mathbb{K} = \mathbb{C}$ every self-adjoint linear operator $T : V \to V$ is diagonalizable.

(e) On a finite-dimensional inner product space over $\mathbb{K} = \mathbb{R}$ every self-adjoint linear operator $T : V \to V$ is diagonalizable.

(f) A linear projection $P : V \to V$ (with $P^2 = P$) on an inner product space is always diagonalizable but need not be orthogonally diagonalizable.

2. Explain how the answers to true/false questions 1(a)–1(c) above are affected if we replace the words "*diagonalizing basis*" by "*orthonormal diagonalizing basis*."

3. Consider $T : V \to V$ on a finite-dimensional inner product space.

(a) How are the eigenvalues of T and T^* related?

(b) How are the dimensions of the eigenspaces related?

(c) What other relations hold between the eigenspaces of T and those of T^*?

Do your answers depend on whether $\mathbb{K} = \mathbb{R}$ or $\mathbb{C}$?

If $A \in \mathrm{M}(n, \mathbb{K})$ with $\mathbb{K} = \mathbb{R}$ or $\mathbb{C}$ is a *self-adjoint* matrix, so that

$$A^* = A \quad (\text{where } (A^*)_{ij} = \overline{A_{ji}} \text{ for all } i, j),$$

and if $V = \mathbb{K}^n$ is equipped with the standard inner product, we have seen that there is a basis in $\mathbb{K}^n$ that diagonalizes the corresponding multiplication operator $L_A(\mathbf{x}) = A \cdot \mathbf{x}$ on V.

4. Now suppose that A is only *symmetric*, with

$$A^{\mathsf{T}} = A \quad (\text{where } (A^{\mathsf{T}})_{ij} = A_{ji} \text{ for all } i, j).$$

When $\mathbb{K} = \mathbb{R}$ symmetry, $A^{\mathsf{T}} = A$ is the same thing as self-adjointness, and suffices to guarantee diagonalizability of L_A (and even orthogonal diagonalizability). What happens when $\mathbb{K} = \mathbb{C}$?

QUESTION. If A is a symmetric $n \times n$ matrix with complex entries, does the $\mathbb{C}$-linear operator $L_A : \mathbb{C}^n \to \mathbb{C}^n$ always have a diagonalizing basis? Prove or give a counterexample.

NOTE: We are not asking for an *orthonormal* diagonalizing basis in $\mathbb{C}^n$.

HINT: The 2×2 case is decisive. Furthermore, if A has distinct eigenvalues in $\mathbb{C}$ it *is* diagonalizable and can't serve as a counterexample. What conditions must the coefficients in a symmetric $A = [a, b; b, c]$ satisfy if the characteristic polynomial $p_A(x) = \det(A - xI)$ fails to have distinct roots in $\mathbb{C}$?

Section 6.5. Reflections, Rotations, and Rigid Motions on $\mathbb{R}^n$

In this section a rotation by angle $\theta = 0$ is still a rotation.

1. TRUE/FALSE QUESTIONS: ("True" if the statement is *always* true.)

(a) The product $R_1 \circ R_2$ of two reflections across hyperplanes through the origin in $\mathbb{R}^n$ is another reflection of the same kind.

(b) The product $\rho_{\theta_1} \circ \rho_{\theta_2}$ of two rotations about the origin in $\mathbb{R}^2$ is another rotation ρ_θ (by what angle?).

(c) An invertible linear operator $T : \mathbb{R}^2 \to \mathbb{R}^2$ with $\det(T) = \pm 1$ is either: (i) A rotation ρ_θ by some angle θ about the origin, or (ii) a reflection r_ℓ across some line ℓ through the origin.

(d) Every nonsingular linear operator $T : \mathbb{R}^3 \to \mathbb{R}^3$ is either: (i) A rotation $R_{\ell,\theta}$ by an angle θ about some axis ℓ through the origin, or (ii) a reflection r_E across some plane E containing the origin.

2. Use Euler's theorem to explain why the product $R_{\ell_1,\theta_1} \circ R_{\ell_2,\theta_2}$ of two rotations about axes through the origin in $\mathbb{R}^3$ is either the identity map I or a rotation R_{ℓ_3,θ_3} by some angle θ_3 about some third axis ℓ_3 through the origin.

HINT: Use that fact that $\mathrm{O}(n)$ is a matrix group and that $L_A \circ L_B = L_{AB}$.

REMARK: This is not easy to prove without benefit of Euler's theorem and some group-theoretic sleight-of-hand. For a challenge: determine the new axis ℓ_3 and the rotation angle θ_3. As a hint: ℓ_3 is a scalar multiple of the calculus cross product $\mathbf{f}_1 \times \mathbf{f}_2$ of vectors that determine the oriented axes ℓ_1, ℓ_2. This fact is of some importance in angular momentum calculations in classical and quantum mechanics.

3. In $\mathbb{R}^2$:

(a) Prove that the product $r_{\ell_1} \circ r_{\ell_2}$ of two reflections across lines that go through the origin is a rotation ρ_θ about the origin by some angle θ.

(b) If the lines ℓ_1, ℓ_2 make angles θ_1, θ_2 with the $+x$-axis, what is the rotation angle θ in (a)?

4. If we only know that $A^{\mathrm{T}}A = I_{n\times n}$ for some $A \in \mathrm{M}(n, \mathbb{R})$, explain why A must be an *orthogonal matrix* in $\mathrm{O}(n)$ with a two-sided inverse, so that

$$A^{\mathrm{T}}A = I_{n\times n} = AA^{\mathrm{T}} \quad (\text{hence } A^{\mathrm{T}} = A^{-1}).$$

What does orthogonality imply about the possible values of $\det(A)$?

5. If $A \in \mathrm{M}(3, \mathbb{R})$ is an orthogonal matrix (one with $A^{\mathrm{T}}A = I$) and $\det(A) = -1$, prove that the multiplication operator $L_A(\mathbf{x}) = A \cdot \mathbf{x}$ on $\mathbb{R}^3$ is a reflection across some 2-dimensional hyperplane E through the origin (as in Example 6.79).

REMARKS: Thus orthogonal linear operators on $\mathbb{R}^3$, those determined by orthogonal matrices $A \in \mathrm{O}(3)$, are either rotations or reflections across 2-dimensional hyperplanes, according to whether $\det(A) = +1$ or -1. A similar result holds when $n = 2$. We have $A \in \mathrm{O}(n) \Rightarrow \det(A) = \pm 1$, so when $n = 3$ all possibilites for orthogonal transformations L_A ($A \in \mathrm{O}(3)$) are accounted for, but things get complicated in higher dimensions. The structure of orthogonal linear operators on $\mathbb{R}^n$ for $n \geq 4$ is discussed in *Linear Algebra II* in Chapter 3, "Tensors and Multilinear Forms."

Section 6.6. Spectral Theorem on Vector and Inner Product Spaces

In this section all inner product spaces are finite-dimensional over $\mathbb{K} = \mathbb{C}$ or $\mathbb{R}$.

1. TRUE/FALSE QUESTIONS: ("True" if the statement is *always* true.)

(a) A self-adjoint linear operator $T : V \to V$ on an inner product space over $\mathbb{K} = \mathbb{R}$ or $\mathbb{C}$ always has a spectral decomposition

$$T = \sum_{\lambda_k \in \mathrm{sp}(T)} \lambda_k P_{\lambda_k}.$$

(Does it matter whether $\mathbb{K} = \mathbb{R}$ or $\mathbb{C}$?).

(b) If $T : V \to V$ is a normal operator on an inner product space, it has a spectral decomposition $T = \sum_{\lambda \in \mathrm{sp}(T)} \lambda P_\lambda$, and the spectral projections P_λ appearing in it are all self-adjoint.

(c) If $T : V \to V$ is a diagonalizable linear operator on an inner product space, each projection in its spectral decomposition $T = \sum_{\lambda_k \in \mathrm{sp}(T)} \lambda_k P_{\lambda_k}$ is a polynomial combination of powers of T, so $P_{\lambda_k} = f_{\lambda_k}(T)$ for some f_{λ_k} in $\mathbb{K}[x]$. (Does it matter whether $\mathbb{K} = \mathbb{R}$ or $\mathbb{C}$?)
(d) If $T : V \to V$ is a normal operator on an inner product space over $\mathbb{C}$, then $T^* = f(T)$ for some polynomial $f(x) \in \mathbb{C}[x]$.

NOTE: This is not true if T is merely diagonalizable.

2. Consider these combinations of linear operators on an inner product space V over $\mathbb{C}$:

(i) $T_1 + T_2$ (ii) $T_1 \circ T_2$
(iii) λT where $\lambda \in \mathbb{C}$ (iv) λT where $\lambda \in \mathbb{R}$

Answer the following questions.
(a) If T_1, T_2 are self-adjoint, which of these combinations are always self-adjoint?
(b) If T_1, T_2 are skew-adjoint, so $T^* = -T$, which of these combinations are always skew-adjoint?
(c) If T_1, T_2 are unitary, which of these combinations are always unitary?
(d) If T_1, T_2 are diagonalizable, which of these combinations are always diagonalizable?

3. Let $V = \mathbb{C}^2$ with the usual euclidean inner product, and $T : V \to V$ the linear operator whose matrix is

$$[T]_{\mathfrak{X}\mathfrak{X}} = \begin{pmatrix} 3 & -1 \\ 1 & 4 \end{pmatrix}$$

with respect to the standard orthonormal basis $\mathfrak{X} = \{e_1, e_2\}$.
(a) Determine the spectrum $\mathrm{sp}_{\mathbb{C}}(T) = \{\lambda_1, \lambda_2\}$ and show that T is diagonalizable by finding an explicit basis $\mathfrak{Y} = \{f_1, f_2\}$ of eigenvectors. Is T orthogonally diagonalizable?
(b) Find a transition matrix $S \in \mathrm{M}(2, \mathbb{C})$ where $[T]_{\mathfrak{Y}\mathfrak{Y}} = S\,[T]_{\mathfrak{X}\mathfrak{X}} S^{-1}$ is diagonal. Is S a *unitary* matrix ($U^* = U^{-1}$)?
(c) Find the matrices $[P_{\lambda_k}]_{\mathfrak{X}\mathfrak{X}}$ and $[P_{\lambda_k}]_{\mathfrak{Y}\mathfrak{Y}}$ that describe the projections appearing in the spectral decomposition $T = \sum_k \lambda_k P_{\lambda_k}$ with respect to the standard basis $\mathfrak{X}$ and the diagonalizing basis $\mathfrak{Y}$.
(d) Calculate the spectral decomposition(s) of e^{tT} ($t \in \mathbb{R}$) with respect to the bases $\mathfrak{X}$ and $\mathfrak{Y}$.

HINT: You need only find one of the matrices in (c) since $P_{\lambda_1} + P_{\lambda_2} = I$.

4. In general, we say that a linear operator $T : V \to V$ on a vector space over $\mathbb{K} = \mathbb{R}$ or $\mathbb{C}$ has a *spectral decomposition* if

$$T = \sum_{\lambda \in \mathrm{sp}(T)} \lambda P_\lambda$$

where the P_λ are linear projection operators (ranges not necessarily pairwise orthogonal) such that

$$\text{(i) } P_\lambda^2 = P_\lambda \neq 0 \qquad \text{(ii) } \sum_\lambda P_\lambda = I \qquad \text{(iii) } P_\lambda P_\mu = 0 \text{ if } \lambda \neq \mu$$

(a) Show that the ranges $V_\lambda = P_\lambda(V)$ of the projections in this decomposition are precisely the eigenspaces $E_\lambda(T)$ and that $V = \bigoplus_{\lambda \in \mathrm{sp}(T)} V_\lambda$.

(b) Explain why all spectral projections P_λ commute with T.

(c) Show that an arbitrary operator $S : V \to V$ commutes with $T \Leftrightarrow S$ commutes with each spectral projection P_λ.

HINT: Implication ($\Leftarrow$) follows from the Spectral Theorem; for ($\Rightarrow$) start by proving the assertion made in Additional Exercise 1(c).

5. Let $T : V \to V$ be diagonalizable with spectral decomposition $T = \sum_{\lambda \in \mathrm{sp}(T)} \lambda P_\lambda$.

(a) Is the adjoint T^* also diagonalizable?

(b) If so, is its spectral decomposition given by

$$T^* = \sum_{\lambda \in \mathrm{sp}(T)} \bar{\lambda} P_\lambda,$$

with the same spectral projections P_λ that appeared in T?

HINTS: What if T is a single linear projection that is not self-adjoint, so $R(T)$ and $K(T)$ not orthogonal subspaces)?

6. Same as the previous problem, but with $T : V \to V$ *orthogonally diagonalizable.*

Section 6.7. Positive Operators and Polar Decomposition

1. TRUE/FALSE QUESTIONS: (“True” if the statement is *always* true.)

(a) An invertible linear operator $T : V \to V$ has a polar decomposition, even if V is an inner product space over $\mathbb{R}$.

NOTE: We have (*unitary*) = (*real-orthogonal*) on real inner product spaces. Adjoints T^* are computed the same way with respect to the given inner product whether $\mathbb{K} = \mathbb{R}$ or $\mathbb{C}$.

(b) The singular values of any linear operator on a finite-dimensional inner product space are also eigenvalues for the operator T.

(c) If $T : V \to V$ is a diagonalizable linear operator on a finite-dimensional inner product space V over $\mathbb{C}$, then T is positive definite if $\mathrm{sp}(T) \subseteq [0, +\infty) + i0$.

(d) If $T \geq 0$ and $\lambda \geq 0$ in $\mathbb{R}$, then $E_\lambda(\sqrt{T}) = E_{\sqrt{\lambda}}(T)$.

2. Determine polar decompositions for the operators L_A on $\mathbb{C}^2$ or $\mathbb{C}^3$ when

$$\text{(a) } A = \begin{pmatrix} 1 & 2 \\ 3 & -2 \end{pmatrix} \qquad \text{(b) } A = \begin{pmatrix} 20 & 4 & 0 \\ 0 & 0 & 1 \\ 4 & 20 & 0 \end{pmatrix}$$

Find the singular values in each case.

3. For a linear operator $T : V \to W$ between finite-dimensional inner product spaces, both T^*T and TT^* are self-adjoint and positive, and hence have unique positive square roots.

(a) Can you see any reason why one might want to define the absolute value of T to be $|T| = \sqrt{T^*T}$ rather than $|T| = \sqrt{TT^*}$?

(b) If $T : V \to W$ is invertible and we take the "wrong" square root $|T|' = \sqrt{TT^*}$ instead of $\sqrt{T^*T}$, show that this leads to a "right-handed" polar decomposition $T = |T|' \cdot U'$.

4. In Example 6.103 and Exercise 6.104 we found the spectral decomposition of a linear operator on $V = \mathbb{C}^2$ whose matrix with respect to the standard orthonormal basis was

$$A = [T]_{\mathfrak{X}\mathfrak{X}} = \begin{pmatrix} 7 & \sqrt{3} \\ \sqrt{3} & 5 \end{pmatrix}.$$

Using that discussion as a guide, find the spectral decompositions of the self-adjoint operators T^*T and TT^*. Then compute

$$\text{(a) } \sqrt{T^*T} \qquad \text{(b) } \sqrt{TT^*}$$

Are these the same operator on $\mathbb{C}^2$? Do they have the same square roots?

Index

Published Titles in This Series

29 **Frederick P. Greenleaf and Sophie Marques,** Linear Algebra I, 2019
28 **László Erdős and Horng-Tzer Yau,** A Dynamical Approach to Random Matrix Theory, 2017
27 **S. R. S. Varadhan,** Large Deviations, 2016
26 **Jerome K. Percus and Stephen Childress,** Mathematical Models in Developmental Biology, 2015
25 **Kurt O. Friedrichs,** Mathematical Methods of Electromagnetic Theory, 2014
24 **Christof Schütte and Marco Sarich,** Metastability and Markov State Models in Molecular Dynamics, 2013
23 **Jerome K. Percus,** Mathematical Methods in Immunology, 2011
22 **Frank C. Hoppensteadt,** Mathematical Methods for Analysis of a Complex Disease, 2011
21 **Frank C. Hoppensteadt,** Quasi-Static State Analysis of Differential, Difference, Integral, and Gradient Systems, 2010
20 **Pierpaolo Esposito, Nassif Ghoussoub, and Yujin Guo,** Mathematical Analysis of Partial Differential Equations Modeling Electrostatic MEMS, 2010
19 **Stephen Childress,** An Introduction to Theoretical Fluid Mechanics, 2009
18 **Percy Deift and Dimitri Gioev,** Random Matrix Theory: Invariant Ensembles and Universality, 2009
17 **Ping Zhang,** Wigner Measure and Semiclassical Limits of Nonlinear Schrödinger Equations, 2008
16 **S. R. S. Varadhan,** Stochastic Processes, 2007
15 **Emil Artin,** Algebra with Galois Theory, 2007
14 **Peter D. Lax,** Hyperbolic Partial Differential Equations, 2006
13 **Oliver Bühler,** A Brief Introduction to Classical, Statistical, and Quantum Mechanics, 2006
12 **Jürgen Moser and Eduard J. Zehnder,** Notes on Dynamical Systems, 2005
11 **V. S. Varadarajan,** Supersymmetry for Mathematicians: An Introduction, 2004
10 **Thierry Cazenave,** Semilinear Schrödinger Equations, 2003
9 **Andrew Majda,** Introduction to PDEs and Waves for the Atmosphere and Ocean, 2003
8 **Fedor Bogomolov and Tihomir Petrov,** Algebraic Curves and One-Dimensional Fields, 2002
7 **S. R. S. Varadhan,** Probability Theory, 2001
6 **Louis Nirenberg,** Topics in Nonlinear Functional Analysis, 2001
5 **Emmanuel Hebey,** Nonlinear Analysis on Manifolds: Sobolev Spaces and Inequalities, 2000
3 **Percy Deift,** Orthogonal Polynomials and Random Matrices: A Riemann-Hilbert Approach, 2000
2 **Jalal Shatah and Michael Struwe,** Geometric Wave Equations, 2000
1 **Qing Han and Fanghua Lin,** Elliptic Partial Differential Equations, Second Edition, 2011